ISBN 978-0-266-14667-4
PIBN 10042956

1 MONTH OF
FREE
READING

at
www.ForgottenBooks.com

By purchasing this book you are eligible for one month membership to ForgottenBooks.com, giving you unlimited access to our entire collection of over 1,000,000 titles via our web site and mobile apps.

To claim your free month visit: www.forgottenbooks.com/free42956

HAMILTON'S ARITHMETICS

SECOND BOOK

BY

SAMUEL HAMILTON, Ph.D.

AUTHOR OF "THE RECITATION," AND SUPERINTENDENT
OF SCHOOLS, ALLEGHENY COUNTY, PA.

NEW JERSEY EDITION

AMERICAN BOOK COMPANY

NEW YORK   CINCINNATI   CHICAGO

PREFACE

THIS book is intended to cover the work of the fifth and sixth years. It is based on, and closely follows, the Course of Study issued by the Department of Public Instruction of the State of New Jersey. It is divided into four parts, corresponding to the four half-year periods in the fifth and sixth years.

The *first half* of the *fifth grade* is devoted chiefly to the treatment of common fractions. The *second half* of the *fifth grade* takes up the simpler processes in decimals. The *first half* of the *sixth grade* completes decimals and begins percentage. The *second half* of the *sixth grade* continues percentage, with its simpler applications, and gives an easy treatment of interest.

In addition, each grade provides for a thorough review of all that has preceded it. Much practice is also given in denominate numbers and in mensuration, in scale drawing, and in practical problems relating to the common affairs of life, such as marketing, traveling, furnishing a house, keeping simple accounts, etc.

Attention is invited to the following features:

1. The prominence given to drill intended to give skill, and the frequency of systematic reviews.

2. The easy steps in gradation.

3. The interesting character of the problems, drawn from the child's activities at home, at school, and at play, and from his relations to community life.

8

4. The close relation of business problems to real conditions.

5. The abundance of exercises for oral drill and the encouragement of pupils to make mental estimates before beginning the written work.

6. The emphasis placed on correct interpretation of problems and on choosing the most economical methods for their solution.

7. The appeal made to observation as a stimulus to mathematical thought.

SAMUEL HAMILTON.

CONTENTS

FIFTH GRADE

SIXTH GRADE

FIFTH GRADE — FIRST HALF

NOTATION AND NUMERATION

ARABIC NUMBERS

1. In 364, for what does the 4 stand? the 6? the 3? 364 = 3 hundreds, 6 tens, 4 ones.

2. Read 2475. For what does each figure stand?

Each figure has a value that depends upon its **place** in the number. The first place beginning at the right of a number is called **ones'** place; the second, **tens'** place; the third, **hundreds'** place; the fourth, **thousands'** place, etc.

3. Show that the left-hand 2 in the number 222 has a value, because of its place, that is 10 times the value of the middle 2, and 100 times the value of the right-hand 2.

When a number contains more than three figures, it is separated into groups, or **periods**, of three figures each, beginning at the right. The names of the places in each period are the same. The name of the first period is **units**; of the second period, **thousands**; of the third period, **millions**.

The following table shows the arrangement of these periods, and the three orders of figures in each period:

MILLIONS' PERIOD			THOUSANDS' PERIOD			UNITS' PERIOD		
Hundred-millions	Ten-millions	Millions	Hundred-thousands	Ten-thousands	Thousands	Hundreds	Tens	Ones
1	0	0,	6	4	1,	3	7	6

This number is read, "100 million, 641 thousand, 376."

Beginning at the right, mark off into periods, and read:

4. 2439	**7.** 758650	**10.** 405682	**13.** 1578563
5. 3750	**8.** 775570	**11.** 140560	**14.** 1556893
6. 4005	**9.** 205605	**12.** 685000	**15.** 1003704

Observe that *naughts* are enumerated but never read; that *and* is never read between integers; and that the *first period* is simply read but not named. Thus, 376 is read three hundred seventy-six.

Write :

16. 6 thousand 3 hundred 75.

17. 10 thousand 4 hundred 6.

18. 150 thousand 5 hundred 25.

19. 5 million 825 thousand 5 hundred 4.

20. 5 million 25 thousand 3.

21. 2 million 22 thousand 60.

22. 6 million 27 thousand 9.

23. 5 million 5 thousand 5.

24. 8 million 8 thousand 85.

25. Ten thousand ten hundred ten.

26. Five million two hundred fifty.

27. Six million six thousand six.

28. Sixty million one hundred fifteen thousand five hundred seventy-nine.

29. Forty million four.

30. One million.

31. Ninety-nine million six hundred thousand nine hundred.

32. Eight million two hundred twenty-eight thousand two hundred one.

33. Ninety-eight million one hundred thousand nine hundred fifty-four.

DOLLARS AND CENTS

10 cents	= 1 dime
10 dimes	= 1 dollar
100 cents	= 1 dollar

Since 10 dimes equal a dollar, a dime is $\frac{1}{10}$ of a dollar, and since 100 cents equal a dollar, a cent is $\frac{1}{100}$ of a dollar. *Tenths* and *hundredths* of a dollar are sometimes called **decimal parts of a dollar.**

1. What decimal part of a dollar are 2 dimes? 6 dimes? 9 dimes?

2. What decimal part of a dollar are 2 cents? 5 cents? 10 cents? 25 cents?

Dollars and cents are written thus: $3.07, $3.25.

Cents are written thus: $.01, $.02, $.03, $.25, $.67.

NOTE. — 1 cent may be written either $0.01 or $.01; 25 cents may be written either $0.25 or $.25, etc. The naught preceding the decimal point does not affect the result, and is sometimes written to show more prominently that cents and not dollars are represented.

The point separating the dollars and cents is called the **decimal point.**

Write in figures:

3. 5 dollars and 8 cents; 107 dollars and 7 cents.

4. 9 dollars and 10 cents; 250 dollars and 5 cents.

5. 107 dollars and 75 cents; 300 dollars and 10 cents.

NOTE. — Always read the word *and* between dollars and cents. Thus, $6.09 is read six dollars *and* nine cents.

Read; then write from dictation:

	a	*b*	*c*	*d*	*e*
6.	\$5.09	\$5.05	\$3.00	\$4.00	\$ 12.09
7.	\$6.00	\$0.07	\$1.02	\$ .98	\$ 20.08
8.	\$8.47	\$9.99	\$1.00	\$1.65	\$254.45
9.	\$ .05	\$ .99	\$1.01	\$ .75	\$201.05
10.	\$7.09	\$ .37	\$1.74	\$ .81	\$150.50

ROMAN NUMBERS

1. Without looking at the clock face, write the Roman numbers showing the twelve hours.

Note.— The Romans used these numbers for placing inscriptions on monuments, and for dividing books into chapters. We use them for the same purposes and on the clock face.

2. Memorize:

Letters	I	V	X	L	C	D	M
Values	1	5	10	50	100	500	1000

I. *When a letter is followed by the same letter or by one of less value, the values of the letters are to be added.* Thus, XX = 20; XI = 11; XXX = 30.

3. Write in Roman numbers: 13, 20, 30, 31, 32, 33.

4. Write in Roman numbers: 16, 17, 18, 26, 27, 28, 36, 37, 38, 12, 21, 22, 23.

II. *When a letter is followed by another of greater value, the value of the smaller is to be subtracted from that of the greater.* Thus, IV = 4; IX = 9; XL = 40; XC = 90.

5. Write in Roman numbers: 40, 41, 42, 43, 45, 46, 47, 48, 90, 91.

III. *When a letter is placed between two letters of greater value, the value of the smaller is to be subtracted from the sum of the other two.* Thus, XIV = 14; XIX = 19; XXIX = 29; XLIX = 49.

6. Write in Roman numbers: 14, 24, 34, 44, 54, 64, 74, 84, 19, 29, 39, 49, 59, 69, 79, 89.

IV. *The letter* C *represents* 100; D *represents* 500; M *represents* 1000. Thus, XC = 90; XCV = 95; MD = 1500; DC = 600; MDCCCCXIII = 1913.

7. A book has twenty chapters. Write them in their order.

Read:

8. XXI	LXV	XLIV	LXXX	LVI	C
9. XXIV	LXXI	XCVII	M	LXI	D
10. XXVIII	XLIII	LXII	MC	LXX	MI

Write in Roman numbers:

11. 15	66	33	100	400	700
12. 25	88	49	200	500	800
13. 75	99	55	300	600	900

FUNDAMENTAL OPERATIONS

ADDITION

Addition is the process of uniting two or more numbers to form one number.

The **addends** are the numbers to be added.

The **sum** or **amount** is the result of addition.

To the Teacher. — Rapid counting by 2's to 9's should be practiced daily.

Oral and Written Work

Answer at sight :

1.	8 9	28 9	48 9	58 9	68 9		6.	18 7	28 7	58 7	68 7	98 7
2.	19 6	39 6	49 6	69 6	89 6		7.	28 6	38 6	58 6	78 6	88 6
3.	18 8	28 8	38 8	98 8	88 8		8.	17 9	57 9	87 9	77 9	97 9
4.	19 5	39 5	59 5	49 5	79 5		9.	16 7	26 7	76 7	96 7	46 7
5.	35 6	95 6	65 6	85 6	45 6		10.	44 9	74 9	84 9	94 9	64 9

Note. — Drill frequently on addition by endings, as in Exs. 1–10.

Add, observing the groups that make 5, 10, 15, 20, etc. :

11.		12.	13.	14.	15.	16.	17.
4		7	8	7	8	3	2
6	10	8	3	4	9	9	6
7		4	7	6	3	8	4
8	10	7	6	5	5	2	8
5		8	4	7	4	7	4
9		1	2	2	6	8	6
8		0	1	6	7	5	5
2	10	9	7	5	9	4	7
5		3	5	4	8	2	8
7	20	8	6	5	9	1	1
8		9	4	6	6	3	9

Write from dictation; then add and test, observing groups:

18.	19.	20.	21.	22.
541	862	720	839	564
462 }8	321	287 }10	463	268 }12
376	456 }15	823	726 }15	374
814 }10	289	541 }10	339	827 }10
296	315	569	187	693

23.	24.	25.	26.	27.
325	4250	20150	3984	52306
237	629	107	4	20006
63	47	360	296	30750
104	2307	26347	307	5600
9	234	6066	6785	170
937	276	205	4397	52879
428	999	304	6123	30562
632	708	591	4187	41028

Add:

28.	29.	30.	31.	32.
35701	575	5609	9401	8026
6531	1039	41	672	643
320	9901	60	56	29
43	601	853	803	660
7501	8010	731	10	9032
4507	301	550	9083	8009

33.	34.	35.	36.
$80.00	$329.60	$200.65	$3245.07
1.05	300.06	30.00	4706.50
6.09	6.09	976.38	5079.06
87.63	709.36	40.79	5904.00

Like numbers are numbers that express the same kind of unit; as, 3 ft. and 5 ft., or 3 and 5; but not 3 ft. and 5 ¢. *Only like numbers can be added or subtracted.*

Time yourself in adding these examples. Then try again and see whether you can beat your first record.

37.	39.	41.	43.
$25.36	$163.75	$243.15	$121.48
43.72	275.84	65.74	83.62
96.81	486.39	182.33	275.14
39.47	928.75	34.62	· 8.73
62.58	265.73	215.73	16.28

38.	40.	42.	44.
$795.90	$678.65	$450.00	$354.00
784.00	875.60	45.76	280.01
450.25	490.00	46.01	45.06
350.00	348.45	9.98	98.00
450.05	246.79	.87	.98
365.00	400.00	1.23	12.12

45.	47.	49.	51.
$456.00	$500.00	$ 10.80	$480.24
580.00	75.25	450.00	24.00
452.98	.87	98.98	35.08
60.00	29.07	50.00	10.15
21.12	108.90	2.50	48.90
29.88	982.13	8.20	62.15

46.	48.	50.	52.
478010	457890	847950	748590
389101	149503	384940	138391
375910	485958	478849	273785
100000	484950	479847	588976
364805	876395	353535	456789
374898	234715	666666	758984

SUBTRACTION

Subtraction is the process of finding the difference between two numbers, or of taking one number from another.

The **minuend** is the number from which we subtract; the **subtrahend** is the number subtracted; and the **difference** or **remainder** is the answer.

Oral Work

Drill thoroughly on these combinations in subtraction:

	a	b	c	d	e	f	g	h	i	j	k	l
1.	9	15	12	16	11	13	12	9	8	17	15	11
	4	7	6	7	4	9	3	3	2	8	4	3
2.	7	10	10	17	18	13	15	16	13	14	17	15
	2	7	6	9	9	7	6	9	8	9	5	9

Subtract at sight:

	a	b	c	d	e	f	g	h	i	j	k	l
3.	11	16	15	12	8	9	13	10	11	14	12	14
	9	8	8	5	3	6	6	4	6	7	4	6

	a	b	c	d	e	f	g	h	i	j	k	l
4.	15	13	11	10	15	14	18	13	17	13	15	16
	11	4	8	9	5	8	16	11	13	12	10	14

5. Subtract 8 from 24, 34, 54, 74; then subtract 9 from each of these numbers; then 6; then 5; then 7.

6. Subtract 9 from 32, 42, 72, 92; then subtract 8 from each of these numbers; then 6; then 7; then 5; then 4; then 3.

7. Subtract 8 from 43, 53, 63, 73; then subtract 7 from each of these numbers; then 9; then 6; then 5; then 4.

8. Subtract 9 from 21, 31, 41, 91; then subtract 7 from each of these numbers; then 8; then 6; then 5; then 4; then 3; then 2.

9. Subtract 9 from 55, 45, 65, 95; then subtract 8 from each of these numbers; then 7; then 6.

10. Subtract 8 from 66, 86, 76, 96; then 7 from each of these numbers; then 9.

11. Subtract 9 from 37, 47, 77, 87; then subtract 8.

12. Subtract 9 from 28, 58, 68, 78; then subtract 8.

Give difference at sight. Thus, $64 - 30 = 34$; $34 - 6 = 28$. Simply say, 34, 28.

	a	b	c	d	e	f	g	h
13.	64	53	72	54	71	57	61	83
	36	16	53	36	39	38	47	37

	a	b	c	d	e	f	g	h
14.	53¢	$56	91	84 pt.	83 lb.	74 pk.	76 ft.
	27¢	$37	57	55 pt.	54 lb.	37 pk.	49 ft.

To **test** subtraction, add the remainder to the subtrahend.

Give differences at sight:

15. $43 - 28$	**18.** $86 - 43$	**21.** $31 - 14$	**24.** $86 - 79$
16. $65 - 32$	**19.** $71 - 18$	**22.** $26 - 19$	**25.** $96 - 93$
17. $94 - 77$	**20.** $69 - 27$	**23.** $37 - 24$	**26.** $18 - 11$

27. When the minuend and the difference are given, how may the subtrahend be found? When the difference and the subtrahend are given, how may the minuend be found?

Written Work

Take each number below from 1000, giving 1 minute to each column:

1. 225	**5.** 216	**9.** 725	**13.** 715	**17.** 375
2. 314	**6.** 500	**10.** 946	**14.** 800	**18.** 814
3. 625	**7.** 499	**11.** 328	**15.** 125	**19.** 731
4. 374	**8.** 795	**12.** 613	**16.** 625	**20.** 656

21. Take each of the numbers in examples 1–20 from 10,000.

Subtract, timing your work. Then try again and see whether you can beat your record. Check results by adding.

22. 607008	**25.** 180260	**28.** 800647
448789	98775	98749
23. 756008	**26.** 100907	**29.** 870009
398497	49899	698058
24. 640006	**27.** 230900	**30.** 906700
290809	97897	798897

Subtract, timing yourself. Then try to beat your record.

	a	*b*	*c*	*d*	*e*	*f*	*g*	*h*
31.	44	69	213	761	298	723	842	841
	39	27	124	148	149	476	364	563

32.	36005 − 19096	42.	630209 − 189768
33.	90000 − 27938	43.	620005 − 246937
34.	23000 − 17500	44.	610034 − 263805
35.	$4629.70 − $3675.84	45.	$2473.87 − $629.75
36.	$1475.558 − $539.47	46.	$9000.45 − $4167.23
37.	$3000.73 − $2036.75	47.	$6343.75 − $900.84
38.	$9143.65 − $6183.69	48.	$9143.92 − $6287.75
39.	$24000.47 − $6937.64	49.	$4816.75 − $2407.84
40.	$2039.05 − $1729.89	50.	$94367.48 − $21697.83
41.	$9400.37 − $2869.94	51.	$21485.86 − $11475.97

MULTIPLICATION

Multiplication is the process of taking one number as many times as there are units in another.

The **multiplicand** is the number multiplied ; the **multiplier** is the number showing how many times the multiplicand is taken; and the **product** is the result of multiplication.

Oral Work

1. $6+6+6+6=?$ In what short way can you find the answer? Multiplication is a *short* form of *addition*.

2. $3 \times \$7 = \$21$. What number is multiplied? By what number is it multiplied? What is the name of the answer?

3. Read: 5×4 ft. ; 6×3 doz.; 9×2 yd.: 8×7¢.

When the multiplier is written *before* the sign ×, as in 4×8 bu., the sign is read "*times*." Thus, 4×8 bu. is read 4 *times* 8 bushels. When the multiplier is written *after* the sign, the sign is read " *multiplied by*." Thus, 4 ft. × 8 is read 4 ft. *multiplied by* 8.

4. Does the expression " 6 apples times 4 " make sense? The multiplier cannot name things, as, feet, hours, etc., since it indicates the number of times the multiplicand is to be taken.

An **abstract number** is a number used without reference to a particular thing ; as, 5, 6, 50.

A **concrete number** is a number used with reference to a particular thing; as, 5 trees, 6 dollars.

The multiplier is always regarded as an *abstract number*, since it shows how many times the multiplicand is taken. The product is always like the multiplicand.

5. Compare the product of 8 × 6 with the product of 6 × 8; the product of 2 × 14 with the product of 14 × 2. *Either factor may be regarded as the multiplier.*

NOTE. — The multiplicand may be either concrete or abstract. When it is concrete, the product has the same name as the multiplicand.

Multiplying and dividing by 10, 20, 100, etc.

1. Multiply 42 by 10. Annex a naught to the right of 42. Is there any difference in the products? Which of these two methods of multiplying is the *shorter?* Multiply 42 by 100. Annex two naughts to the right of 42.

2. Multiply each of the following numbers by 10 ; by 100 :

43　26　75　96　283　694　786　813　465　710

Written Work

1. Multiply 323 by 325.　　　　　　　　SHORT FORM

Multiplicand,	323		323
Multiplier,	325		325
1st partial product,	1615 =	5 × 323	1615
2d partial product,	6460 =	20 × 323	646
3d partial product,	96900 =	300 × 323	969
Entire product,	104975 =	325 × 323	104975

To **test** multiplication, use the multiplicand for a multiplier and perform the multiplication again; or, divide the product by the multiplier.

2. How much will 48 chairs cost at $1.25 each?

$$
\begin{array}{r}
\$\ 1.25 \\
48 \\
\hline
10\ 00 \\
50\ 0 \\
\hline
\$60.00
\end{array}
$$

Multiply as before, and mark off from the right, in the product, two places for cents.

3. 45×63 bu. $= ?$ **16.** $42 \times 89 = ?$ **29.** $47 \times \$\ 5.67 = ?$

4. 29×87 ft. $= ?$ **17.** $50 \times 78 = ?$ **30.** $78 \times \$\ 3.50 = ?$

5. 46×215 doz. $= ?$ **18.** $45 \times 38 = ?$ **31.** $43 \times \$\ 5.69 = ?$

6. $78 \times \$326 = ?$ **19.** $45 \times 67 = ?$ **32.** $75 \times \$\ 8.97 = ?$

7. $86 \times \$293 = ?$ **20.** $48 \times 67 = ?$ **33.** $82 \times \$49.85 = ?$

8. $91 \times \$145 = ?$ **21.** $67 \times 34 = ?$ **34.** $98 \times \$67.80 = ?$

9. $97 \times 609 = ?$ **22.** $47 \times 200 = ?$ **35.** $65 \times \$99.94 = ?$

10. $85 \times 987 = ?$ **23.** $67 \times 450 = ?$ **36.** $76 \times \$87.87 = ?$

11. $68 \times 694 = ?$ **24.** $54 \times 709 = ?$ **37.** $78 \times \$66.05 = ?$

12. $65 \times 45 = ?$ **25.** $75 \times 908 = ?$ **38.** $46 \times \$68.07 = ?$

13. $78 \times 56 = ?$ **26.** $56 \times 109 = ?$ **39.** $25 \times \$60.80 = ?$

14. $78 \times 90 = ?$ **27.** $47 \times 305 = ?$ **40.** $32 \times \$56.93 = ?$

15. $50 \times 52 = ?$ **28.** $44 \times 333 = ?$ **41.** $45 \times \$40.98 = ?$

42.
$$
\begin{array}{r}
85 \\
670 \\
\hline
5950 \\
510 \\
\hline
56950
\end{array}
$$

43.
$$
\begin{array}{r}
6754 \\
608 \\
\hline
54032 \\
40524 \\
\hline
4106432
\end{array}
$$

44.
$$
\begin{array}{r}
\$60.70 \\
80\ 06 \\
\hline
364\ 20 \\
48560 \\
\hline
\$485964.20
\end{array}
$$

Find products and test:

45. 809 × 3750	**60.** 678 × 8190	**75.** 500 × 6780
46. 370 × 2009	**61.** 547 × 7800	**76.** 450 × 6001
47. 400 × 3098	**62.** 101 × 8901	**77.** 780 × 6791
48. 209 × 6708	**63.** 560 × 9000	**78.** 560 × 6401
49. 609 × 8078	**64.** 798 × 9008	**79.** 101 × 1000
50. 458 × 6009	**65.** 567 × 7980	**80.** 109 × 2300
51. 650 × 8079	**66.** 440 × 7860	**81.** 708 × 8001
52. 407 × 7900	**67.** 450 × 7009	**82.** 650 × 5160
53. 608 × 8004	**68.** 579 × 6009	**83.** 765 × 7180
54. 440 × 7980	**69.** 478 × 4890	**84.** 832 × 9001
55. 765 × 8798	**70.** 679 × 8009	**85.** 689 × 5010
56. 508 × 2347	**71.** 579 × 1001	**86.** 789 × 6870
57. 278 × 8750	**72.** 809 × 9001	**87.** 340 × 1009
58. 476 × 2609	**73.** 581 × 8001	**88.** 678 × 5689
59. 805 × 3672	**74.** 591 × 5001	**89.** 671 × 1009

Multiply and test:

90. 6425	**a.** 245	
91. 1024	**b.** 344	Form 100 problems by multi-
92. 8720	**c.** 564	plying each multiplicand by
93. 9652	**d.** 746	each of the multipliers; thus:
94. 8665	**e.** 804	90 *a.* 245 × 6425 = ?
95. 7894	by · **f.** 961	90 *b.* 344 × 6425 = ?
96. 8465	**g.** 869	97 *i.* 968 × 7695 = ?
97. 7695	**h.** 796	Write, solve, and test each
98. 8425	**i.** 968	problem in $1\frac{1}{2}$ minutes.
99. 9476	**j.** 898	

DIVISION

Division is the process of finding how many times one number contains another, or of separating a number into equal parts.

The **dividend** is the number divided; the **divisor** is the number by which we divide; and the **quotient** is the result of division.

The **remainder** is the part of the dividend remaining when the division is not exact.

Oral Work

1. What is the sign of division? Division is indicated in three ways; thus, $15 \div 5$, $\frac{15}{5}$, $5)15$.

2. In problem 1, which number is the dividend? which is the divisor?

If the dividend and divisor are concrete, they must have the same name. The quotient is then abstract. Thus, $7 (divisor) is contained in $21 (dividend) 3 times (quotient).

When the divisor is abstract and the dividend concrete, the quotient has the same name as the dividend. Thus, $21 ÷ 7 = $3, or $\frac{1}{7}$ of $21 = $3.

When we consider that $7 is contained 3 times in $21, the problem differs from the separation of $21 into 7 equal *parts*. The latter kind of division is called **partition**.

3. Find $\frac{1}{8}$ of 120; 160; 200; 400; 480; 960.

4. How do you prove division? Illustrate.

Give the quotients at sight:

5. $160 \div 4$	**10.** $720 \div 24$	**15.** $750 \div 15$	**20.** $880 \div 4$
6. $280 \div 7$	**11.** $900 \div 15$	**16.** $400 \div 4$	**21.** $1200 \div 20$
7. $960 \div 8$	**12.** $500 \div 2$	**17.** $360 \div 6$	**22.** $7200 \div 60$
8. $1080 \div 12$	**13.** $900 \div 45$	**18.** $360 \div 5$	**23.** $9000 \div 60$
9. $900 \div 5$	**14.** $100 \div 25$	**19.** $900 \div 3$	**24.** $3600 \div 30$

Dividing by 10, 100, 1000, etc.

Oral Work

1. Divide 40 by 10. Cut off a naught from the right of 40. Is there any difference in the quotients? Which of these two methods of dividing is the *shorter?*

2. Divide 2436 by 100. $1|00|24|36$
$24\frac{36}{100}$

3. Divide each of the following numbers by 10; by 100; by 1000: 3000; 46,000; 273,000; 619,000; 81,400; 2000; 8626; 46,153; 814,256.

Written Work

1. Divide 81906 by 34. Notice in the third division that 30 is smaller than the divisor. Place naught in the quotient, write 6 to the right of 30, and divide 306 by 34.

```
         2409, quotient
    34)81906, dividend
       68
       ───
       139
       136
       ───
        306
        306
        ───
```

To test division, multiply the quotient by the divisor. If there is a remainder, add it to the product.

Divide and test:

2. 218645 ÷ 44	**9.** 432107 ÷ 31	**16.** 95625 by 43
3. 218465 ÷ 95	**10.** 654321 ÷ 52	**17.** 62181 by 63
4. 836219 ÷ 23	**11.** 978001 ÷ 73	**18.** 63595 by 79
5. 346924 ÷ 63	**12.** 803402 ÷ 84	**19.** 736840 by 65
6. 637185 ÷ 88	**13.** 549802 ÷ 95	**20.** 406090 by 65
7. 46221 ÷ 21	**14.** 400001 ÷ 89	**21.** 13965 by 133
8. 28497 ÷ 21	**15.** 309008 ÷ 97	**22.** 16023 by 147

23. 678001 by 105	**30.** $70004 by 172	**37.** 765432 by 509
24. 700000 by 220	**31.** $74029 by 181	**38.** 678900 by 678
25. 850020 by 307	**32.** $26686 by 878	**39.** 394201 by 727
26. 449091 by 145	**33.** $41324 by 492	**40.** 400009 by 836
27. 330789 by 232	**34.** $87912 by 578	**41.** 801020 by 905
28. 276548 by 321	**35.** $35973 by 709	**42.** 100987 by 744
29. 983217 by 144	**36.** $20044 by 908	**43.** 254016 by 823

Divide and test :

44. 6464341	*a.* 268	Form 100 problems by dividing each dividend by each of the divisors; thus :
45. 7846760	*b.* 354	
46. 5864548	*c.* 676	
47. 8645341	*d.* 758	44 *a.* 6464341 ÷ 268 = ?
48. 9624872	*e.* 865	44 *b.* 6464341 ÷ 354 = ?
49. 7784100 by	*f.* 984	49 *c.* 7784100 ÷ 676 = ?
50. 6810404	*g.* 789	Write, solve, and test each problem in 2 minutes.
51. 7904025	*h.* 897	
52. 4867045	*i.* 509	
53. 3234567	*j.* 890	

Written Work

Division of dollars and cents.

1. If 20 bu. apples cost $18.00, find the cost per bushel.

$ 0.90, cost of 1 bu.
20)$18.00

Find the cost of 1 if :

3. 12 yd. cloth cost $13.44

4. 16 books cost $13.60

5. 15 yd. ribbon cost $1.35

2. At 8¢ per quart, how many quarts of berries can be bought for $1.84?

23, no. of quarts.
8¢)184¢

Find the number if :

6. Pads at 8¢ cost $3.20

7. Caps at 72¢ cost $10.80

8. Plates at 18¢ cost $8.10

FACTORS AND MULTIPLES

The numbers which multiplied together make any *product* are called the **factors** of the product. Thus, in $2 \times 3 = 6$, 2 and 3 are the factors of 6.

Oral Work

1. Name two factors that produce 8, 10, 12, 14, 16, 18, 20.

2. The product of two factors is 24, and one of the factors is 3. What is the other factor?

	Factor	Product	Other Factor			Factor	Product	Other Factor
3.	2	20	?		**9.**	10	40	?
4.	5	50	?		**10.**	15	80	?
5.	8	24	?		**11.**	12	60	?
6.	9	27	?		**12.**	11	77	?
7.	7	35	?		**13.**	13	52	?
8.	4	28	?		**14.**	17	68	?

An **integer** or an **integral number** is a whole number.

When we say that 8 is a multiple of both 2 and 4, we mean that both 2 and 4 are contained in 8 an *integral number* of times. When we say that 20 is a multiple of both 4 and 5, we mean that both 4 and 5 are exactly contained in 20.

A multiple of a number is a number that contains two or more numbers an integral number of times.

15. Find the factors of 18, as, 9×2, or 6×3. Since 18 exactly contains 9 and 2 and 6 and 3, it is a multiple of 9 and 2 and of 6 and 3.

The number 24 is a multiple of what two numbers? the number 36? 48? 16? 12?

The *dividend* in division is always the product of two factors, — the divisor and the quotient. Thus, in $48 \div 6 = 8$, 48 is the product of the factors 6 and 8.

Find the unknown terms:

	Divisor	Dividend	Quotient		Divisor	Dividend	Quotient
16.	12	144	?	26.	84	7	?
17.	10	100	?	27.	?	135	45
18.	12	600	?	28.	?	125	5
19.	12	180	?	29.	?	60	12
20.	?	288	12	30.	?	81	9
21.	?	144	12	31.	?	150	2
22.	?	450	15	32.	21	?	24
23.	?	500	10	33.	36	?	27
24.	15	?	18	34.	48	?	42
25.	20	?	40	35.	56	?	72

Written Work

	Divisor	Dividend	Quotient		Divisor	Dividend	Quotient
1.	25	650	?	11.	35	2205	?
2.	23	736	?	12.	?	2548	52
3.	51	918	?	13.	49	2793	?
4.	?	544	32	14.	36	?	46
5.	?	812	28	15.	?	2263	73
6.	24	672	?	16.	64	2816	?
7.	33	?	29	17.	37	?	39
8.	39	?	18	18.	?	1775	25
9.	16	?	38	19.	48	?	52
10.	27	?	34	20.	49	3136	?

COMMON FRACTIONS

To THE TEACHER. — The review of the work of the Fourth Grade should continue while the work in fractions is being introduced.

Halves, fourths, and eighths.

Oral Work

1 unit $= \frac{2}{2}$.   1 unit $= \frac{4}{4}$.   1 unit $= \frac{8}{8}$.

$$\frac{1}{2} \quad = \quad \frac{2}{4} \quad = \quad \frac{4}{8}.$$

1. What is $\frac{1}{2}$ of 8? $\frac{1}{2}$ of 16? $\frac{1}{2}$ of 24?

2. What is $\frac{1}{4}$ of 8? $\frac{1}{4}$ of 16? $\frac{1}{4}$ of 24? $\frac{3}{4}$ of 24?

3. What is $\frac{1}{8}$ of 8? $\frac{1}{8}$ of 16? $\frac{1}{8}$ of 24? $\frac{3}{8}$ of 24?

4. The first circle is divided into how many equal parts? What is each part called?

5. The second circle is divided into how many equal parts? What is each part called?

6. The third circle is divided into how many equal parts? What is each part called?

7. $\frac{1}{2}$ of the first circle $= \frac{?}{4}$ of the second circle. $\frac{1}{2}$ of the first circle $= \frac{?}{8}$ of the third circle.

8. How many halves of a circle are there in a whole circle? how many fourths of a circle? how many eighths of a circle?

9. Change $\frac{1}{2}$ to fourths; thus, $\frac{1}{2} = \frac{2}{4}$. Change $\frac{3}{4}$ to fourths.

10. Change $\frac{1}{2}$ to eighths; $\frac{3}{4}$ to eighths.

11. How many units are there in $\frac{4}{4}$? $\frac{4}{4}$? $\frac{8}{8}$?

12. $\frac{1}{2}$ of the first circle $+ \frac{1}{2}$ of the first circle = how many times the first circle? Then $\frac{1}{2} + \frac{1}{2} =$ how many?

13. $\frac{1}{4}$ of the second circle $+ \frac{1}{4}$ of the second circle = what part of the second circle? Then $\frac{1}{4} + \frac{1}{4} =$ how many?

14. $\frac{2}{4}$ of the second circle $+ \frac{2}{4}$ of the second circle = how many times the second circle? Then $\frac{2}{4} + \frac{2}{4} =$ how many?

15. $\frac{2}{8}$ of the third circle = what part of the second circle?

16. $\frac{4}{8}$ of the third circle = $\frac{1}{2}$ of the same circle.

17. $\frac{1}{4} = \frac{?}{8}$; $\frac{1}{2} = \frac{?}{8} = \frac{?}{4}$; $\frac{4}{8} = \frac{?}{2} = \frac{?}{4}$.

18. $\frac{1}{2}$ of an orange $= \frac{?}{4} = \frac{?}{8}$ of the same orange.

19. Four boys each have $\$\frac{1}{4}$. How many dollars have they?

20. $\$\frac{1}{4} + \$\frac{1}{4} = \$\frac{2}{4}$; $\$\frac{1}{2} + \$\frac{1}{4} = \$\frac{3}{4}$; $\frac{1}{8}$ day $+ \frac{1}{2}$ day $= \frac{?}{8}$ day.

21. $\$\frac{3}{4} + \$\frac{3}{4} = \$\frac{6}{4}$; $\$\frac{6}{4} =$ how many dollars?

22. $\frac{6}{4}$ days $= —$ days; $\frac{6}{4}$ days = 1 day and $\frac{?}{4}$ days.

23. $\$\frac{1}{2} + \$\frac{1}{2} + \$\frac{1}{2} + \$\frac{1}{2} = \$\frac{?}{2}$.

24. Draw, in order of their size, $\frac{1}{2}$ of the first circle; $\frac{3}{4}$ of the second circle, and $\frac{8}{8}$ of the third circle.

A **unit** is any one thing; as, 1, 1$\not\in$, 1 boy, 1 horse.

A **fraction** is one or more of the equal parts of a unit.

The **denominator** of a fraction, which is written below the line, shows into how many equal parts the unit is divided.

The **numerator** of a fraction, which is written above the line, shows how many equal parts of the fraction are taken.

The **terms** of a fraction are the numerator and the denominator.

25. Write in figures one third; one fourth; five eighths. How many figures are needed to express a common fraction? In the fraction $\frac{3}{4}$, what does the 4 show? the 3?

26. Read $\frac{1}{2}$; $\frac{1}{3}$; $\frac{2}{3}$; $\frac{3}{4}$; $\frac{3}{8}$; $\frac{5}{8}$; $\frac{7}{8}$.

27. Write seven eighths; thirteen sixteenths.

One half is written $\frac{1}{2}$; *one fourth,* $\frac{1}{4}$; *three fourths,* $\frac{3}{4}$; *three eighths,* $\frac{3}{8}$, etc.

Fractions are said to be **equivalent** when they have the same value. Thus, $\frac{1}{2}$, $\frac{2}{4}$, and $\frac{4}{8}$ are equivalent fractions.

A number made up of a whole number and a fraction is a **mixed number**; as, $4\frac{1}{2}$, $3\frac{7}{8}$, $5\frac{1}{4}$.

Written Work

1. Add $\frac{1}{2}$ and $\frac{3}{8}$.

$\frac{1}{2} = \frac{4}{8}$
$\frac{3}{8} = \frac{3}{8}$
$\frac{7}{8}$, sum.

Change $\frac{1}{2}$ to eighths. The sum of $\frac{4}{8}$ and $\frac{3}{8}$ is $\frac{7}{8}$.

2. Add $2\frac{1}{2}$ and $1\frac{3}{8}$.

$2\frac{1}{2}$
$1\frac{3}{8}$
$3\frac{7}{8}$, sum.

First add the fractions as above. Then add the whole numbers. The sum is $3\frac{7}{8}$.

3. From $\frac{1}{2}$ subtract $\frac{3}{8}$.

$\frac{1}{2} = \frac{4}{8}$
$\frac{3}{8} = \frac{3}{8}$
$\frac{1}{8}$, difference.

Change $\frac{1}{2}$ to eighths. The difference between $\frac{4}{8}$ and $\frac{3}{8}$ is $\frac{1}{8}$.

4. From $12\frac{1}{2}$ subtract $2\frac{3}{8}$.

$12\frac{1}{2} = 12\frac{4}{8}$
$2\frac{3}{8} = \ \ 2\frac{3}{8}$
$10\frac{1}{8}$, difference.

The difference between the fractions is $\frac{1}{8}$. The difference between the whole numbers is 10. The difference is therefore $10\frac{1}{8}$.

5. $\frac{1}{2}+\frac{1}{4}=?$ $\frac{1}{4}+\frac{1}{8}=?$ $\frac{1}{2}+\frac{1}{4}+\frac{1}{8}=?$

6. $\frac{1}{2}+\frac{1}{2}=\frac{2}{2}=1.$ $\frac{1}{2}+\frac{3}{4}=\frac{5}{4}=1\frac{1}{4}.$ $\frac{3}{4}+\frac{2}{4}+\frac{3}{4}=\frac{8}{4}=?$

7. $\frac{1}{4}+\frac{3}{8}=?$ $\frac{3}{8}+\frac{5}{4}=?$ $\frac{1}{2}+\frac{3}{8}+\frac{5}{4}=?$

8. $\frac{1}{2}-\frac{1}{8}=?$ $\frac{1}{2}-\frac{1}{4}=?$ $\frac{3}{4}-\frac{1}{8}=?$ $\frac{3}{4}-\frac{3}{8}=?$

Add:

9. $3\frac{1}{2}$	**10.** $7\frac{1}{4}$	**11.** $7\frac{1}{4}$	**12.** $9\frac{1}{8}$	**13.** $5\frac{1}{4}$.
$3\frac{1}{8}$	$8\frac{1}{2}$	$9\frac{3}{8}$	$6\frac{1}{2}$	$12\frac{1}{4}$

Subtract; then add:

14. $10\frac{3}{8}$	**17.** $12\frac{1}{2}$	**20.** $27\frac{3}{4}$	**23.** $19\frac{1}{4}$	**26.** $36\frac{1}{2}$
$5\frac{1}{4}$	$6\frac{1}{8}$	$8\frac{3}{8}$	$6\frac{1}{8}$	$16\frac{3}{8}$

15. $62\frac{5}{8}$	**18.** $63\frac{3}{4}$	**21.** $26\frac{3}{4}$	**24.** $18\frac{3}{4}$	**27.** $40\frac{3}{4}$
$31\frac{3}{8}$	$39\frac{1}{4}$	$24\frac{3}{8}$	$9\frac{3}{8}$	$20\frac{1}{8}$

16. $5\frac{1}{2}$	**19.** $9\frac{3}{4}$	**22.** $9\frac{1}{2}$	**25.** $5\frac{1}{4}$	**28.** $8\frac{3}{4}$
$3\frac{1}{4}$	$8\frac{1}{2}$	$3\frac{1}{2}$	4	$2\frac{1}{2}$

Add:

29. $3\frac{1}{2}$	**30.** $67\frac{1}{4}$	**31.** $65\frac{1}{2}$	**32.** $25\frac{3}{8}$	**33.** $56\frac{1}{4}$
$4\frac{1}{4}$	$6\frac{1}{2}$	$7\frac{3}{8}$	$8\frac{1}{8}$	$6\frac{1}{8}$

34. Ruth bought $\frac{1}{2}$ yd. of red ribbon and $\frac{1}{4}$ yd. of blue ribbon. How much did she buy all together?

35. Marian had $1\frac{1}{8}$ yd. of linen and used $\frac{1}{2}$ yd. How much had she left?

36. John had a string $\frac{3}{4}$ yd. long from which he cut $\frac{5}{8}$ yd. How long was the part remaining?

37. If I have $2\frac{1}{4}$ lb. of pepper, and use $\frac{3}{8}$ lb., how much have I left?

Halves, sixths, and twelfths.

Oral Work

1 unit = $\frac{2}{2}$.　　1 unit = $\frac{6}{6}$.　　1 unit = $\frac{12}{12}$.

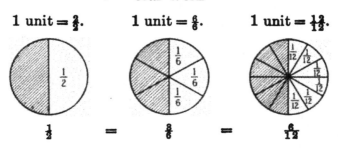

$\frac{1}{2}$　=　$\frac{3}{6}$　=　$\frac{6}{12}$

1. What is $\frac{1}{2}$ of 12? $\frac{1}{2}$ of 24? $\frac{1}{2}$ of 36?

2. What is $\frac{1}{6}$ of 12? $\frac{1}{6}$ of 24? $\frac{1}{6}$ of 36? $\frac{5}{6}$ of 36?

3. What is $\frac{1}{12}$ of 12? $\frac{1}{12}$ of 24? $\frac{1}{12}$ of 36? $\frac{5}{12}$ of 36?

4. $\frac{1}{2} = \frac{?}{12}$; $\frac{6}{12} = \frac{?}{2}$; $\frac{3}{6} = \frac{?}{2}$; $\frac{1}{2} = \frac{?}{6}$; $\frac{6}{12} = \frac{?}{6}$; $\frac{1}{6} + \frac{6}{6} + \frac{2}{6} = $ — ones.

5. $\frac{2}{6}$ of a circle equals $\frac{?}{12}$ of the same circle.

Draw and cut out of paper circles or oblongs to show these relations.

6. Which is larger, $\frac{1}{2}$ of an apple or $\frac{5}{6}$ of the same apple? how much larger?

7. $\frac{1}{2} = \frac{?}{12}$; $\frac{1}{6} = \frac{?}{12}$; $\frac{5}{6} = \frac{?}{12}$; $\frac{24}{12} = $ — ones.

8. Show by paper-cutting that $\frac{1}{6}$ of a circle equals $\frac{2}{12}$ of the same circle.

9. $\frac{2}{2}$ of a circle equals $\frac{?}{12}$ of the same circle.

10. Show by oblongs that $\frac{1}{2}$ of any oblong equals $\frac{3}{6}$ of the same oblong or $\frac{6}{12}$ of the same oblong.

11. Show by paper-cutting that $\frac{1}{2}$ of any unit is less in size than $\frac{5}{6}$ of the same unit.

12. How many halves are there in an orange? how many sixths? how many twelfths?

13. Show that if $\frac{4}{6}$ of an apple is taken away, $\frac{1}{6}$, or $\frac{2}{12}$, remains.

14. Arrange the following in order of their size. Which are less than 1? equal to 1? greater than 1?

$$\frac{1}{2}, \frac{1}{6}, \frac{7}{12}, \frac{3}{2}, \frac{6}{6}, \frac{13}{12}, \frac{5}{2}, \frac{12}{6}, \frac{4}{2}, \frac{24}{12}.$$

Written Work

1. Add $\frac{5}{6}$ and $\frac{6}{12}$.

$\frac{5}{6} = \frac{10}{12}$

$\frac{6}{12} = \frac{6}{12}$

$\frac{16}{12} = 1\frac{8}{12}$, or $1\frac{1}{3}$, sum.

Change the fraction $\frac{5}{6}$ to twelfths. The sum of $\frac{10}{12}$ and $\frac{6}{12}$ is $\frac{16}{12}$, or $1\frac{4}{12}$, or $1\frac{1}{3}$.

2. Add $3\frac{5}{6}$ and $2\frac{6}{12}$.

$3\frac{5}{6} = 3\frac{10}{12}$

$2\frac{6}{12} = 2\frac{6}{12}$

$6\frac{1}{3}$, sum.

First add the fractions as above. $\frac{16}{12}$ $= 1\frac{1}{3}$. Write $\frac{1}{3}$ and carry the 1. The sum of the whole numbers is $1 + 2 + 3$, or 6. The sum is therefore $6\frac{1}{3}$.

3. From $\frac{5}{6}$ subtract $\frac{6}{12}$.

$\frac{5}{6} = \frac{10}{12}$

$\frac{6}{12} = \frac{6}{12}$

$\frac{4}{12}$, difference.

Change $\frac{5}{6}$ to twelfths. The difference between $\frac{10}{12}$ and $\frac{6}{12}$ is $\frac{4}{12}$.

4. From $3\frac{5}{6}$ subtract $2\frac{6}{12}$.

$3\frac{5}{6} = 3\frac{10}{12}$

$2\frac{6}{12} = 2\frac{6}{12}$

$1\frac{4}{12}$, difference.

The difference between the fractions is $\frac{4}{12}$. The difference between the whole numbers is 1. The sum is therefore $1\frac{4}{12}$.

Add:

5. $\frac{1}{2} + \frac{1}{6}$.   **6.** $\frac{1}{6} + \frac{6}{12}$.   **7.** $1\frac{1}{2} + 2\frac{1}{6}$.   **8.** $3\frac{7}{12} + 1\frac{1}{6}$.

Subtract:

9. $\frac{1}{2} - \frac{1}{6}$.   **10.** $\frac{5}{6} - \frac{6}{12}$.   **11.** $3\frac{7}{12} - 1\frac{1}{2}$.   **12.** $4\frac{5}{6} - 3\frac{1}{2}$.

First add; then subtract:

13. $19\frac{6}{12}$ **14.** $12\frac{11}{12}$ **15.** $16\frac{9}{12}$ **16.** $15\frac{7}{12}$ **17.** $12\frac{5}{6}$

$\ 6\frac{1}{6}$ $\ 6\frac{5}{6}$ $\ 5\frac{1}{2}$ $\ 10\frac{1}{2}$ $\ 10\frac{5}{12}$

18. $11\frac{1}{2}$ **19.** $6\frac{1}{2}$ **20.** $41\frac{5}{8}$ **21.** $16\frac{1}{2}$ **22.** $15\frac{7}{12}$

    $4\frac{1}{6}$ $5\frac{1}{12}$ $16\frac{7}{12}$ $12\frac{1}{6}$ $5\frac{1}{2}$

23. Julia bought $\frac{1}{12}$ doz. pearl buttons and $\frac{5}{6}$ doz. bone buttons. What part of a dozen did she buy? How many buttons?

24. James worked $3\frac{5}{6}$ hr. on Saturday and $1\frac{1}{2}$ hr. on Monday. How many hours did he work both days? How many more hours did he work on Saturday than on Monday?

25. Frank's marks for the month average $91\frac{1}{6}$, and John's average $89\frac{1}{2}$. Find the difference in their averages.

Thirds, sixths, and ninths.

Oral Work

1 unit $= \frac{3}{3}$.   1 unit $= \frac{6}{6}$.   1 unit $= \frac{9}{9}$.

$\frac{1}{3}$ $=$ $\frac{2}{6}$ $=$ $\frac{3}{9}$.

1. What is $\frac{1}{3}$ of 6? $\frac{1}{3}$ of 12? $\frac{1}{3}$ of 18? $\frac{2}{3}$ of 18?

2. What is $\frac{1}{6}$ of 6? $\frac{1}{6}$ of 12? $\frac{1}{6}$ of 18? $\frac{5}{6}$ of 18?

3. What is $\frac{1}{9}$ of 9? $\frac{1}{9}$ of 18? $\frac{1}{9}$ of 27? $\frac{4}{9}$ of 27?

4. Into how many parts is the first circle divided? the second circle? the third circle?

5. $\frac{1}{3}$ of a circle $= \frac{?}{6}$ of the circle $= \frac{?}{9}$ of the circle.

6. $\frac{2}{3}$ of a circle $= \frac{?}{6}$ of the circle $= \frac{?}{9}$ of the circle.

7. $\frac{3}{3}$ of a circle $= \frac{?}{6}$ of the circle $= \frac{?}{9}$ of the circle.

8. $\frac{1}{3} + \frac{2}{3} = \frac{?}{6}$; $\frac{1}{3} + \frac{2}{3} = \frac{?}{9}$; $\frac{1}{3} + \frac{2}{3} = \frac{?}{6}$.

9. $\frac{1}{3}$ of an hour $+ \frac{1}{3}$ of an hour $= \frac{?}{6}$ of an hour.

10. $\frac{2}{3}$ of a day $+ \frac{2}{3}$ of a day $= \frac{?}{6}$ of a day.

11. $\frac{1}{3} + \frac{1}{6} = -$; $\frac{1}{3} + \frac{1}{6} = -$; $\frac{1}{3} + \frac{1}{6} = -$; $\frac{1}{4} + \frac{1}{3} = -$.

12. Draw an oblong and show that $\frac{1}{3}$ of the oblong $= \frac{2}{6}$ of the oblong $= \frac{3}{9}$ of the oblong.

13. $\frac{1}{3} = \frac{?}{6}$; $\frac{2}{3} = \frac{?}{6}$; $\frac{2}{3} = \frac{?}{9}$; $\frac{2}{3} = \frac{?}{9}$.

14. Draw squares and show that $\frac{1}{3} = \frac{2}{6}$; that $\frac{2}{3} = \frac{4}{6}$; that $\frac{1}{3} = \frac{3}{9}$; that $\frac{2}{3} = \frac{6}{9}$.

15. How many thirds equal one unit? how many sixths? how many ninths?

16. $\frac{6}{6} =$ how many units? $\frac{12}{6} =$ how many units? $\frac{18}{6} =$ how many units? $\frac{9}{9} =$ how many units?

Fractional equivalents of the yard and the foot.

Oral Work

One foot.   *One yard.*

1. A foot is what part of a yard?

2. 2 feet are what part of a yard?

3. Into how many parts is the yard divided?

4. How many feet equal $\frac{1}{3}$ of a yard? $\frac{2}{3}$ of a yard? $\frac{3}{3}$ of a yard?

5. Measure a yard on the blackboard. Divide the yard into feet. Divide a foot into inches.

6. How many inches equal $\frac{1}{2}$ of a foot? $\frac{2}{3}$ of a foot? $\frac{3}{4}$ of a foot? $1\frac{1}{2}$ ft.? $2\frac{1}{2}$ ft.?

7. 6 in. are what part of a foot? of 2 ft.? of a yard?

8. 4 in. are what part of a foot? of 2 ft.?

9. $1\frac{1}{2}$ ft. $+ 1\frac{1}{2}$ ft. $= —$ ft.

10. $\frac{1}{2} = \frac{?}{4}$; $\frac{1}{4} + \frac{1}{4} = \frac{?}{4}$; $\frac{1}{2} = \frac{?}{6}$; $\frac{1}{2} + \frac{2}{6} = \frac{?}{6}$.

11. $1\frac{1}{4}$ ft. $+ \frac{1}{2}$ ft. $=$ how many feet? $1\frac{1}{4} + \frac{1}{2} = —$.

12. $\frac{1}{3}$ yd. $+ \frac{1}{2}$ yd. $+ \frac{1}{6}$ yd. $= —$ yd. $= —$ ft.

13. $2\frac{1}{3}$ ft. $+ 3\frac{2}{3}$ ft. $= —$ ft. $= —$ yd.

14. $\frac{3}{4}$ ft. $+ \frac{1}{4}$ ft. $= —$ ft.; $\frac{3}{4}$ ft. $+ \frac{3}{4}$ ft. $+ \frac{2}{4}$ ft. $= —$ ft.

15. $\frac{4}{2}$ ft. $= —$ ft.; $\frac{6}{3}$ ft. $= —$ ft.; $\frac{6}{4}$ ft. $= —$ ft.

Written Work

Add:

16. $2\frac{1}{2}$ in.	**17.** $5\frac{1}{2}$ yd.	**18.** $\frac{1}{6}$ ft.	**19.** $2\frac{1}{4}$ ft.
$3\frac{1}{2}$ in.	$8\frac{1}{4}$ yd.	$2\frac{1}{4}$ ft.	$8\frac{1}{6}$ ft.

Subtract:

20. $3\frac{3}{4}$ ft.	**21.** $7\frac{1}{3}$ yd.	**22.** $3\frac{1}{4}$ ft.	**23.** $20\frac{1}{4}$ ft.
$2\frac{2}{4}$ ft.	$6\frac{1}{6}$ yd.	$2\frac{1}{6}$ ft.	$15\frac{1}{6}$ ft.

Review — Written Work

Add:

1. $11\frac{1}{8}$	**4.** $3\frac{1}{8}$	**7.** $29\frac{1}{2}$	**10.** $97\frac{3}{8}$	**13.** $10\frac{5}{6}$
$5\frac{1}{8}$	$2\frac{5}{8}$	$6\frac{1}{4}$	$\frac{2}{4}$	$\frac{1}{3}$
$2\frac{1}{2}$	$12\frac{1}{6}$	$18\frac{1}{2}$	$\frac{1}{8}$	$4\frac{1}{6}$

2. $6\frac{2}{8}$	**5.** $11\frac{1}{8}$	**8.** $17\frac{3}{4}$	**11.** $8\frac{1}{8}$	**14.** $80\frac{9}{12}$
$4\frac{1}{2}$	$5\frac{6}{8}$	$5\frac{3}{8}$	$3\frac{3}{4}$	$4\frac{1}{2}$
$2\frac{1}{2}$	$3\frac{1}{8}$	$4\frac{1}{4}$	$\frac{1}{2}$	$70\frac{1}{6}$

3. $6\frac{1}{2}$	**6.** $40\frac{1}{4}$	**9.** $90\frac{1}{6}$	**12.** $7\frac{1}{2}$	**15.** $24\frac{1}{8}$
$12\frac{3}{4}$	$16\frac{1}{2}$	$2\frac{1}{3}$	$4\frac{7}{8}$	$5\frac{2}{9}$
$2\frac{1}{4}$	$5\frac{1}{3}$	$7\frac{3}{8}$	$2\frac{1}{2}$	$3\frac{7}{9}$

Subtract:

16. $25\frac{2}{3}$	**21.** $87\frac{2}{4}$	**26.** $80\frac{11}{12}$	**31.** $57\frac{7}{9}$
$17\frac{1}{6}$	$14\frac{1}{2}$	$16\frac{5}{6}$	$16\frac{2}{3}$

17. $16\frac{7}{12}$	**22.** $25\frac{1}{4}$	**27.** $45\frac{5}{6}$	**32.** $14\frac{7}{8}$
$4\frac{1}{2}$	$16\frac{1}{8}$	$16\frac{1}{2}$	$10\frac{1}{2}$

18. $19\frac{1}{2}$	**23.** $37\frac{7}{8}$	**28.** $17\frac{3}{4}$	**33.** $13\frac{3}{8}$
$16\frac{1}{6}$	$16\frac{3}{4}$	$16\frac{3}{8}$	$4\frac{2}{3}$

19. $8\frac{7}{9}$	**24.** $15\frac{5}{6}$	**29.** $49\frac{3}{4}$	**34.** $6\frac{3}{4}$
$5\frac{3}{8}$	$13\frac{1}{2}$	20	$4\frac{1}{2}$

20. $13\frac{7}{8}$	**25.** $20\frac{3}{8}$	**30.** $16\frac{7}{8}$	**35.** $27\frac{5}{8}$
10	$16\frac{1}{6}$	$12\frac{1}{4}$	$1\frac{1}{2}$

36. $3\frac{1}{2}+2\frac{1}{4}=?$ 　 $3\frac{1}{6}+2\frac{2}{3}-\frac{1}{9}=?$ 　 $3\frac{1}{2}+2\frac{2}{8}-\frac{1}{4}=?$

Fifths, tenths, and fifteenths.

Oral Work

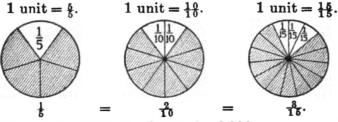

1 unit $=\frac{5}{5}$. 　　　 1 unit $=\frac{10}{10}$. 　　　 1 unit $=\frac{15}{15}$.

$\frac{1}{5}$ 　　 $=$ 　　 $\frac{2}{10}$ 　　 $=$ 　　 $\frac{8}{15}$.

1. What is $\frac{2}{5}$ of 25? $\frac{3}{10}$ of 20? $\frac{2}{15}$ of 30?

2. Into how many parts is the first circle divided? the second circle? the third circle?

3. Observe the parts of each circle that are not shaded. $\frac{1}{5}=\frac{?}{10}=\frac{?}{15}$.

4. Then $\frac{2}{5} = \frac{?}{10} = \frac{?}{15}$; $\frac{3}{5} = \frac{?}{10} = \frac{?}{15}$; $\frac{4}{5} = \frac{?}{10} = \frac{?}{15}$.

5. Each of five boys had $\frac{1}{5}$ of a dollar. How many dollars did they have?

6. $\frac{2}{5}$ of a circle $+ \frac{3}{5}$ of the same circle $= \frac{?}{5}$ of the circle. Then $\frac{2}{5} + \frac{3}{5} = \frac{?}{5}$; $\frac{4}{5} - \frac{2}{5} = \frac{?}{5}$; $\frac{4}{5} - \frac{3}{5} = \frac{?}{5}$.

7. How many parts of a unit are there in $\frac{1}{5} + \frac{1}{5} + \frac{1}{5}$? in $\frac{2}{5} + \frac{2}{5}$? in $\frac{2}{10} + \frac{5}{10} + \frac{3}{10}$? in $\frac{3}{15} + \frac{8}{15} + \frac{3}{15}$? in $\frac{8}{15} - \frac{3}{15}$?

8. $\frac{1}{5} + \frac{1}{10} = \frac{?}{10}$; $\frac{1}{5} + \frac{1}{15} = \frac{?}{15}$; $\frac{1}{5} + \frac{1}{5} = \frac{?}{15}$.

9. $\frac{5}{5} + \frac{5}{5} = \frac{?}{5}$. Then $\frac{10}{5} =$ how many units?

10. $\frac{8}{5} =$ how many units and $\frac{?}{5}$ remaining?

11. $\frac{12}{5} =$ how many units and $\frac{?}{5}$ remaining?

12. $\frac{15}{10} =$ how many units and $\frac{5}{10}$ remaining?

13. Change to units and parts of units: $\frac{8}{5}$, $\frac{8}{5}$, $\frac{12}{5}$, $\frac{12}{10}$.

Written Work

Add :

1. $2\frac{1}{5}$
    $3\frac{1}{10}$

2. $11\frac{1}{5}$
    $5\frac{3}{10}$

3. $25\frac{3}{5}$ mi.
    $4\frac{1}{10}$ mi.

4. $5\frac{8}{15}$
    $4\frac{1}{5}$

5. $24\frac{1}{5}$ mi.
    $4\frac{1}{15}$ mi.
    $5\frac{1}{5}$ mi.

6. $23\frac{1}{5}$
    $7\frac{3}{10}$
    $31\frac{1}{15}$

7. $50\frac{1}{5}$
    $35\frac{8}{15}$
    $4\frac{4}{5}$

8. $24\frac{1}{5}$ da.
    $3\frac{3}{10}$ da.
    $4\frac{1}{5}$ da.

Subtract :

9. $3\frac{1}{5}$
    $2\frac{1}{10}$

10. $25\frac{1}{5}$ hr.
    $13\frac{1}{15}$ hr.

11. $14\frac{2}{5}$
    $10\frac{1}{10}$

12. $78\frac{2}{5}$ min.
    $42\frac{1}{15}$ min.

13. James works at a store $1\frac{1}{10}$ hours each day ; John works $2\frac{1}{5}$ hours ; and Frank $1\frac{3}{10}$ hours. How many hours do they work all together?

14. Jane walks to school in $\frac{1}{3}$ hr. and Martha in $\frac{1}{4}$ hr. How much longer does it take Jane than Martha? how many minutes longer?

15. Nine eggs are what part of a dozen? 9 inches is what part of a foot?

16. May earned $\$\frac{3}{10}$; Henry earned $\$\frac{2}{5}$; and Rose earned $\$\frac{1}{5}$. What part of a dollar did they earn together? how many cents?

Add:

17. $3\frac{2}{5}$ $2\frac{1}{15}$	**19.** $2\frac{1}{5}$ $3\frac{7}{10}$	**21.** $1\frac{7}{10}$ $2\frac{1}{5}$	**23.** $3\frac{7}{15}$ $2\frac{4}{5}$
18. $3\frac{4}{15}$ $3\frac{2}{5}$	**20.** $5\frac{2}{5}$ $7\frac{3}{10}$	**22.** $3\frac{1}{15}$ $4\frac{1}{5}$	**24.** $3\frac{9}{10}$ $2\frac{1}{5}$

Subtract:

25. $5\frac{4}{5}$ $2\frac{1}{10}$	**26.** $3\frac{3}{10}$ $1\frac{1}{5}$	**27.** $2\frac{9}{10}$ $1\frac{3}{5}$	**28.** $3\frac{7}{15}$ $2\frac{1}{5}$

Halves, fourths, and sixteenths.

Oral Work

1 unit $= \frac{2}{2}$.　　1 unit $= \frac{4}{4}$.　　1 unit $= \frac{16}{16}$.

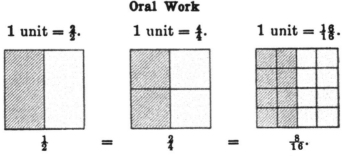

$\frac{1}{2}$　$=$　$\frac{2}{4}$　$=$　$\frac{8}{16}$.

1. How do these three units compare in size?

2. Into how many parts is the first square divided? the second square? the third square?

3. $\frac{1}{2}$ of the first square = — fourths of the second square = — sixteenths of the third square.

4. $\frac{2}{2}$ = — unit; $\frac{4}{4}$ = — unit; $\frac{8}{8}$ = — unit; $\frac{16}{16}$ = — unit.

5. $\frac{1}{2} = \frac{?}{4} = \frac{?}{8} = \frac{?}{16}$.   **6.** $\frac{1}{4} = \frac{?}{8} = \frac{?}{16}$; $\frac{1}{8} = \frac{?}{16}$.

7. $\frac{2}{4} = \frac{?}{8} = \frac{?}{16}$; $\frac{3}{4} = \frac{?}{8} = \frac{?}{16}$.

8. $\frac{2}{8} = \frac{?}{16} = \frac{?}{4}$; $\frac{4}{8} = \frac{?}{16} = \frac{?}{2}$; $\frac{6}{8} = \frac{?}{16} = \frac{?}{4}$.

9. $\frac{1}{2} + \frac{1}{4} + \frac{1}{8} = \frac{?}{8}$; $\frac{1}{4} + \frac{3}{8} + \frac{3}{16} = \frac{?}{16}$; $\frac{9}{16} - \frac{1}{4} = \frac{?}{16}$.

Written Work

Add:

10. $3\frac{1}{16}$ ft. **11.** $\$16\frac{1}{2}$ **12.** $12\frac{1}{4}$ **13.** $10\frac{3}{16}$
$5\frac{1}{4}$ ft. $20\frac{3}{4}$ $14\frac{1}{4}$ $5\frac{1}{2}$
$2\frac{1}{2}$ ft. $17\frac{1}{16}$ $10\frac{5}{16}$ $6\frac{3}{4}$

Subtract:

14. $\$12\frac{1}{2}$ **15.** $23\frac{3}{4}$ yd.   **16.** $13\frac{1}{2}$ mi.   **17.** $68\frac{3}{4}$
$8\frac{1}{16}$ $18\frac{3}{16}$ yd. $9\frac{3}{16}$ mi. $52\frac{5}{16}$

18. A flower bed is $4\frac{1}{2}$ ft. long and $3\frac{1}{4}$ ft. wide. Find the distance around it.

19. The school ground is in the form of a square, $13\frac{1}{2}$ rd. on a side. Find the distance in rods around it.

20. If I cut $\frac{1}{2}$ yd. of ribbon from $3\frac{3}{4}$ yd., how much ribbon have I left?

21. A stick $5\frac{3}{16}$ in. long is broken in two pieces. One piece is $3\frac{1}{8}$ in. long. How long is the other piece?

22. I pay $\$3\frac{1}{2}$ for a chair, and $\$4\frac{3}{4}$ for a table. What is my total bill?

Thirds, sixths, and twelfths.

Oral Work

$$1 \text{ unit} = \tfrac{3}{3} = \tfrac{6}{6} = \tfrac{12}{12}.$$

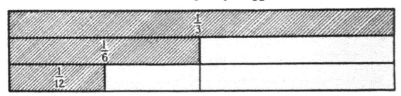

1. Show by the diagram how many sixths are equal to one third; how many twelfths are equal to one sixth.

2. $\frac{1}{3} = \frac{2}{6} = \frac{?}{12}$; $\frac{2}{3} = \frac{?}{6} = \frac{?}{12}$.

3. Cut a paper oblong and fold it into thirds; into sixths; into twelfths.

4. $\frac{1}{3}$ hour $= \frac{?}{12}$ of an hour; $\frac{1}{6}$ hour $= \frac{?}{12}$ of an hour.

5. Change $\frac{1}{3}$ and $\frac{1}{6}$ to twelfths. Change $\frac{4}{12}$ to thirds.

6. Cut three oblongs of the same size out of paper. Fold the first oblong into thirds; the second oblong into sixths; and the third oblong into twelfths.

7. Change $\frac{2}{3}$ and $\frac{5}{6}$ to twelfths. Change $\frac{6}{12}$ to sixths.

8. $\frac{4}{12}$ of an oblong equal how many thirds of the oblong?

9. $\frac{12}{12}$ equal how many units; $\frac{12}{6}$ equal how many units?

Add:

10. $\frac{1}{3}$, $\frac{1}{6}$, $\frac{1}{12}$	**15.** $3\frac{2}{3}$, $4\frac{1}{6}$	**20.** $3\frac{1}{3}$, $3\frac{1}{6}$, $2\frac{1}{12}$
11. $\frac{2}{3}$, $\frac{1}{6}$, $\frac{1}{12}$	**16.** $2\frac{2}{3}$, $3\frac{1}{12}$	**21.** $2\frac{1}{3}$, $2\frac{1}{3}$, $2\frac{5}{6}$
12. $\frac{2}{3}$, $\frac{1}{12}$, $\frac{5}{12}$	**17.** $4\frac{1}{6}$, $5\frac{1}{12}$	**22.** $2\frac{1}{12}$, $2\frac{1}{3}$, $\frac{1}{3}$
13. $\frac{1}{3}$, $\frac{5}{6}$, $\frac{1}{12}$	**18.** $5\frac{1}{12}$, $6\frac{2}{3}$	**23.** $3\frac{1}{3}$, $2\frac{7}{12}$, $5\frac{5}{12}$
14. $\frac{1}{3}$, $\frac{7}{12}$, $\frac{5}{6}$	**19.** $7\frac{1}{6}$, $8\frac{2}{3}$	**24.** $3\frac{1}{3}$, $4\frac{1}{6}$, $5\frac{1}{12}$

Written Work

1. Change to units and parts of units: $\frac{8}{4}$, $\frac{9}{8}$, $\frac{10}{8}$, $\frac{17}{6}$, $\frac{15}{8}$, $\frac{8}{5}$, $\frac{18}{10}$, $\frac{19}{16}$, $\frac{20}{18}$.

Add:

2. $29\frac{1}{2}$	4. $7\frac{1}{2}$	6. $39\frac{1}{8}$	8. $5\frac{1}{4}$
$32\frac{1}{4}$	$10\frac{3}{4}$	$42\frac{5}{12}$	$6\frac{5}{8}$
$45\frac{5}{12}$	$25\frac{7}{8}$	$23\frac{3}{4}$	$12\frac{2}{3}$

3. $27\frac{3}{4}$ ft.	5. $15\frac{1}{4}$ mi.	7. $14\frac{1}{8}$ bu.	9. $12\frac{1}{4}$ da.
$45\frac{7}{8}$ ft.	$29\frac{4}{15}$ mi.	$19\frac{1}{4}$ bu.	$10\frac{1}{4}$ da.
$25\frac{1}{4}$ ft.	$31\frac{1}{10}$ mi.	$16\frac{5}{8}$ bu.	$13\frac{1}{12}$ da.

Subtract:

10. $8\frac{1}{8}$	12. $17\frac{1}{8}$	14. $14\frac{3}{4}$	16. $32\frac{5}{12}$
$6\frac{1}{12}$	$15\frac{1}{12}$	$12\frac{1}{12}$	$30\frac{1}{2}$

11. $7\frac{3}{4}$ in.	13. $9\frac{3}{4}$ bu.	15. $10\frac{1}{4}$ lb.	17. $15\frac{11}{12}$
$5\frac{5}{12}$ in.	$7\frac{1}{12}$ bu.	$6\frac{1}{12}$ lb.	$5\frac{3}{8}$

18. The floor of a room is $13\frac{5}{6}$ ft. long and $12\frac{1}{4}$ ft. wide. Find the distance around the room. How much longer is the room than it is wide?

19. Mary's hair ribbon is $15\frac{1}{8}$ in. long and $1\frac{1}{8}$ in. wide. Find the difference between the length and the width of the ribbon.

20. James worked after school $1\frac{5}{6}$ hr. Monday, $1\frac{1}{8}$ hr. Tuesday, $1\frac{8}{12}$ hr. Wednesday, 2 hr. Thursday, and $1\frac{5}{8}$ hr. Friday. On Saturday he worked $10\frac{1}{4}$ hr. How many hours did he work during the week?

21. Morgan lives $1\frac{3}{8}$ mi. from school, and Frank $1\frac{1}{4}$ mi. in the same direction. In going to school, Morgan walks how much farther than Frank? Find the distance both walk in going to and coming from school in a day.

Fourths, eighths, and sixteenths.

Oral Work

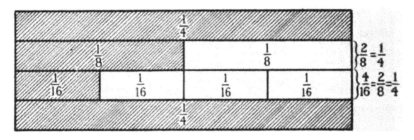

1. Show by the diagram how many eighths are equal to one fourth; how many sixteenths are equal to one eighth; how many sixteenths are equal to one fourth.

2. $\frac{2}{8} = \frac{?}{16}$; $\frac{2}{8} = \frac{?}{4}$; $\frac{4}{16} = \frac{?}{4}$; $\frac{4}{16} = \frac{?}{8}$?

3. Any unit can be divided into how many halves? 3ds? 4ths? 5ths? 6ths? 7ths? 8ths? 10ths? 16ths? 24ths? etc.?

4. Add $\frac{1}{4}$ and $\frac{1}{8}$; $\frac{1}{8}$ and $\frac{1}{16}$. From $\frac{7}{8}$ take $\frac{1}{4}$.

5. From $\frac{9}{16}$ take $\frac{1}{8}$; $\frac{1}{4}$; $\frac{1}{2}$; $\frac{3}{8}$.

6. $\frac{5}{4}$ means that a unit ($\frac{4}{4}$) and a part of a unit ($\frac{1}{4}$) have been added. What does $\frac{9}{8}$ mean? $\frac{17}{16}$?

Written Work

Add:

1. $18\frac{1}{4}$ in.	**2.** $15\frac{1}{16}$ bu.	**3.** $19\frac{1}{4}$	**4.** $40\frac{7}{8}$
$20\frac{1}{8}$ in.	$27\frac{1}{8}$ bu.	$32\frac{3}{16}$	$30\frac{1}{16}$
$39\frac{1}{16}$ in.	$41\frac{3}{4}$ bu.	$20\frac{1}{8}$	$18\frac{1}{4}$

Subtract:

5. $9\frac{1}{8}$ yd.	**6.** $14\frac{1}{4}$ da.	**7.** $28\frac{11}{16}$	**8.** $39\frac{7}{16}$
$7\frac{1}{16}$ yd.	$7\frac{1}{8}$ da.	$13\frac{3}{8}$	$8\frac{1}{4}$

REDUCTION OF FRACTIONS

1. Notice in the diagram on p. 42 that $\frac{1}{4} = \frac{4}{16}$. By what number are both numerator and denominator of $\frac{1}{4}$ *multiplied* to change it to $\frac{4}{16}$? Is there any difference in *value* between $\frac{1}{4}$ and $\frac{4}{16}$? Notice that the terms in $\frac{4}{16}$ are larger or *higher* than in $\frac{1}{4}$. The change of $\frac{1}{4}$ to the equal fraction $\frac{4}{16}$ is called **changing or reducing $\frac{1}{4}$ to higher terms.**

2. By what number must both terms of $\frac{4}{16}$ be *divided* to change $\frac{4}{16}$ to $\frac{1}{4}$? Is there any difference in *value* between $\frac{4}{16}$ and $\frac{1}{4}$? Which fraction has the **lower terms**? The change of $\frac{4}{16}$ to $\frac{1}{4}$ is called **reducing $\frac{4}{16}$ to lower terms.**

3. Notice that in the diagram $\frac{4}{16} = \frac{2}{8} = \frac{1}{4}$. When $\frac{4}{16}$ is changed to $\frac{2}{8}$ it is reduced to *lower* terms but not to its *lowest* terms, since $\frac{2}{8}$ can be changed to still lower terms, $\frac{1}{4}$. Can $\frac{1}{4}$ be reduced to still lower terms? The change of $\frac{4}{16}$ to $\frac{1}{4}$ is called **reducing $\frac{4}{16}$ to its lowest terms.**

4. By what number must both terms of $\frac{1}{2}$ be multiplied to change it to the equal fraction $\frac{8}{8}$? By what number must both terms of $\frac{8}{10}$ be divided to change it to the equal fraction $\frac{4}{5}$? Is $\frac{4}{5}$ in its lowest terms?

Multiplying or dividing both terms of a fraction by the same number does not change its value.

5. Read in lowest terms at sight :

$\frac{2}{4}$; $\frac{8}{12}$; $\frac{6}{9}$; $\frac{2}{4}$; $\frac{12}{16}$; $\frac{10}{16}$; $\frac{7}{9}$; $\frac{2}{6}$; $\frac{9}{12}$; $\frac{9}{15}$; $\frac{6}{8}$; $\frac{8}{10}$; $\frac{4}{6}$.

6. $\frac{8}{9}$; $\frac{6}{10}$; $\frac{7}{8}$; $\frac{4}{12}$; $\frac{8}{16}$; $\frac{6}{14}$; $\frac{7}{14}$; $\frac{4}{10}$; $\frac{9}{10}$; $\frac{12}{16}$; $\frac{4}{16}$; $\frac{10}{12}$.

7. Observe that the fractions we have been studying show that *any unit* may be divided into any number of equal parts and any part may be used as a *unit of measure*. Just as we say 10¢ means 10 × 1¢, the *unit of measure*, so $\frac{3}{4}$ means 3 × $\frac{1}{4}$, the *unit of measure*.

Oral Work

A fraction is changed to **higher** terms when it is changed to a fraction having a *larger* denominator.

Thus, $\frac{1}{2} = \frac{2}{4} = \frac{4}{8} = \frac{8}{16}$, etc.

A fraction is changed to **lower** terms when it is changed to a fraction having a *smaller* denominator.

Thus, $\frac{8}{16} = \frac{4}{8} = \frac{2}{4} = \frac{1}{2}$.

Give the equivalents called for and explain the steps in the work.   Thus, dividing both terms of $\frac{8}{12}$ by 4 gives $\frac{2}{3}$; multiplying both terms of $\frac{2}{3}$ by 4 gives $\frac{8}{12}$.

1. $\frac{1}{4} = \frac{?}{8} = \frac{?}{16} = \frac{?}{24}$.   2. $\frac{2}{3} = \frac{?}{6} = \frac{?}{9} = \frac{?}{12} = \frac{?}{18}$.

3. $\frac{3}{4} = \frac{?}{8} = \frac{?}{12} = \frac{?}{16} = \frac{?}{20} = \frac{?}{24} = \frac{?}{28} = \frac{?}{32}$.

Change:

4. $\frac{2}{9}$ to thirds   9. $\frac{6}{16}$ to eighths

5. $\frac{4}{12}$ to thirds   10. $\frac{8}{16}$ to thirds

6. $\frac{5}{10}$ to halves   11. $\frac{4}{10}$ to fifths

7. $\frac{3}{18}$ to sixths   12. $\frac{6}{12}$ to halves

8. $\frac{4}{16}$ to fourths   13. $\frac{8}{12}$ to sixths

14. Change $\frac{1}{2}$, $\frac{1}{4}$, $\frac{2}{3}$, $\frac{7}{8}$, $\frac{5}{6}$, $\frac{7}{12}$, $\frac{5}{8}$, $\frac{5}{12}$, each to twenty-fourths.

15. Change $\frac{10}{12}$, $\frac{16}{18}$, $\frac{16}{24}$, $\frac{10}{30}$, $\frac{12}{24}$, $\frac{24}{36}$, each to sixths.

Change:

16. $\frac{3}{6}$ to halves   22. $\frac{4}{8}$ to fourths

17. $\frac{8}{12}$ to fourths   23. $\frac{2}{16}$ to eighths

18. $\frac{2}{20}$ to tenths   24. $\frac{14}{20}$ to tenths

19. $\frac{2}{8}$ to fourths   25. $\frac{2}{30}$ to fifteenths

20. $\frac{5}{25}$ to fifths   26. $\frac{9}{48}$ to sixteenths

21. $\frac{4}{12}$ to sixths   27. $\frac{4}{8}$ to fourths

Oral and Written Work

1. Divide 10, 12, 24, 36, 38 and 50, each by 2. Divide other numbers ending in 2, 4, 6, 8, or 0 by 2.

A number is divisible by 2, if the ones' figure is 2, 4, 6, 8, or 0.

Change to lowest terms:

2. $\frac{8}{12}$, $\frac{6}{24}$, $\frac{20}{24}$, $\frac{24}{32}$, $\frac{22}{40}$, $\frac{28}{40}$, $\frac{16}{30}$, $\frac{17}{20}$, $\frac{9}{10}$.

3. Change $\frac{36}{40}$ to lowest terms. Thus, $\frac{36}{40} = \frac{18}{20} = \frac{9}{10}$. Or, if you notice that 4 divides both 36 and 40, you may divide these numbers by 4 and say at once $\frac{36}{40} = \frac{9}{10}$.

4. Divide 15, 25, 40, 125, 150, each by 5. What is the ones' figure in each dividend? Divide other numbers ending in 5, or 0, by 5.

A number is divisible by 5, if its ones' figure is 5, or 0.

Read in lowest terms:

5. $\frac{15}{20}$, $\frac{10}{40}$, $\frac{25}{40}$, $\frac{30}{35}$, $\frac{45}{60}$, $\frac{20}{40}$, $\frac{50}{150}$, $\frac{15}{60}$.

6. Change to lowest terms:

$\frac{24}{54}$, $\frac{15}{36}$, $\frac{11}{42}$, $\frac{55}{135}$, $\frac{27}{42}$, $\frac{105}{120}$, $\frac{36}{36}$, $\frac{120}{160}$, $\frac{125}{175}$, $\frac{72}{144}$, $\frac{48}{66}$, $\frac{75}{150}$.

A number is divisible by 3 if the sum of its digits, or figures, is divisible by 3.

A number is divisible by 9 if the sum of its digits is divisible by 9.

7. Memorize all the numbers to 63 that are divisible by 9; by 3.

Change to lowest terms:

	a	b	c	d	e	f
8.	$\frac{81}{144}$	$\frac{121}{144}$	$\frac{108}{132}$	$\frac{102}{120}$	$\frac{99}{132}$	$\frac{72}{108}$.
9.	$\frac{60}{144}$	$\frac{90}{135}$	$\frac{105}{150}$	$\frac{144}{288}$	$\frac{144}{156}$	$\frac{121}{132}$.
10.	$\frac{100}{225}$	$\frac{75}{115}$	$\frac{220}{400}$	$\frac{80}{128}$	$\frac{216}{264}$	$\frac{156}{168}$.

Changing improper fractions to whole or mixed numbers.

Oral Work

1. Name the whole numbers and the fractions in 5, $\frac{15}{2}$, $\frac{7}{4}$ ft., $\frac{64}{6}$ lb., $20, $\frac{25}{6}$ yd., 10 oz., $\frac{22}{7}$ in.

2. In changing $\frac{7}{4}$ to integers and fractions, think first how many fourths it takes to make one whole unit.   Then $\frac{7}{4}$ = how many whole units and $\frac{3}{4}$ remaining?

A fraction equal to or greater than a whole unit is an **improper fraction.**

3. Which of the following fractions are equal to 1? Which are greater than 1?   Which are less than 1?

$\frac{1}{2}$, $\frac{2}{2}$, $\frac{3}{2}$, $\frac{4}{2}$; $\frac{1}{4}$, $\frac{3}{4}$, $\frac{4}{4}$, $\frac{7}{4}$; $\frac{1}{5}$, $\frac{3}{5}$, $\frac{5}{5}$, $\frac{9}{5}$; $\frac{8}{8}$, $\frac{9}{8}$, $\frac{9}{8}$, $\frac{7}{8}$; $\frac{17}{16}$, $\frac{7}{16}$, $\frac{15}{16}$, $\frac{19}{16}$.

Change to mixed numbers or integers:

4. $\frac{45}{8}$	19. $\frac{86}{7}$	34. $\frac{75}{12}$
5. $\frac{25}{4}$	20. $\frac{65}{7}$ oz.	35. $\frac{75}{7}$ mi.
6. $\frac{37}{9}$	21. $\$\frac{75}{6}$	36. $\frac{93}{8}$ rd.
7. $\frac{46}{5}$	22. $\frac{99}{12}$ lb.	37. $\frac{64}{8}$ bu.
8. $\frac{29}{3}$	23. $\frac{143}{12}$ hr.	38. $\frac{110}{9}$ in.
9. $\frac{65}{8}$	24. $\frac{224}{11}$ min.	39. $\frac{115}{10}$ A.
10. $\frac{75}{9}$	25. $\frac{120}{15}$ da.	40. $\frac{65}{4}$ lb.
11. $\frac{65}{7}$	26. $\frac{49}{5}$ sec.	41. $\frac{125}{10}$ min.
12. $\frac{56}{8}$	27. $\frac{89}{12}$ hr.	42. $\frac{110}{2}$ oz.
13. $\frac{102}{8}$	28. $\frac{89}{16}$ min.	43. $\frac{80}{18}$ lb.
14. $\frac{79}{7}$	29. $\frac{87}{8}$ min.	44. $\frac{109}{12}$ da.
15. $\frac{45}{3}$	30. $\frac{100}{16}$ bu.	45. $\frac{53}{4}$ in.
16. $\frac{95}{16}$	31. $\frac{195}{8}$ oz.	46. $\frac{102}{16}$ lb.
17. $\frac{64}{16}$	32. $\frac{99}{12}$ in.	47. $\frac{87}{10}$ min.
18. $\frac{99}{12}$	33. $\frac{84}{8}$ mi.	48. $\frac{107}{8}$ bu.

Changing whole or mixed numbers to improper fractions.

Oral Work

1. Change 1, 2, 3, 4, 5, 6, 7, 8, 9, 10, each to the fractional unit 4ths; to 5ths; to 6ths; to 7ths. Thus, $1 = \frac{4}{4}$, $2 = \frac{8}{4}$, etc.

2. Change $1\frac{1}{2}$, $1\frac{1}{4}$, $2\frac{2}{5}$, $3\frac{5}{7}$, $2\frac{7}{8}$, $8\frac{8}{10}$, $9\frac{6}{11}$, $10\frac{5}{12}$, each to the fractional units indicated by the fraction of the mixed numbers. Thus, $1\frac{1}{2} = \frac{3}{2}$.

3. What kind of fractions are $\frac{3}{2}$, $\frac{7}{4}$, $\frac{18}{6}$?

Read as improper fractions at sight:

	a	b	c	d	e	f	g	h	i
4.	$4\frac{3}{4}$	$6\frac{7}{8}$	$2\frac{4}{5}$	$3\frac{2}{8}$	$9\frac{1}{10}$	$4\frac{7}{12}$	$6\frac{3}{8}$	$10\frac{5}{8}$	$9\frac{3}{4}$
5.	$4\frac{2}{3}$	$5\frac{3}{4}$	$5\frac{2}{6}$	$2\frac{4}{5}$	$4\frac{8}{11}$	$2\frac{1}{10}$	$2\frac{1}{8}$	$4\frac{6}{7}$	$5\frac{4}{5}$
6.	$4\frac{3}{8}$	$3\frac{7}{12}$	$9\frac{1}{4}$	$8\frac{5}{6}$	$7\frac{2}{9}$	$9\frac{3}{10}$	$7\frac{5}{12}$	$4\frac{3}{10}$	$6\frac{2}{3}$

Written Work

1. Change $11\frac{7}{15}$ to an improper fraction.

$$11\frac{7}{15}$$
$$\underline{\quad 15 \quad}$$
$$\overline{165}$$
$$\underline{\quad\; 7 \quad}$$
$$\overline{172}$$
$$\overline{15}$$

Since 1 unit $= \frac{15}{15}$, 11 units $= 11 \times \frac{15}{15} = \frac{165}{15}$; $\frac{165}{15} + \frac{7}{15} = \frac{172}{15}$.

Change to improper fractions:

2. $15\frac{2}{3}$	**8.** $51\frac{2}{11}$	**14.** $85\frac{5}{16}$	**20.** $42\frac{11}{25}$
3. $17\frac{3}{4}$	**9.** $60\frac{5}{12}$	**15.** $92\frac{2}{3}$	**21.** $71\frac{9}{15}$
4. $18\frac{5}{6}$	**10.** $45\frac{6}{13}$	**16.** $85\frac{7}{12}$	**22.** $83\frac{7}{8}$
5. $25\frac{5}{7}$	**11.** $56\frac{9}{14}$	**17.** $48\frac{19}{20}$	**23.** $65\frac{17}{20}$
6. $35\frac{7}{8}$	**12.** $75\frac{8}{15}$	**18.** $78\frac{3}{16}$	**24.** $28\frac{23}{25}$
7. $42\frac{7}{9}$	**13.** $40\frac{9}{10}$	**19.** $80\frac{11}{50}$	**25.** $47\frac{7}{8}$

26. $75\frac{18}{20}$	**30.** $82\frac{3}{4}$	**34.** $41\frac{11}{16}$	**38.** $64\frac{11}{14}$
27. $37\frac{4}{5}$	**31.** $50\frac{18}{20}$	**35.** $26\frac{7}{16}$	**39.** $98\frac{11}{60}$
28. $46\frac{7}{12}$	**32.** $25\frac{7}{12}$	**36.** $61\frac{8}{25}$	**40.** $54\frac{5}{8}$
29. $54\frac{7}{9}$	**33.** $34\frac{5}{8}$	**37.** $43\frac{11}{12}$	**41.** $37\frac{11}{16}$

Changing to similar fractions; least common denominator.

Oral Work

Fractions like $\frac{1}{6}$, $\frac{2}{6}$, and $\frac{4}{6}$, which have the same denominator, are said to have a **common denominator** (c. d.).

Similar fractions are fractions that have a common denominator.

1. Notice that $\frac{1}{2}$ and $\frac{1}{3}$ may be changed to similar fractions by changing them to fractions having the common denominator, 6, or 12, or 18, or 24, etc.

$$\text{Thus, } \frac{1}{2} = \frac{3}{6}; \quad \frac{1}{3} = \frac{2}{6}.$$
$$\text{Or, } \frac{1}{2} = \frac{6}{12}; \quad \frac{1}{3} = \frac{4}{12}.$$
$$\text{Or, } \frac{1}{2} = \frac{9}{18}; \quad \frac{1}{3} = \frac{6}{18}.$$

Is any denominator less than 6 a common denominator of $\frac{1}{2}$ and $\frac{1}{3}$? Then what is the **least common denominator** (l. c. d.) of $\frac{1}{2}$ and $\frac{1}{3}$?

2. Since 4 is the least number that contains 2 and 4, 4 is the least common denominator of $\frac{1}{2}$ and $\frac{1}{4}$. For the same reason, the least common denominator of $\frac{1}{2}$ and $\frac{1}{5}$ is 10; of $\frac{1}{4}$ and $\frac{1}{8}$ is 8; of $\frac{2}{3}$ and $\frac{2}{4}$ is 12; of $\frac{3}{5}$ and $\frac{1}{10}$ is 10, etc.

3. Name some common denominators of $\frac{1}{4}$ and $\frac{1}{6}$. Name their least common denominator.

4. What is the least common denominator of $\frac{1}{2}$, $\frac{1}{4}$, and $\frac{1}{8}$; of $\frac{1}{2}$, $\frac{1}{6}$, and $\frac{1}{12}$? What is the least common denominator of $\frac{1}{2}$, $\frac{1}{3}$, and $\frac{1}{9}$? (Notice that 6 is not contained in 9. Hence the least common denominator is 18.)

Fill in the following blanks:

NOTE. — When you can not find the l. c. d. by inspection, try multiples of the largest denominator until you find one in which each of the other denominators is exactly contained. Thus, in 3, 9, 15, try 15, 30, 45.

DENOMINATORS	LEAST COMMON DENOMINATOR	DENOMINATORS	LEAST COMMON DENOMINATOR
5. 3, 5, 10	——	**13.** 4, 3, 12	——
6. 2, 6, 8	——	**14.** 16, 3, 2	——
7. 3, 4, 2	——	**15.** 12, 5, 4	——
8. 3, 9, 6	——	**16.** 14, 2, 7	——
9. 2, 5, 3	——	**17.** 14, 2, 3	——
10. 3, 8, 4	——	**18.** 11, 12	——
11. 3, 9, 15	——	**19.** 7, 9	——
12. 5, 4, 10	——	**20.** 11, 4, 2	——

Change to similar fractions:

21. $\frac{1}{4}, \frac{1}{8}$ **25.** $\frac{5}{6}, \frac{2}{3}$ **29.** $\frac{5}{12}, \frac{7}{8}$ **33.** $\frac{5}{6}, \frac{7}{9}$

22. $\frac{3}{4}, \frac{5}{6}$ **26.** $\frac{5}{9}, \frac{5}{18}$ **30.** $\frac{5}{8}, \frac{7}{9}$ **34.** $\frac{11}{15}, \frac{4}{5}$

23. $\frac{7}{8}, \frac{1}{2}$ **27.** $\frac{3}{4}, \frac{7}{8}$ **31.** $\frac{5}{6}, \frac{8}{8}$ **35.** $\frac{3}{10}, \frac{5}{6}$

24. $\frac{7}{10}, \frac{4}{5}$ **28.** $\frac{5}{6}, \frac{4}{9}$ **32.** $\frac{5}{6}, \frac{5}{12}$ **36.** $\frac{4}{15}, \frac{3}{10}$

Written Work

1. Change $\frac{3}{4}$, $\frac{7}{10}$, and $\frac{2}{5}$ to similar fractions having the least common denominator.

l. c. d. = 20

$$\frac{3 \times 5}{4 \times 5} = \frac{15}{20}$$

You see at once that the l. c. d. is 20.

$$\frac{2 \times 4}{5 \times 4} = \frac{8}{20}$$

To change $\frac{1}{4}$ to 20ths, divide the required denominator, 20 by the given denominator, 4. Multiply both terms by the quotient, 5. Proceed in the same way with the other fractions.

$$\frac{7 \times 2}{10 \times 2} = \frac{14}{20}$$

Change to similar fractions having the l. c. d.:

2. $\frac{1}{2}, \frac{1}{3}$	9. $\frac{1}{2}, \frac{1}{3}, \frac{1}{4}$	16. $\frac{9}{16}, \frac{7}{20}$	23. $\frac{7}{12}, \frac{5}{8}, \frac{3}{4}$
3. $\frac{1}{4}, \frac{1}{6}$	10. $\frac{1}{6}, \frac{2}{3}, \frac{3}{4}$	17. $\frac{5}{9}, \frac{5}{12}$	24. $\frac{1}{12}, \frac{3}{4}, \frac{2}{3}$
4. $\frac{1}{2}, \frac{1}{4}$	11. $\frac{1}{2}, \frac{3}{8}, \frac{5}{6}$	18. $\frac{5}{6}, \frac{1}{4}, \frac{5}{12}$	25. $\frac{1}{20}, \frac{1}{2}, \frac{3}{5}$
5. $\frac{2}{3}, \frac{3}{4}$	12. $\frac{3}{4}, \frac{2}{5}, \frac{2}{3}$	19. $\frac{3}{10}, \frac{2}{5}, \frac{1}{2}$	26. $\frac{5}{6}, \frac{3}{4}, \frac{1}{3}$
6. $\frac{5}{6}, \frac{3}{8}$	13. $\frac{7}{8}, \frac{4}{5}, \frac{3}{10}$	20. $\frac{5}{16}, \frac{7}{8}, \frac{7}{32}$	27. $\frac{3}{8}, \frac{1}{4}, \frac{1}{2}$
7. $\frac{9}{10}, \frac{3}{5}$	14. $\frac{5}{12}, \frac{5}{6}, \frac{2}{3}$	21. $\frac{3}{16}, \frac{3}{4}, \frac{5}{8}$	28. $\frac{5}{6}, \frac{3}{4}, \frac{1}{5}$
8. $\frac{5}{8}, \frac{7}{12}$	15. $\frac{1}{4}, \frac{1}{3}, \frac{7}{9}$	22. $\frac{5}{13}, \frac{3}{4}, \frac{9}{26}$	29. $\frac{1}{8}, \frac{4}{5}, \frac{5}{16}$

ADDITION OF FRACTIONS

Oral Work

1. Can you add $\frac{2}{6}$ and $\frac{1}{6}$? Can you add $\frac{1}{5}$ and $\frac{1}{5}$ without change? What change must be made in $\frac{1}{2}$ and $\frac{1}{5}$ before they can be added? in $\frac{2}{3}$ and $\frac{1}{4}$?

2. $\frac{1}{5} = \frac{?}{10}$; $\frac{2}{5} = \frac{?}{10}$; $\frac{3}{5} = \frac{?}{10}$; $\frac{5}{5} = \frac{?}{10}$.

3. $\frac{1}{2} = \frac{?}{10}$; $\frac{1}{2} + \frac{1}{5} = \frac{?}{10}$; $\frac{1}{2} + \frac{1}{3} = \frac{?}{6}$; $\frac{1}{2} + \frac{1}{3} = \frac{?}{8}$.

4. Can you add $\frac{1}{2}$ and $\frac{1}{5}$ without change? Change both fractions to tenths. Can they then be added?

5. Can you add $\frac{1}{2}$ and $\frac{1}{3}$ without change? Change both fractions to sixths. Can they then be added?

6. When $\frac{1}{2}$ and $\frac{1}{4}$ are to be added, to what similar fractions should they be changed?

7. What are the denominators of the fractions in example 4? To what like or common denominators did you change both fractions?

8. What are the denominators of the fractions in example 5? To what denominator did you change the fraction $\frac{1}{2}$? $\frac{1}{3}$? Why?

9. After two or more fractions are changed to like or common denominators, that is, after they have been made similar, what is the *second step in adding them?*

10. Add $\frac{1}{2}$, $\frac{1}{3}$, $\frac{1}{6}$; $\frac{1}{3}$, $\frac{1}{4}$, $\frac{1}{12}$.

11. Observe that in problem 10, $\frac{1}{2} + \frac{1}{3} + \frac{1}{6} = \frac{6}{6}$, or 1, and that $\frac{1}{3} + \frac{1}{4} + \frac{1}{12} = \frac{8}{12}$, or $\frac{2}{3}$.

12. What is the *third step* in adding fractions?

Why is the first step not necessary in the following?

13. $\frac{2}{8} + \frac{1}{8}$ **15.** $\frac{1}{6} + \frac{3}{6}$ **17.** $\frac{1}{8} + \frac{3}{8} + \frac{5}{8} + \frac{7}{8}$

14. $\frac{4}{9} + \frac{5}{9}$ **16.** $\frac{3}{7} + \frac{4}{7}$ **18.** $\frac{1}{10} + \frac{3}{10} + \frac{5}{10} + \frac{9}{10}$

Written Work

1. Add $\frac{3}{4}$ and $\frac{5}{9}$.

$36 = $ l. c. d.

$\frac{3}{4} \times \frac{9}{9} = \frac{27}{36}$

$\frac{5}{9} \times \frac{4}{4} = \frac{20}{36}$

$\frac{3}{4} + \frac{5}{9} = \frac{47}{36}$, or $1\frac{11}{36}$

The l. c. d. of $\frac{3}{4}$ and $\frac{5}{9}$ is 36; $\frac{3}{4} = \frac{27}{36}$ and $\frac{5}{9} = \frac{20}{36}$; $\frac{27}{36} + \frac{20}{36} = \frac{47}{36}$, or $1\frac{11}{36}$.

Observe the three steps in adding fractions:

1. If necessary, make the fractions similar, that is, change them to a common denominator.

2. Write the sum of the numerators over the common denominator.

3. Change the sum to its simplest form.

Add:

2. $\frac{3}{4}$, $\frac{7}{9}$ **8.** $\frac{2}{11}$, $\frac{4}{6}$ **14.** $\frac{2}{3}$, $\frac{7}{8}$, $\frac{3}{4}$

3. $\frac{7}{9}$, $\frac{3}{8}$ **9.** $\frac{9}{10}$, $\frac{7}{8}$ **15.** $\frac{4}{6}$, $\frac{6}{7}$

4. $\frac{6}{7}$, $\frac{4}{5}$ **10.** $\frac{2}{3}$, $\frac{3}{4}$, $\frac{1}{2}$ **16.** $\frac{2}{7}$, $\frac{1}{3}$, $\frac{3}{4}$

5. $\frac{4}{5}$, $\frac{5}{8}$ **11.** $\frac{2}{3}$, $\frac{6}{7}$, $\frac{5}{8}$ **17.** $\frac{4}{5}$, $\frac{1}{2}$, $\frac{3}{8}$

6. $\frac{5}{10}$, $\frac{3}{8}$ **12.** $\frac{6}{7}$, $\frac{1}{9}$ **18.** $\frac{7}{8}$, $\frac{7}{18}$, $\frac{8}{9}$

7. $\frac{5}{6}$, $\frac{7}{9}$ **13.** $\frac{3}{4}$, $\frac{4}{5}$, $\frac{1}{2}$ **19.** $\frac{11}{12}$, $\frac{4}{5}$, $\frac{9}{10}$

Adding mixed numbers.

Written Work

1. Add $8\frac{2}{3}$ and $12\frac{3}{5}$.

$$15 = \text{l. c. d.}$$
$$8\frac{2}{3} = 8\frac{10}{15}$$
$$12\frac{3}{5} = 12\frac{9}{15}$$
$$8\frac{2}{3} + 12\frac{3}{5} = 20\frac{19}{15}, \text{ or }$$
$$21\frac{4}{15}$$

The l. c. d. of $\frac{2}{3}$ and $\frac{3}{5}$ is 15.
$\frac{2}{3} = \frac{10}{15}$; $\frac{3}{5} = \frac{9}{15}$. The sum of $\frac{10}{15}$ and $\frac{9}{15}$ is $\frac{19}{15}$, which equals $1\frac{4}{15}$. The 1 is added to the sum of 12 and 8, making 21, which with $\frac{4}{15}$, makes $21\frac{4}{15}$.

Add:

2. $7\frac{2}{3}$
   $8\frac{1}{2}$

3. $150\frac{3}{4}$
   $68\frac{5}{6}$

4. $80\frac{5}{6}$
   $18\frac{2}{3}$

5. $32\frac{1}{2}$
   $60\frac{5}{12}$

6. $175\frac{1}{2}$
   $16\frac{1}{2}$

7. $350\frac{7}{12}$
   $267\frac{3}{4}$

8. $120\frac{3}{4}$
   $261\frac{3}{10}$

9. $135\frac{1}{2}$
   $122\frac{4}{15}$

10. $80\frac{1}{4}$
   $12\frac{5}{24}$

Add:

11. $1\frac{5}{6} + \frac{3}{5} + 2\frac{7}{8}$

12. $10\frac{3}{8} + 12\frac{2}{3} + 5\frac{7}{12}$

13. $2\frac{5}{12} + 5\frac{3}{4} + 9\frac{4}{15}$

14. $1\frac{3}{4} + 7\frac{1}{4} + 8\frac{7}{8}$

15. $9\frac{3}{5} + 16\frac{7}{10} + 5\frac{11}{15}$

16. $4\frac{7}{8} + \frac{2}{3} + 1\frac{3}{16}$

17. $3\frac{1}{2} + 4\frac{2}{3} + 1\frac{7}{12}$

18. $4\frac{2}{3} + 5\frac{7}{10} + 9\frac{5}{18}$

19. $12\frac{2}{3} + \frac{3}{5} + \frac{7}{8} + 3\frac{9}{10}$

20. $2\frac{2}{6} + 1\frac{1}{8} + 3\frac{1}{4}$.

21. A man walked $3\frac{1}{8}$ miles one hour, $3\frac{1}{4}$ miles the second hour, and $2\frac{1}{4}$ miles the third hour. How far did he walk?

22. A farmer sold corn for $\$14\frac{3}{4}$, wheat for $\$37\frac{1}{2}$, and rye for $\$15\frac{1}{10}$. How much did he receive for all?

23. A clerk spent $\$18\frac{1}{2}$ a month for board, $\$9\frac{1}{2}$ for a room, and $\$4\frac{7}{10}$ for clothes. How much did he spend in one month?

24. I sold $\frac{1}{4}$ of an acre of land to one man, $2\frac{2}{5}$ acres to another, and $1\frac{3}{4}$ acres to another. How many acres did I sell?

25. Find the perimeter or distance around a sheet of paper $9\frac{1}{4}$ in. by $5\frac{1}{8}$ in.

26. Mr. Seward traveled $19\frac{1}{2}$ miles on Monday, $21\frac{3}{4}$ miles on Tuesday, and $22\frac{5}{8}$ miles on Wednesday. How far did he travel in three days?

27. John deposited $\$6\frac{1}{5}$ in a savings bank, James $\$7\frac{7}{10}$, Henry $\$9\frac{3}{4}$, and Joseph $\$11\frac{1}{2}$. How much did they all deposit?

28. A merchant bought a barrel of flour for $\$3\frac{3}{8}$, a barrel of apples for $\$3\frac{7}{10}$, and a barrel of sugar for $\$15\frac{1}{4}$. What was the amount of his bill?

29. Four boys weigh, respectively, $90\frac{1}{2}$ pounds, $95\frac{1}{4}$ pounds, $98\frac{7}{8}$ pounds, and $101\frac{9}{16}$ pounds. What is their total weight?

30. A field is $75\frac{3}{4}$ rods long, and $30\frac{7}{8}$ rods wide. Find the distance around the field.

31. The widths of 4 lots are as follows: $30\frac{1}{2}$ feet, $42\frac{5}{10}$ feet, $38\frac{11}{20}$ feet, and $48\frac{7}{12}$ feet. Find the entire width of the lots.

32. Five floors of an office building are each 12 feet high, two floors are each $13\frac{3}{4}$ feet high, and one floor is $18\frac{5}{8}$ feet high. What is the total height of the floors?

33. A vessel sails $402\frac{7}{8}$ miles the first day, $370\frac{9}{10}$ miles the second day, $325\frac{3}{4}$ miles the third day, and is then $309\frac{1}{2}$ miles from New York. Find how far the ship will have sailed when it reaches New York.

34. James Wilson planted 4 rows of potatoes to test different kinds for quality and quantity. The first row dug $5\frac{7}{8}$ bushels; the second, $5\frac{1}{3}$ bushels; the third, $5\frac{3}{4}$ bushels; and the fourth, $4\frac{9}{16}$ bushels. Find the total amount raised in the 4 rows.

35. A man spends $\frac{1}{6}$ of his salary for clothing, $\frac{1}{4}$ for board, and $\frac{1}{10}$ for traveling expenses. What fractional part of his salary does he spend for these purposes?

36. A newsboy earned $\$\frac{2}{3}$ one day, $\$\frac{8}{10}$ another day, and $\$\frac{1}{4}$ a third day. How much did he earn in the 3 days?

37. A stick was broken into two pieces — one $3\frac{3}{4}$ ft. long and the other $1\frac{1}{4}$ ft. long. How long was the whole stick?

38. If a man earns $\$3\frac{2}{5}$ a day, and a boy $\$\frac{1}{4}$ a day, how much do the man and the boy earn together in a day?

39. Find the total cost of the following purchases: flour $\$1\frac{2}{5}$, sugar $\$\frac{1}{4}$, dried beef $\$\frac{8}{10}$, and corned beef $\$\frac{1}{2}$?

40. Four loads of coal weighed as follows: 2 tons, $1\frac{3}{4}$ tons, $2\frac{1}{4}$ tons, and $2\frac{1}{2}$ tons. How much did the four loads weigh?

41. The rainfall in April was $4\frac{1}{10}$ inches, in May $3\frac{3}{4}$ inches, and in June $4\frac{1}{5}$ inches. What was the total rainfall for the 3 months?

42. A farmer drives in one day $12\frac{2}{3}$ miles, then $6\frac{7}{8}$ miles, and then $9\frac{1}{2}$ miles. How far does he drive?

43. The feed for a horse cost $\$5\frac{3}{4}$ per month; for a cow, $\$4\frac{1}{5}$ per month. If a man has 2 horses and 2 cows, how much will it cost to feed them a month?

44. Three lengths of ribbon were cut from a piece — the first, $1\frac{1}{4}$ yd. long, the second, $2\frac{1}{8}$ yd. long, and the third $3\frac{1}{16}$ yd. long. How much ribbon was cut all together from the piece?

SUBTRACTION OF FRACTIONS

Oral Work

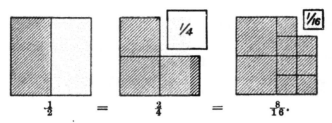

$$\tfrac{1}{2} \quad = \quad \tfrac{2}{4} \quad = \quad \tfrac{8}{16}.$$

1. 1 sq. in. $-\frac{1}{2}$ sq. in. $=$ —— sq. in.

2. $\frac{1}{2}$ sq. in. $-\frac{1}{4}$ sq. in. $=$ —— sq. in.

3. 1 sq. in. $-\frac{1}{4}$ sq. in. $=$ —— sq. in.

4. $\frac{1}{2}$ sq. in. $-\frac{4}{16}$ sq. in. $=$ —— sq. in.

5. $\frac{1}{2}$ sq. in. $-\frac{1}{16}$ sq. in. $\doteq$ —— sq. in.

6. $\frac{3}{4} - \frac{1}{4} = \frac{?}{4}$; $\frac{5}{8} - \frac{1}{8} = \frac{?}{8}$; $\frac{9}{10} - \frac{3}{10} = \frac{?}{10}$.

7. Give answers to the following : $\frac{11}{12} - \frac{5}{12} = ?$ $\frac{11}{18} - \frac{5}{18} = ?$ $\frac{6}{11} - \frac{4}{11} = ?$ $\frac{19}{20} - \frac{3}{20} = ?$ $\frac{24}{25} - \frac{3}{25} = ?$

8. What do you notice about the denominators of the fractions you have subtracted in example 7 ?

In subtraction of fractions, just as in subtraction of whole numbers, the minuend must be *larger* than the subtrahend.

9. When the denominators are alike, what do you subtract ?

10. Could you subtract $\frac{1}{3}$ from $\frac{1}{2}$ without change ? How may these fractions be made similar ?

11. When the denominators are *unlike*, what is the *first step ?* What is the *second step ?* What is the *third step ?*

Subtract :

12. $\$\frac{1}{2} - \$\frac{1}{4}$; $\frac{3}{8}$ ft. $- \frac{1}{4}$ ft. ; $\frac{7}{9}$ yd. $- \frac{2}{3}$ yd. ; $\frac{11}{12}$ ft. $- \frac{7}{8}$ ft.

13. $\frac{9}{10} - \frac{2}{5}$; $\frac{16}{21} - \frac{3}{7}$; $\frac{11}{12} - \frac{2}{3}$; $\frac{4}{5} - \frac{1}{2}$; $\frac{2}{5} - \frac{4}{15}$; $\frac{7}{8} - \frac{1}{4}$.

Subtracting fractions or mixed numbers that are not similar.

Written Work

1. From $\frac{3}{4}$ take $\frac{3}{5}$.

$20 = $ l. c. d.

$\frac{3}{4} \times \frac{5}{5} = \frac{15}{20}$

$\frac{3}{5} \times \frac{4}{4} = \frac{12}{20}$

$\frac{15}{20} - \frac{12}{20} = \frac{3}{20}$

The l. c. d. is 20. $\frac{3}{4} = \frac{15}{20}$; $\frac{3}{5} = \frac{12}{20}$; $\frac{15}{20} - \frac{12}{20}$ $= \frac{3}{20}$.

Observe the three steps in subtracting fractions:

1. If necessary, make the fractions similar; that is, reduce them to a common denominator.

2. Subtract the smaller numerator from the greater and write the result over the common denominator.

3. Change the result to its simplest form.

Find differences :

2. $\frac{1}{2} - \frac{1}{8}$

3. $\frac{1}{4} - \frac{1}{5}$

4. $\frac{1}{3} - \frac{1}{4}$

5. $\frac{1}{2} - \frac{1}{8}$

6. $\frac{1}{3} - \frac{1}{7}$

7. $\frac{1}{8} - \frac{1}{12}$

8. $\frac{1}{2} - \frac{1}{9}$

9. $\frac{1}{3} - \frac{1}{8}$

10. $\frac{1}{6} - \frac{1}{7}$

11. $\frac{1}{6} - \frac{1}{8}$

12. $\frac{1}{8} - \frac{1}{9}$

13. $\frac{1}{4} - \frac{1}{7}$

14. $\frac{1}{5} - \frac{1}{9}$

15. $\frac{1}{2} - \frac{1}{7}$

16. $\frac{1}{3} - \frac{1}{12}$

17. $\frac{1}{4} - \frac{1}{8}$

18. $\frac{1}{5} - \frac{1}{6}$

19. $\frac{1}{4} - \frac{1}{6}$

20. $\frac{1}{5} - \frac{1}{10}$

21. $\frac{1}{6} - \frac{1}{8}$

22. $\frac{3}{4} - \frac{1}{6}$

23. From a piece of cloth containing $\frac{7}{8}$ of a yard, $\frac{3}{4}$ of a yard was sold. What part of a yard remained ?

24. From a city lot containing $\frac{3}{8}$ of an acre, $\frac{5}{16}$ of an acre was sold. What part of an acre remained ?

25. A man traveled $\frac{1}{4}$ of a certain distance the first hour, and $\frac{1}{5}$ of the distance the second hour. What part of the distance farther did he travel the first hour than the second ?

26. From 7 take $\frac{2}{3}$.

$$7 = 6\frac{3}{3}$$
$$\frac{2}{3} = \frac{2}{3}$$
$$7 - \frac{2}{3} = 6\frac{1}{3}$$

Change 7 to 6 and $\frac{3}{3}$. Subtracting $\frac{2}{3}$ from $\frac{3}{3}$ gives $\frac{1}{3}$, which annexed to 6 gives $6\frac{1}{3}$.

Find differences:

27.	$3 - \frac{1}{2}$	**34.**	$9 - \frac{5}{6}$	**41.**	$125 - \frac{1}{18}$
28.	$12 - \frac{3}{8}$	**35.**	$3 - \frac{4}{7}$	**42.**	$10 - \frac{4}{5}$
29.	$22 - \frac{11}{12}$	**36.**	$100 - \frac{9}{10}$	**43.**	$11 - \frac{3}{4}$
30.	$7 - \frac{1}{8}$	**37.**	$18 - \frac{14}{15}$	**44.**	$40 - \frac{3}{8}$
31.	$133 - \frac{8}{9}$	**38.**	$28 - \frac{11}{14}$	**45.**	$7 - \frac{8}{11}$
32.	$44 - \frac{3}{25}$	**39.**	$55 - \frac{13}{16}$	**46.**	$51 - \frac{7}{10}$
33.	$18 - \frac{7}{8}$	**40.**	$4 - \frac{19}{20}$	**47.**	$48 - \frac{7}{8}$

48. If I have \$2 and spend \$$\frac{4}{5}$ how much have I left?

49. A vessel contained 8 gallons of oil. After $\frac{7}{8}$ of a gallon had leaked out, how much remained?

50. A grocer who had bought 10 bushels of potatoes, sold $\frac{3}{4}$ of a bushel. How many bushels remained?

51. From $12\frac{3}{4}$ take $10\frac{1}{2}$.

$$4 = \text{l. c. d.}$$
$$12\frac{3}{4} = 12\frac{3}{4}$$
$$10\frac{1}{2} = 10\frac{2}{4}$$
$$12\frac{3}{4} - 10\frac{1}{2} = 2\frac{1}{4}$$

Change $\frac{1}{2}$ to fourths. $\frac{1}{2} = \frac{2}{4}$. $\frac{3}{4} - \frac{2}{4} = \frac{1}{4}$; $12 - 10 = 2$. $2 + \frac{1}{4} = 2\frac{1}{4}$.

Find differences:

52.	$4\frac{3}{4} - 3\frac{1}{2}$	**59.**	$79\frac{13}{16} - 26\frac{5}{8}$	**66.**	$100\frac{1}{4} - 52\frac{1}{4}$
53.	$7\frac{2}{3} - 4\frac{1}{2}$	**60.**	$97\frac{17}{21} - 35\frac{4}{7}$	**67.**	$78\frac{2}{3} - 35\frac{5}{6}$
54.	$10\frac{5}{6} - 3\frac{1}{8}$	**61.**	$121\frac{1}{2} - 66\frac{2}{5}$	**68.**	$50\frac{1}{4} - 40\frac{1}{3}$
55.	$10\frac{7}{8} - 2\frac{3}{4}$	**62.**	$80\frac{5}{8} - 14\frac{1}{2}$	**69.**	$124\frac{5}{11} - 112\frac{1}{3}$
56.	$12\frac{7}{9} - 5\frac{2}{3}$	**63.**	$98\frac{8}{9} - 32\frac{1}{3}$	**70.**	$240\frac{11}{16} - 200\frac{2}{3}$
57.	$24\frac{7}{10} - 11\frac{2}{5}$	**64.**	$45\frac{23}{30} - 30\frac{2}{3}$	**71.**	$15\frac{7}{8} - 8\frac{1}{2}$
58.	$31\frac{11}{12} - 18\frac{5}{6}$	**65.**	$25\frac{4}{9} - 12\frac{1}{4}$	**72.**	$117\frac{3}{4} - 15\frac{5}{8}$

Find the value of:

73. $7\frac{1}{4} - 2\frac{5}{8}$.

$$8 = \text{l. c. d.}$$
$$7\frac{1}{4} = 6\frac{5}{4} = 6\frac{10}{8}$$
$$2\frac{5}{8} = \qquad 2\frac{5}{8}$$
$$\overline{7\frac{1}{4} - 2\frac{5}{8} = 4\frac{5}{8}}$$

Since $\frac{5}{8}$ is greater than $\frac{1}{4}$, change $7\frac{1}{4}$ to $6\frac{5}{4}$. The l. c. d. is 8. $\frac{1}{4} = \frac{10}{8}$. Subtract $2\frac{5}{8}$ from $6\frac{10}{8}$. The remainder is $4\frac{5}{8}$.

74. $3\frac{1}{8} - 1\frac{1}{2}$

75. $5 \ - 1\frac{3}{8}$

76. $5\frac{1}{4} - 2\frac{2}{3}$

77. $10\frac{3}{8} - 5\frac{3}{4}$

78. $20\frac{1}{12} - 13\frac{5}{6}$

79. $50\frac{1}{6} - 30\frac{1}{3}$

80. $6\frac{2}{3} - 2\frac{6}{7}$

81. $11\frac{3}{8} - 5\frac{3}{4}$

82. $44\frac{2}{4} - 12\frac{1}{2}$

83. $36\frac{1}{4} - 11\frac{4}{15}$

84. $81\frac{1}{6} - 14\frac{6}{7}$

85. $21\frac{3}{6} - 12\frac{2}{3}$

86. One motorman's trip takes $4\frac{2}{3}$ hours, and another's $2\frac{4}{5}$ hours. How much longer is the first trip than the second?

87. A man bought two suits of clothes, one costing $\$35\frac{1}{2}$ and the other $\$28\frac{3}{4}$. How much more did the first suit cost than the second?

88. James lives $1\frac{9}{20}$ miles from the schoolhouse, and Samuel, $1\frac{1}{5}$ miles away. How much farther does Samuel have to walk to school than James?

89. If a man earns $\$3\frac{2}{3}$ a day, and a boy $\$\frac{3}{4}$ a day, how much more does the man earn in a day than the boy earns?

90. A boy is $4\frac{5}{12}$ feet tall. His sister is $3\frac{3}{8}$ feet tall. How much taller is the boy than his sister?

91. The top of a door is $12\frac{1}{2}$ feet above the ground, and the bottom of it is $4\frac{3}{4}$ feet above the ground. How high is the door?

92. From a barrel containing $51\frac{1}{2}$ gallons of oil, $17\frac{1}{2}$ gallons were sold in one day, and $25\frac{1}{2}$ gallons another day. How many gallons remained unsold?

93. A station agent who was paid $\$60$ per month spent in one month $\$12\frac{2}{5}$ for groceries, $\$7\frac{7}{10}$ for meat, and $\$15\frac{1}{4}$ for other expenses. How much did he save?

REVIEW

Find the value of :

1. $\frac{2}{3} + \frac{3}{4} - \frac{5}{6}$

2. $3\frac{1}{2} + 2\frac{1}{4} - 1\frac{5}{8}$

3. $5\frac{1}{3} + 4\frac{5}{6} - 4\frac{5}{12}$

4. $1\frac{5}{8} + 2\frac{1}{12} - 1\frac{5}{6}$

5. $9\frac{9}{15} - 4\frac{2}{3} + 3\frac{7}{10}$

6. $12\frac{1}{4} - 5\frac{1}{2} + 2\frac{2}{9}$

7. $\frac{3}{4} - \frac{1}{2} + \frac{7}{8} + \frac{17}{18}$

8. $20 + \frac{1}{16} + \frac{1}{12} - 2\frac{1}{4}$

9. $\frac{1}{3} + \frac{1}{4} + \frac{5}{6} - \frac{7}{12}$

10. $18\frac{11}{24} - 10\frac{3}{8} + 5\frac{1}{3} + 2\frac{5}{6}$

11. $15 - 3\frac{2}{3} + 4\frac{1}{6}$

12. $40 + 60 + 30\frac{11}{13}$

13. $19\frac{27}{32} - 11\frac{5}{8} + 1\frac{5}{16}$

14. $7\frac{7}{9} + 4\frac{1}{3} + 11\frac{5}{18}$

15. $3\frac{5}{12} + 9\frac{5}{6} - 7\frac{7}{24}$

16. $22\frac{2}{11} + 7\frac{1}{3} - 20\frac{10}{33}$

17. $8\frac{1}{4} + 22\frac{1}{4} + 43\frac{3}{8}$

18. $225\frac{5}{8} + 132\frac{5}{12} + 80\frac{11}{16}$

19. $12\frac{3}{8} + 19\frac{7}{8} - 27\frac{5}{8}$

20. $19\frac{1}{6} + 11\frac{3}{8} + 14\frac{1}{3}$

21. Arthur bought a hat and gave in payment $10. If he received in change $7\frac{3}{4}$, how much did the hat cost?

22. The treasurer of a literary society received $492\frac{1}{2}$. He spent for light and heat $50\frac{3}{4}$, for new books $77\frac{1}{4}$, for a lecturer's expenses $26\frac{1}{2}$, and for music for an entertainment $80. How much remained in his hands?

23. A teacher's salary was $95 per month. He spent $20 for board, $7\frac{1}{2}$ for room, $9\frac{3}{5}$ for clothes, and $2\frac{1}{10}$ for other expenses. How much did he save out of his month's salary?

24. A dealer bought $61\frac{3}{4}$ bushels of apples from one man; $127\frac{1}{2}$ bushels from another; and $89\frac{3}{4}$ bushels from another. How much did they cost at 75¢ per bushel?

25. A student spends $\frac{1}{4}$ of the day in study, $\frac{1}{8}$ in recitations, $\frac{1}{12}$ at his meals, and $\frac{1}{8}$ in recreation and exercise. What part of the day has he left for sleep?

26. A boy walked around a field $60\frac{1}{2}$ rods long and 40 rods wide. How far did he walk?

27. William had \$$45\frac{1}{2}$; he gave his sister \$$37\frac{4}{5}$. How much had he remaining?

28. Mr. Mellon owned $\frac{1}{8}$ of an oil well. Later he bought $\frac{7}{16}$ of it. What part of the well did he not own?

29. A clerk earns \$40 a month and spends for board \$$13\frac{1}{2}$, for clothes \$$7\frac{3}{4}$, and for other expenses \$$14\frac{4}{5}$. How much of his month's salary does he save?

30. From a piece of ribbon containing 10 yards were sold $\frac{1}{2}$ of a yard, $1\frac{1}{4}$ yards, $\frac{3}{4}$ of a yard, $\frac{5}{8}$ of a yard, $2\frac{3}{4}$ yards, and 3 yards. How many yards remained unsold?

31. A mail carrier works 8 hours per day. In making three deliveries he works $1\frac{1}{8}$ hours, $2\frac{1}{3}$ hours, and $2\frac{3}{4}$ hours. How much time has he for a fourth delivery?

32. $\frac{2}{5}$ of a pole is in the ground, $\frac{3}{8}$ of it in the water, and the rest in the air. What part of the pole is in the air?

33. Harry caught 3 fish. The first weighed $\frac{1}{2}$ of a pound, the second $\frac{7}{8}$ of a pound, and the third $1\frac{1}{4}$ pounds. How much did the 3 fish weigh?

34. A real estate agent bought $16\frac{1}{2}$ acres of land. He sold $2\frac{1}{4}$ acres to one man, and $5\frac{7}{8}$ acres to another man. How many acres had he remaining?

35. The sum of two fractions is $\frac{7}{8}$, and one of the fractions is $\frac{3}{5}$. What is the other fraction?

36. The McJunkin dairy sold milk as follows: Monday, $65\frac{3}{8}$ gallons; Tuesday, $60\frac{1}{2}$ gallons; Wednesday, $71\frac{3}{4}$ gallons; Thursday, 69 gallons; Friday, $67\frac{7}{8}$ gallons; and Saturday, $90\frac{1}{4}$ gallons. What were the total sales for the week?

37. A salesman traveled in 4 days as follows: Monday, $122\frac{1}{2}$ miles; Tuesday, $187\frac{3}{4}$ miles; Wednesday, $93\frac{1}{8}$ miles; and Thursday, $207\frac{3}{16}$ miles. How many miles did he travel in all?

38. If a boat will carry safely 800 pounds, how many pounds of provisions can be carried when three men whose weights are, respectively, $165\frac{1}{2}$ pounds, $182\frac{3}{4}$ pounds, and 208 pounds, are in the boat?

39. In 3 days in June the sun shone in New York $14\frac{3}{8}$ hours, $14\frac{3}{20}$ hours, and $14\frac{19}{80}$ hours. How many hours of sunshine were there in these 3 days? How many hours without sunshine were there?

40. Raymond walks $1\frac{7}{8}$ miles to school, and George walks $\frac{3}{4}$ of a mile. How far do they both walk, and how much farther does Raymond walk than George?

41. A playground is $16\frac{1}{4}$ yards wide, and $25\frac{2}{3}$ yards long. Find the distance around it. How much greater is the playground in length than in width?

42. Measure the distance around your school ground in feet and fractions of a foot; then in yards and fractions of a yard; then secure a tape line and measure it in rods and fractions of a rod. Find its *perimeter* in each of these three units of measure.

43. With a yardstick let each pupil measure in feet and fractions of a foot the length and the width of the different rooms in each one's house, first finding the perimeter of each room, and then the difference between the length and the width of each room.

44. The rainfall in three different months was as follows: $3\frac{7}{10}$ inches; $4\frac{2}{5}$ inches; $2\frac{3}{10}$ inches. Find mentally the rainfall for the three months, and the difference between the greatest rainfall and the rainfall for each of the other months.

45. Mrs. Thompson's baby gained 2 oz. the first week, 3 oz. the second week, 4 oz. the third week, and 8 oz. the fourth week. Express each week's gain as a fractional part of a pound and find how many pounds the baby gained.

46. Make up two problems involving addition and subtraction of fractions, and two involving addition and subtraction of mixed numbers.

47. A western farmer sold three loads of alfalfa in one summer. The first weighed $2\frac{3}{4}$ tons; the second, $2\frac{7}{8}$ tons; and the third, $1\frac{3}{4}$ tons. Find the entire number of tons sold.

48. James's father cut ice on his mill pond three times during the winter. The first cutting was $6\frac{5}{8}$ inches, the second cutting was $8\frac{3}{4}$ inches, and the third cutting was $5\frac{9}{16}$ inches. Find the entire thickness of the ice for the three cuttings.

49. In problem 48 find the difference between the greatest thickness and each of the other two thicknesses.

50. Donald is 4 feet 8 inches tall, and Arline is 3 feet 9 inches tall. Express their heights in feet as mixed numbers. How much taller is Donald than Arline?

51. The weather man reported 3 hours and 20 minutes sunshine on Monday; 2 hours and 40 minutes sunshine on Tuesday; and 5 hours and 15 minutes sunshine on Wednesday. Express the time of sunshine of each day as a fractional part of a day, and find the total time of sunshine for the three days.

52. Ben ran 100 yards in $9\frac{8}{10}$ sec. and Vincent ran the same distance in $11\frac{1}{4}$ sec. How much did Ben beat Vincent's time?

53. George weighs 70 lb. 8 oz. and James weighs 98 lb. 4 oz. Express their weights as mixed numbers and find the sum and the difference of their weights.

54. Bertha grew $\frac{3}{4}$ of an inch in three months, Paul $\frac{5}{8}$ of an inch, and Susan $\frac{5}{16}$ of an inch. Express the difference between Bertha's growth and each of the other two.

55. Mary bought three remnants of ribbon. The first was $3\frac{3}{4}$ yards; the second, $\frac{7}{8}$ of a yard, and the third, $2\frac{1}{8}$ yards. How much ribbon did Mary buy?

First add each number in *a* to the numbers on the same line in *b, c, d, e, f, g*. Thus, in 56, add $2\frac{4}{5}$, first to $3\frac{5}{8}$, then to $4\frac{3}{4}$, then to $8\frac{5}{8}$, etc. In 57 add $4\frac{5}{8}$, first to $6\frac{2}{3}$, then to $5\frac{7}{8}$, etc. Then subtract each number in column *a* from the numbers on the same line in *b, c, d, e, f, g*. Thus, in 56, subtract $2\frac{4}{5}$, first from $3\frac{5}{8}$, then from $4\frac{3}{4}$, then from $8\frac{5}{8}$ etc. In 57 subtract $4\frac{5}{8}$, first from $6\frac{2}{3}$, then from $5\frac{7}{8}$, etc.

	a	*b*	*c*	*d*	*e*	*f*	*g*
56.	$2\frac{4}{5}$	$3\frac{5}{8}$	$4\frac{3}{4}$	$8\frac{5}{8}$	$5\frac{1}{2}$	$7\frac{1}{4}$	$9\frac{3}{8}$
57.	$4\frac{5}{8}$	$6\frac{2}{3}$	$5\frac{7}{8}$	$5\frac{1}{8}$	$8\frac{1}{4}$	$9\frac{5}{8}$	$9\frac{1}{4}$
58.	$8\frac{3}{10}$	$11\frac{4}{5}$	$12\frac{7}{10}$	$15\frac{4}{7}$	$22\frac{5}{12}$	$32\frac{5}{16}$	$25\frac{11}{12}$
59.	$9\frac{1}{4}$	$40\frac{5}{8}$	$45\frac{8}{16}$	$48\frac{9}{10}$	$14\frac{8}{10}$	$60\frac{5}{8}$	$65\frac{4}{15}$
60.	$8\frac{1}{10}$	$75\frac{8}{15}$	$80\frac{5}{13}$	$59\frac{8}{16}$	$77\frac{3}{8}$	$68\frac{5}{9}$	$98\frac{4}{11}$

TEST EXERCISES IN ADDITION AND SUBTRACTION OF FRACTIONS

Oral Work

To the Teacher. — Test pupils' ability to do this work accurately and neatly in a given time. Let them do the work a second time, and try to beat the first record.

Add:

	a	*b*	*c*	*d*	*e*	*f*	*g*	*h*	*i*	*j*
1.	$\frac{5}{8}$	$\frac{8}{4}$	$\frac{15}{10}$	$\frac{8}{9}$	$\frac{11}{16}$	$\frac{9}{10}$	$\frac{12}{13}$	$\frac{7}{11}$	$\frac{9}{14}$	$\frac{10}{16}$
	$\frac{7}{8}$	$\frac{10}{4}$	$\frac{11}{10}$	$\frac{4}{9}$	$\frac{14}{16}$	$\frac{7}{10}$	$\frac{10}{13}$	$\frac{9}{11}$	$\frac{8}{14}$	$\frac{20}{16}$
2.	$\frac{4}{5}$	$\frac{5}{2}$	$\frac{8}{4}$	$\frac{9}{8}$	$\frac{8}{6}$	$\frac{10}{9}$	$\frac{9}{7}$	$\frac{5}{12}$	$\frac{5}{13}$	$\frac{25}{16}$
	$\frac{15}{5}$	$\frac{6}{2}$	$\frac{4}{4}$	$\frac{8}{3}$	$\frac{5}{6}$	$\frac{16}{9}$	$\frac{12}{7}$	$\frac{14}{12}$	$\frac{9}{13}$	$\frac{22}{16}$

Written Work

Add:

3.	$3\frac{2}{6}$ $5\frac{3}{4}$	**6.**	$6\frac{1}{6}$ $7\frac{4}{8}$	**9.**	$8\frac{2}{3}$ $5\frac{1}{3}$	**12.**	$3\frac{1}{2}$ $8\frac{3}{4}$	**15.**	$4\frac{1}{8}$ $5\frac{2}{7}$
4.	$3\frac{4}{5}$ $2\frac{3}{10}$	**7.**	$7\frac{5}{8}$ $4\frac{1}{12}$	**10.**	$6\frac{5}{6}$ $2\frac{5}{12}$	**13.**	$12\frac{1}{8}$ $15\frac{3}{4}$	**16.**	$7\frac{7}{8}$ $2\frac{1}{2}$
5.	$25\frac{5}{6}$ $21\frac{3}{10}$	**8.**	$63\frac{7}{12}$ $50\frac{3}{10}$	**11.**	$61\frac{5}{18}$ $27\frac{4}{15}$	**14.**	$98\frac{3}{16}$ $15\frac{11}{24}$	**17.**	$6\frac{7}{8}$ $3\frac{3}{16}$

Subtract, and test results by adding:

18.	$24\frac{2}{3}$ $14\frac{1}{4}$	**20.**	$40\frac{2}{3}$ $12\frac{1}{10}$	**22.**	$44\frac{7}{8}$ $29\frac{3}{4}$	**24.**	$14\frac{11}{12}$ $9\frac{3}{10}$	**26.**	$27\frac{5}{8}$ $19\frac{7}{16}$
19.	$91\frac{13}{16}$ $47\frac{1}{3}$	**21.**	$36\frac{2}{3}$ $18\frac{3}{4}$	**23.**	$15\frac{2}{3}$ $9\frac{1}{10}$	**25.**	$40\frac{7}{8}$ $20\frac{3}{16}$	**27.**	$18\frac{7}{12}$ $11\frac{1}{3}$

28. James worked $6\frac{3}{4}$ hours after school in one week, Henry worked $4\frac{1}{8}$ hours, and Frank worked $2\frac{1}{2}$ hours. Find the number of hours the three boys were employed during the week.

29. Mary's hair ribbon contains $\frac{3}{4}$ of a yard; Ruth's hair ribbon contains $\frac{7}{8}$ of a yard; and Martha's sash contains $2\frac{1}{4}$ yards. How many yards of ribbon are there in all three pieces?

30. John is $7\frac{3}{4}$ years old, and his father is 30 years and 6 months old. How much older is John's father than John?

31. If a rug is $6\frac{3}{8}$ feet long, and $4\frac{5}{12}$ feet wide, what is the difference between its length and its width?

MULTIPLICATION OF FRACTIONS

Multiplying a fraction by a whole number.

Oral Work

  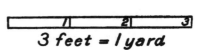

3 feet = 1 yard

1. Into how many halves is the square divided?
2. Two times $\frac{1}{2}$ the square = —— square.
3. Two times $\frac{1}{2}$ the circle = —— circle.
4. Two times $\$\frac{1}{2}$ = —— dollars.
5. Five times $\$\frac{1}{2}$ = —— dollars.
6. Into how many feet is the yard divided?
7. What is one of these parts called?
8. What are two parts called?
9. $2 \times \frac{1}{3}$ of a yard = —— yard.
10. $6 \times \frac{1}{3}$ of a yard = —— yards.
11. $6 \times \frac{1}{2}$ of a circle = —— circles.
12. $12 \times \frac{1}{2}$ of a circle = —— circles.
13. $4 \times \frac{1}{2}$ is the same as $\frac{1}{2} + \frac{1}{2} + \frac{1}{2} + \frac{1}{2}$.
14. $6 \times \frac{1}{2} = \frac{6}{2}$, or 3. Therefore, to multiply $\frac{1}{2}$ by 6, we say 6 times $\frac{1}{2} = \frac{6}{2}$, or 3.

Give products:

15. $8 \times \frac{1}{2}$	20. $9 \times \frac{2}{3}$	25. $7 \times \frac{4}{9}$	30. $3 \times \frac{7}{11}$
16. $12 \times \frac{1}{4}$	21. $8 \times \frac{3}{4}$	26. $12 \times \frac{8}{9}$	31. $12 \times \frac{4}{7}$
17. $6 \times \frac{2}{3}$	22. $9 \times \frac{9}{10}$	27. $4 \times \frac{7}{11}$	32. $5 \times \frac{7}{8}$
18. $7 \times \frac{3}{7}$	23. $6 \times \frac{4}{5}$	28. $8 \times \frac{3}{7}$	33. $6 \times \frac{4}{5}$
19. $10 \times \frac{3}{5}$	24. $8 \times \frac{8}{9}$	29. $11 \times \frac{7}{8}$	34. $6 \times \frac{8}{9}$

Multiplying a mixed number by a whole number.

Oral Work

1. Find the cost of 6 eggs at $3\frac{1}{2}$ cents apiece.

2. Find the cost of 4 qt. of oil at $4\frac{1}{2}$ cents per quart.

3. How much will 8 lb. of rice cost at $6\frac{1}{2}$ cents per pound?

4. Find the cost of 12 lb. of sugar at $6\frac{1}{2}$ cents per pound.

5. When berries are sold for $8\frac{1}{2}$ cents per basket, find the cost of 6 baskets.

6. A man earns $\$1\frac{3}{4}$ per day. How much does he earn in 6 days?

7. When apples are sold for $\$1\frac{1}{4}$ per bushel, find the cost of 8 bushels.

8. Mary pays $\$\frac{3}{4}$ for music lessons and takes two lessons a week. How much do her music lessons cost her in 4 weeks?

9. John makes $\frac{2}{3}$ of a cent on each paper and averages 40 papers a day. Find his profit.

10. We pay $8\frac{1}{2}$¢ a quart for milk. How much is our milk bill in a week, if we use 2 qt. per day?

11. By buying 25 cents worth of street car tickets, each ticket costs me $4\frac{1}{6}$ cents. Find the cost of 12 tickets.

12. A storekeeper makes $1\frac{3}{4}$ cents on each can of corn. How much does he make on 18 cans?

13. John works for $5\frac{1}{2}$ cents per hour. If he works 8 hours a day, how much does he earn in 2 days?

14. Mary uses $1\frac{3}{4}$ yd. of ribbon for a bow. How much does it take for 9 such bows?

15. Find the cost of a dozen eggs at $3\frac{1}{3}$ cents apiece.

16. I pay \$5¾ for a boy's suit. Find the cost of 4 such suits.

17. 3 baskets of cherries cost 25 cents. How much will 12 baskets cost?

SUGGESTION. — 12 baskets will cost how many times the cost of 3 baskets?

Written Work

1. Find $5 \times 2\frac{2}{3}$.

$2\frac{2}{3}$
5

$\frac{10}{3} = 3\frac{1}{3}$ This means that $5 \times \frac{2}{3}$ is to be added to 5×2.
10 $5 \times \frac{2}{3} = \frac{10}{3}$, or $3\frac{1}{3}$; $5 \times 2 = 10$; $3\frac{1}{3} + 10 = 13\frac{1}{3}$.

$13\frac{1}{3}$

Find the value of:

2. $8 \times 4\frac{1}{2}$	**11.** $10 \times 2\frac{2}{3}$	**20.** $125 \times 18\frac{3}{4}$
3. $10 \times 4\frac{1}{2}$	**12.** $12 \times 3\frac{3}{4}$	**21.** $72 \times 2\frac{4}{5}$
4. $9 \times 2\frac{2}{3}$	**13.** $20 \times 5\frac{3}{5}$	**22.** $100 \times 14\frac{7}{10}$
5. $12 \times 3\frac{3}{4}$	**14.** $45 \times 12\frac{7}{8}$	**23.** $132 \times 5\frac{5}{9}$
6. $11 \times 3\frac{3}{11}$	**15.** $120 \times 22\frac{3}{8}$	**24.** $168 \times 10\frac{5}{11}$
7. $14 \times 2\frac{4}{7}$	**16.** $154 \times 11\frac{6}{7}$	**25.** $20 \times 18\frac{4}{5}$
8. $12 \times 8\frac{1}{2}$	**17.** $96 \times 6\frac{2}{3}$	**26.** $90 \times 15\frac{5}{6}$
9. $6 \times 8\frac{1}{2}$	**18.** $144 \times 9\frac{1}{4}$	**27.** $50 \times 16\frac{1}{4}$
10. $5 \times 3\frac{2}{5}$	**19.** $80 \times 4\frac{1}{2}$	**28.** $200 \times 15\frac{3}{20}$

29. A book dealer purchases 125 books at wholesale at $1\frac{1}{5}$ each. Find the cost.

30. The car fare from Pittsburgh to Chicago on the Ft. Wayne R.R. is \$10½. Find the amount received from the sale of 50 tickets.

31. A newsdealer buys 300 papers at $1\frac{1}{4}$ cents each and sells them at 2 cents each. Find the cost and the gain.

32. A huckster buys 20 dozen bananas at 10 cents per dozen and sells them at the rate of $2\frac{1}{2}$ cents each. Find his gain.

33. A fruit dealer buys a barrel of apples for $\$4\frac{1}{2}$. The barrel contains 240 apples. He sells one half of them at the rate of 2 for 5 cents and the remainder at the rate of 3 for 5 cents. Find his profit.

Suggestion.—At 2 for 5¢, 120 apples will cost 60 × 5¢; at 3 for 5¢, 120 apples will cost 40 × 5¢.

Finding fractional parts of whole numbers.

Oral Work

1. $\frac{1}{2}$ of 2 units = ——unit.

2. $\frac{1}{2}$ of $2 = ——dollar.

We have learned that we may divide a unit into any number of parts and then take any number of these parts. Thus, $\frac{3}{4}$ of $60 means that $60 (60 units) is divided into 4 equal parts of $15 each and that 3 of these parts, or $45, are taken.

Since $\frac{1}{4}$ of $60 = $15, $\frac{3}{4}$ of $60 = 3 × $15, or $45.

Find the following:

3. $\frac{3}{4}$ of $24	**7.** $\frac{5}{9}$ of 36 minutes	**11.** $\frac{7}{9}$ of 72 cents
4. $\frac{9}{10}$ of 60 horses	**8.** $\frac{11}{12}$ of $144	**12.** $\frac{9}{11}$ of 99
5. $\frac{2}{3}$ of 36 days	**9.** $\frac{6}{13}$ of 78 miles	**13.** $\frac{4}{15}$ of 75
6. $\frac{4}{5}$ of 20 hours	**10.** $\frac{5}{11}$ of 44 rods	**14.** $\frac{3}{7}$ of 84

15. To find $\frac{1}{3}$ of 4, into how many parts do you divide 4? How do you find one of these parts? Divide 4 by 3.

16. Explain what you mean by $\frac{1}{5}$ of a number; by $\frac{3}{4}$ of a number; by $\frac{2}{5}$ of a number.

17. Find $\frac{2}{3}$ of 4.

Solution.—$\frac{1}{3}$ of $4 = \frac{4}{3}$, and $\frac{2}{3}$ of $4 = 2 \times \frac{4}{3}$, or $\frac{8}{3}$; $\frac{8}{3} = 2\frac{2}{3}$.

Find the following:

18. $\frac{2}{5}$ of 7	**24.** $\frac{9}{10}$ of 12	**30.** $\frac{3}{4}$ of 20
19. $\frac{3}{4}$ of 9	**25.** $\frac{11}{15}$ of 8	**31.** $\frac{5}{6}$ of 6
20. $\frac{7}{8}$ of 12	**26.** $\frac{7}{11}$ of 8	**32.** $\frac{4}{5}$ of 9
21. $\frac{5}{6}$ of 8	**27.** $\frac{8}{9}$ of 9	**33.** $\frac{3}{8}$ of 11
22. $\frac{9}{10}$ of 10	**28.** $\frac{4}{11}$ of 12	**34.** $\frac{7}{9}$ of 10
23. $\frac{1}{4}$ of 8	**29.** $\frac{5}{6}$ of 7	**35.** $\frac{5}{6}$ of 12

36. Multiply $\frac{3}{4}$ by 8. Which number is the multiplier? Which number is the multiplicand? Observe that $\frac{3}{4}$ of 8 gives the same result.

If the sign of multiplication is written after a fractional multiplier, it may be read "of." Thus, $\frac{3}{4} \times 12$ may be read "$\frac{3}{4}$ of 12."

Read the following problems:

37. $\frac{2}{3} \times 9$ **38.** $\frac{3}{4} \times 6$ **39.** $\frac{7}{8} \times 4$ **40.** $\frac{5}{7} \times 6$

Written Work

1. A merchant owing $1200 gave his check for $\frac{3}{4}$ of the amount. For how much did he write his check?

2. Three men own 2500 acres of land. The first owns $\frac{2}{5}$ of it, the second $\frac{3}{5}$ of the remainder, and the third the remainder. How many acres does each own?

3. If a laborer works $\frac{4}{5}$ of the days in a common year, how many days does he work?

4. A student's expenses at college are $480 per year. If board and tuition cost $\frac{2}{3}$ of that amount, how much is spent for room rent, books, etc.?

5. A contractor agrees to erect a building for $24,570. Labor costs $\frac{1}{3}$ of the amount, material $\frac{2}{5}$ of the remainder. Find his profit.

6. $\frac{4}{5}$ of the entire enrollment of 14,720 in school are girls. Find the number of girls and the number of boys.

7. $\frac{7}{16}$ of $960 is paid in a year for rent.   Find the monthly rent bill.

8. An automobile cost $3456 and the expenses and repairs for one year were $\frac{3}{16}$ of the cost.   Find the expenses.

9. A western farmer bought a farm of 160 acres at $25 an acre.   He erected a house costing $\frac{3}{5}$ as much as the land, and a barn costing $\frac{1}{2}$ as much as the house.   Find the total cost of the property.

Multiplying a whole number by a mixed number.

Oral Work

1. Multiply 12 by $6\frac{3}{4}$; 10 by $7\frac{2}{5}$.

2. $7\frac{1}{4}$ times 8 hours are how many hours?

3. How much do $2\frac{3}{4}$ pounds of candy cost at 40 cents a pound?

4. I bought $4\frac{7}{8}$ yards of ribbon at 40 cents a yard. How much did it cost?

5. A boy walks 3 miles in an hour.   How far can he walk at the same rate in $2\frac{5}{6}$ hours?

6. James is 6 years old.   His mother is $4\frac{5}{6}$ times as old. How old is she?

7. How much will $10\frac{3}{4}$ pounds of meat cost at 16 cents a pound?

8. A man bought $7\frac{1}{2}$ gallons of oil at 12¢ a gallon. How much did he pay for it?

9. When gas costs 25 cents per thousand feet, what is my bill for $10\frac{2}{3}$ thousand feet?

10. If a lot cost $200 and a house 8⅝ times as much, how much did the house cost?

11. A man worked 20¾ days in a month for $2 a day. How much did he earn?

12. How many inches equal 9¾ feet?

13. At 60 cents a bushel, how much will 2½ bushels of wheat cost?

14. I bought 1¾ dozen collars at $2 per dozen. How much did they cost?

15. At 12 cents a pound, how much will 15¾ pounds of raisins cost?

16. How much will 5⅛ bushels of raspberries cost at $2 a bushel?

17. If a plumber is paid 75 cents per hour, how much does he receive in 3⅔ hours?

18. How far will an automobile travel in 2½ hours if it travels 18 miles in one hour?

19. If the freight from New York to Albany on a ton of merchandise is 33 cents, how much will it be on 5 8/11 tons?

20. A gallon of water weighs 8 pounds. How much do 10¾ gallons weigh?

Written Work

1. Multiply 18 by 14⅔.

$$
\begin{array}{r}
18 \\
14\frac{2}{3} \\
\hline
12 = \tfrac{2}{3} \text{ of } 18 \\
72 \\
18 \\
\hline
264
\end{array}
$$

14⅔ times 18 means that ⅔ of 18 is to be added to 14 × 18. ⅔ of 18 = 12, which added to 14 × 18 = 264.

Find products:

2. $7\frac{1}{2} \times 6$	**16.** $20\frac{1}{20} \times 100$	**30.** $116\frac{3}{8} \times 54$
3. $15\frac{1}{8} \times 9$	**17.** $42\frac{1}{11} \times 55$	**31.** $112\frac{3}{10} \times 50$
4. $27\frac{1}{6} \times 12$	**18.** $64\frac{1}{13} \times 89$	**32.** $88\frac{3}{7} \times 28$
5. $120\frac{1}{8} \times 40$	**19.** $72\frac{1}{14} \times 42$	**33.** $30\frac{5}{8} \times 160$
6. $216\frac{1}{10} \times 50$	**20.** $102\frac{1}{16} \times 80$	**34.** $19\frac{7}{12} \times 24$
7. $73\frac{1}{8} \times 15$	**21.** $125\frac{1}{24} \times 120$	**35.** $3\frac{7}{15} \times 60$
8. $140\frac{1}{4} \times 28$	**22.** $12\frac{2}{3} \times 9$	**36.** $145\frac{3}{11} \times 55$
9. $100\frac{1}{10} \times 60$	**23.** $14\frac{3}{5} \times 10$	**37.** $48\frac{6}{7} \times 84$
10. $95\frac{1}{5} \times 45$	**24.** $20\frac{3}{4} \times 12$	**38.** $21\frac{3}{4} \times 16$
11. $81\frac{1}{7} \times 21$	**25.** $35\frac{5}{6} \times 18$	**39.** $40\frac{1}{5} \times 25$
12. $120\frac{1}{8} \times 81$	**26.** $95\frac{1}{7} \times 42$	**40.** $121\frac{6}{7} \times 49$
13. $144\frac{1}{2} \times 108$	**27.** $100\frac{3}{5} \times 20$	**41.** $10\frac{3}{8} \times 18$
14. $150\frac{1}{15} \times 60$	**28.** $124\frac{1}{4} \times 120$	**42.** $14\frac{7}{8} \times 24$
15. $180\frac{1}{18} \times 18$	**29.** $65\frac{5}{8} \times 32$	**43.** $20\frac{1}{12} \times 84$

44. If the rate of sailing of a vessel is 18 miles an hour, how far will it sail in $24\frac{1}{2}$ hours?

45. Find the cost of $12\frac{1}{4}$ tons of coal at $6 a ton.

46. Find the cost of $16\frac{1}{2}$ yards of broadcloth at $1.50 a yard.

Note. — The seller usually regards any part of a cent as an additional cent.

47. A farmer sold $5\frac{1}{2}$ acres of land at $40 an acre. How much did he receive for it? If the buyer sold it at $60 an acre, how much did he gain?

48. Find the total cost of $6\frac{3}{4}$ pounds of steak at 28 ¢ a pound, and a $3\frac{1}{4}$ pound chicken at 36 ¢ a pound.

Multiplying a fraction by a fraction.

Oral Work

1. What is $\frac{1}{3}$ of 6 feet? $\frac{2}{3}$ of 6 feet?

2. What is $\frac{1}{3}$ of 6 sevenths? $\frac{2}{3}$ of 6 sevenths?

$\frac{1}{3}$ of $\frac{6}{7}$ means $\frac{1}{3}$ of 6 equal parts of a unit that has been divided into 7 equal parts.

$\frac{1}{3}$ of $\frac{6}{7} = \frac{2}{7}$; and $\frac{2}{3}$ of $\frac{6}{7} = 2$ times $\frac{2}{7}$, or $\frac{4}{7}$.

Find:

3. $\frac{1}{2}$ of $\frac{4}{5}$	6. $\frac{2}{3}$ of $\frac{6}{7}$	9. $\frac{2}{5}$ of $\frac{10}{11}$
4. $\frac{1}{4}$ of $\frac{8}{9}$	7. $\frac{3}{4}$ of $\frac{8}{9}$	10. $\frac{1}{6}$ of $\frac{20}{21}$
5. $\frac{1}{5}$ of $\frac{5}{6}$	8. $\frac{3}{5}$ of $\frac{5}{8}$	11. $\frac{5}{6}$ of $\frac{24}{25}$

12. What is $\frac{2}{3}$ of $\frac{3}{5}$?

$\frac{1}{3}$ of $\frac{3}{5} = \frac{1}{5}$; and $\frac{2}{3}$ of $\frac{3}{5} = 2$ times $\frac{1}{5}$, or $\frac{2}{5}$. Observe that $\frac{2}{3}$ of $\frac{3}{5} = \frac{2}{3} \times \frac{3}{5} = \frac{6}{15}$, or $\frac{2}{5}$.

Multiply the numerators together and the denominators together, and reduce the result to its lowest terms.

Written Work

1. Find $\frac{2}{5}$ of $\frac{4}{7}$. This means $\frac{2}{5} \times \frac{4}{7}$; $\frac{2}{5} \times \frac{4}{7} = \frac{8}{35}$.

Find products:

2. $\frac{4}{5} \times \frac{5}{6}$	5. $\frac{8}{9} \times \frac{5}{8}$	8. $\frac{8}{9} \times \frac{6}{7}$	11. $\frac{3}{11} \times \frac{5}{6}$
3. $\frac{4}{7} \times \frac{8}{9}$	6. $\frac{5}{6} \times \frac{8}{11}$	9. $\frac{1}{5} \times \frac{8}{9}$	12. $\frac{4}{7} \times \frac{6}{7}$
4. $\frac{4}{5} \times \frac{4}{7}$	7. $\frac{6}{7} \times \frac{9}{10}$	10. $\frac{6}{7} \times \frac{4}{5}$	13. $\frac{8}{9} \times \frac{9}{10}$

14. Find $1\frac{1}{2} \times 1\frac{3}{4}$.

Change to improper fractions. Thus, $1\frac{1}{2} \times 1\frac{3}{4} = \frac{3}{2} \times \frac{7}{4} = \frac{21}{8}$, or $2\frac{5}{8}$.

15. $1\frac{2}{3} \times 2\frac{1}{2}$	18. $4\frac{1}{3} \times 2\frac{1}{2}$	21. $4\frac{2}{3} \times 7\frac{1}{4}$	24. $12\frac{1}{2} \times 4\frac{1}{3}$
16. $3\frac{1}{2} \times 2\frac{2}{5}$	19. $1\frac{1}{8} \times 2\frac{1}{4}$	22. $9\frac{2}{3} \times 12\frac{1}{2}$	25. $10\frac{2}{3} \times 3\frac{1}{8}$
17. $3\frac{1}{8} \times 1\frac{1}{3}$	20. $3\frac{3}{4} \times 1\frac{1}{4}$	23. $6\frac{1}{4} \times 9\frac{3}{4}$	26. $12\frac{1}{2} \times 12\frac{1}{2}$

27. Find the cost of $3\frac{1}{2}$ quarts of milk at $8\frac{1}{2} \cancel{c}$ per quart.

28. Find the cost of $8\frac{3}{4}$ yards of gingham at $8\frac{1}{2} \cancel{c}$ a yard.

Use of cancellation in multiplication of fractions.

Written Work

1. Find $\frac{2}{3}$ of $\frac{3}{5}$.

(1) $\frac{2}{3} \times \frac{3}{5} = \frac{6}{15}$, or $\frac{2}{5}$

(2) $\frac{2 \times \cancel{3}}{\cancel{3} \times 5} = \frac{2}{5}$

In changing $\frac{6}{15}$ to $\frac{2}{5}$, both terms of the fraction are divided by 3. Hence, in finding the value of $\frac{2 \times 3}{3 \times 5}$, the work may be shortened by rejecting the factor 3 from both dividend and divisor, as indicated in the second model.

Cancellation is the process of shortening operations by striking out equal factors from both dividend and divisor.

Cancel equal factors from both dividend and divisor, whenever possible.

Thus, $\frac{2}{3} \times 7\frac{1}{2} \times 4 = \frac{2}{\cancel{3}} \times \frac{\cancel{15}^{5}}{\cancel{2}} \times \frac{4}{1} = \frac{20}{1} = 20.$

2. $\frac{5}{8} \times 2\frac{2}{7} \times 2$

3. $1\frac{3}{7} \times 4\frac{3}{8} \times 1\frac{1}{2}$

4. $4 \times 2\frac{1}{2} \times 1\frac{1}{5}$

5. $\frac{7}{9} \times 3 \times 4\frac{1}{2}$

6. $3 \times \frac{2}{3}$ of $\frac{3}{4}$

7. $5\frac{1}{2} \times 3\frac{1}{4} \times 4$

8. $6\frac{1}{8} \times 8 \times 2\frac{1}{7}$

9. $5\frac{1}{4} \times 2\frac{1}{3} \times 4$

10. $\frac{2}{3}$ of $\frac{3}{4} \times 6\frac{1}{2}$

11. $\frac{7}{8} \times 8 \times 12\frac{1}{7}$

12. $\frac{9}{10}$ of $3\frac{3}{4} \times 10$

13. $5\frac{1}{2} \times 2\frac{2}{11} \times 3\frac{1}{8}$

14. $5 \times 3\frac{1}{5} \times 3\frac{3}{8}$

15. $16\frac{1}{5} \times 4\frac{1}{9} \times 3$

16. At $8\frac{3}{4}$ cents per pound, how much is the expressage on a package weighing 64 pounds?

17. At $\$9\frac{5}{6}$ a pair, find the cost of 15 pairs of lace curtains.

18. At the rate of $33\frac{1}{4}$ miles an hour, how far will a train travel in $10\frac{2}{3}$ hours?

19. If a load of coal weighs $5\frac{3}{8}$ tons, find the cost at $\$6\frac{2}{3}$ per ton.

20. When hay is selling at $\$12\frac{1}{4}$ per ton, find the cost of $6\frac{3}{8}$ tons.

PROBLEMS

1. In a certain building, during January, $181\frac{9}{10}$ tons of coal were consumed. If the coal cost $$2\frac{1}{4}$ per ton, what was the total expense for coal?

2. If a soldier is allowed $\frac{3}{8}$ of a pound of meat a day, how much meat will it take to supply 280 men for $6\frac{1}{2}$ days?

3. A coal dealer sold 1850 tons of coal. $\frac{2}{5}$ of his sales were anthracite coal, the rest bituminous. The former he sold for $$6\frac{1}{4}$ per ton, the latter for $$2\frac{1}{4}$ per ton. How much were his total sales worth?

4. A school is open $5\frac{3}{4}$ hours each day. How many hours is it open in a month of 20 school days? in a term of $4\frac{1}{2}$ months?

5. A boy lives $1\frac{3}{8}$ miles from his school and attends 150 days in the term. How many miles does he walk in a term both to and from school?

6. A department store employs 100 cash girls at $$4\frac{3}{4}$ per week and 120 clerks at $$8\frac{3}{4}$ per week. Find the amount paid to all.

7. Find the cost of sewing buttons on 32 suits, at $2\frac{7}{8}$ ¢ a suit.

8. A contractor averages $6\frac{7}{8}$ rods a day in digging a sewer. How long is the sewer if it takes him 39 days to dig it?

9. A rural-mail carrier travels $23\frac{7}{8}$ miles for each delivery. Find the number of miles he travels in making 310 deliveries.

10. An ocean steamer burns on an average $201\frac{5}{16}$ tons of coal in a day. How much coal will it consume in a voyage of 7 days?

DIVISION OF FRACTIONS

Dividing a whole number by a fraction.

Oral Work

1. How many halves are there in this square?

2. How many times is $\frac{1}{2}$ contained in 1?

3. How many times is $\$\frac{1}{2}$ contained in $\$1$? in $\$2$? in $\$4$?

$1 \div \frac{1}{2} = 2.$

4. What is the quotient of 2 *balls* divided by 1 *ball?* of 2 *cents* divided by 1 *cent?* of 2 *halves* divided by 1 *half?* of $\frac{2}{2} \div \frac{1}{2}$? What is the quotient of 4 *cents* divided by 2 *cents?* of 4 *halves* divided by 2 *halves?* of $\frac{4}{2} \div \frac{2}{2}$? of $\frac{8}{2} \div \frac{2}{2}$?

5. How many *fourths* are there in this square? What, then, is the quotient of 1 divided by $\frac{1}{4}$? of 2 divided by $\frac{1}{4}$?

6. How many $\$\frac{1}{4}$ are there in $\$1$? in $\$8$?

7. What is the quotient of 4 *cents* divided by 1 *cent?* of 4 *fourths* divided by 1 *fourth?* of $\frac{4}{4} \div \frac{1}{4}$?

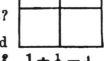

$1 \div \frac{1}{4} = 4.$

8. Since $1 = \frac{4}{4}$, is there any difference between the quotients of $1 \div \frac{1}{4}$ and $\frac{4}{4} \div \frac{1}{4}$? Since $2 = \frac{8}{4}$, is there any difference between the quotients of $2 \div \frac{1}{4}$ and $\frac{8}{4} \div \frac{1}{4}$? Since $3 = \frac{12}{4}$, is there any difference between the quotients of $3 \div \frac{1}{4}$ and $\frac{12}{4} \div \frac{1}{4}$?

9. Explain why $2 \div \frac{1}{2} = \frac{4}{2} \div \frac{1}{2}$; $\frac{2}{3} \div \frac{1}{4} = \frac{8}{12} \div \frac{3}{12}$; $\frac{5}{6} \div \frac{1}{5} = \frac{25}{30} \div \frac{6}{30}$.

Give quotients:

10. $2 \div \frac{1}{2}$	**16.** $5 \div \frac{1}{3}$	**22.** $12 \div \frac{1}{2}$	**28.** $15 \div \frac{1}{2}$
11. $4 \div \frac{1}{2}$	**17.** $6 \div \frac{1}{4}$	**23.** $10 \div \frac{1}{2}$	**29.** $2 \div \frac{1}{8}$
12. $5 \div \frac{1}{2}$	**18.** $6 \div \frac{1}{3}$	**24.** $9 \div \frac{1}{3}$	**30.** $4 \div \frac{1}{8}$
13. $2 \div \frac{1}{8}$	**19.** $8 - \frac{1}{3}$	**25.** $6 \div \frac{1}{5}$	**31.** $5 \div \frac{1}{5}$
14. $4 \div \frac{1}{8}$	**20.** $5 \div \frac{1}{4}$	**26.** $12 \div \frac{1}{4}$	**32.** $16 \div \frac{1}{2}$
15. $3 \div \frac{1}{3}$	**21.** $4 \div \frac{1}{5}$	**27.** $4 \div \frac{1}{8}$	**33.** $10 \div \frac{1}{4}$

Dividing any number by a fraction by inverting the terms of the divisor.

Oral Work

1. How many times is $\frac{1}{2}$ inch contained in 1 inch? How many times is $\frac{1}{3}$ yard contained in 1 yard? $\frac{1}{4}$ foot in 1 foot? Draw figures to illustrate.

When the fraction $\frac{1}{2}$ is changed to $\frac{2}{1}$, the fraction is said to be **inverted**. It then shows how many times the fraction is contained in 1.

2. $1 \div \frac{1}{3} =$ —— $1 \div \frac{1}{4} =$ —— $1 \div \frac{1}{5} =$ ——

3. $1 \div \frac{1}{6} =$ —— $1 \div \frac{1}{7} =$ —— $1 \div \frac{1}{8} =$ ——

Observe that in each of the above problems the quotient equals $1 \times$ *the fraction inverted.*

Thus, $1 \div \frac{1}{3} = 1 \times \frac{3}{1}$, or 3; $1 \div \frac{1}{4} = 1 \times \frac{4}{1}$, or 4; $1 \div \frac{1}{7} = 1 \times \frac{7}{1}$, or 7; $1 \div \frac{1}{5} = 1 \times \frac{5}{1}$, or 5; $1 \div \frac{1}{8} = 1 \times \frac{8}{1}$, or 8.

4. Since $1 \div \frac{1}{3} = 1 \times \frac{3}{1}$, or 3, then $1 \div \frac{2}{3} = 1 \times \frac{3}{2}$, or $1\frac{1}{2}$.

How many times is each of the following fractions contained in 1: $\frac{4}{8}$? $\frac{6}{10}$? $\frac{3}{4}$? $\frac{4}{5}$? $\frac{7}{8}$?

The number of times each of the above fractions is contained in 1 equals the number of times the numerator of each fraction is contained in its denominator.

Find quotients by inverting the divisor and multiplying:

Thus, $3 \div \frac{4}{5} = 3 \times \frac{5}{4} = \frac{15}{4} = 3\frac{3}{4}$.

5.	$1 \div \frac{1}{6}$	12.	$2 \div \frac{2}{5}$	19.	$8 \div \frac{7}{8}$	26.	$15 \div \frac{3}{4}$
6.	$1 \div \frac{2}{8}$	13.	$3 \div \frac{1}{4}$	20.	$16 \div \frac{2}{4}$	27.	$16 \div \frac{2}{3}$
7.	$1 \div \frac{2}{5}$	14.	$4 \div \frac{4}{5}$	21.	$24 \div \frac{4}{5}$	28.	$12 \div \frac{4}{9}$
8.	$1 \div \frac{5}{6}$	15.	$5 \div \frac{5}{6}$	22.	$20 \div \frac{4}{5}$	29.	$8 \div \frac{7}{12}$
9.	$1 \div \frac{3}{5}$	16.	$6 \div \frac{7}{8}$	23.	$15 \div \frac{3}{5}$	30.	$7 \div \frac{8}{15}$
10.	$1 \div \frac{3}{8}$	17.	$9 \div \frac{9}{10}$	24.	$30 \div \frac{2}{3}$	31.	$6 \div \frac{5}{6}$
11.	$1 \div \frac{5}{6}$	18.	$10 \div \frac{4}{9}$	25.	$12 \div \frac{5}{6}$	32.	$9 \div \frac{7}{9}$

*Any number may be **divided** by a fraction by inverting the terms of the divisor and multiplying.*

Written Work

1. Divide $4\frac{1}{2}$ by $\frac{3}{4}$.

$4\frac{1}{2} = \frac{9}{2}$

$\frac{9}{2} \div \frac{3}{4} = \frac{\overset{3}{\cancel{9}}}{\cancel{2}} \times \frac{\overset{2}{\cancel{4}}}{\cancel{3}} = 6.$

Change $4\frac{1}{2}$ to the improper fraction $\frac{9}{2}$. Invert the divisor and multiply, using cancellation. The result is 6.

2. Divide 128 by $\frac{8}{9}$.

$128 \div \frac{8}{9} = \overset{16}{\cancel{128}} \times \frac{9}{\cancel{8}} = 144$

Since $\frac{8}{9}$ is contained in 1, $\frac{9}{8}$ times, it is contained in 128, $128 \times \frac{9}{8}$, or 144 times.

Use cancellation in multiplying.

Find quotients:

3.	$26 \div \frac{1}{3}$	9.	$1\frac{4}{5} \div \frac{9}{10}$	15.	$1\frac{1}{8} \div \frac{3}{4}$
4.	$3\frac{3}{4} \div \frac{2}{3}$	10.	$18 \div \frac{1}{4}$	16.	$2\frac{1}{5} \div \frac{3}{4}$
5.	$25 \div \frac{1}{2}$	11.	$48 \div \frac{2}{3}$	17.	$16 \div \frac{2}{3}$
6.	$2\frac{1}{2} \div \frac{3}{4}$	12.	$12 \div 3\frac{1}{5}$	18.	$3\frac{1}{2} \div \frac{7}{8}$
7.	$2\frac{1}{8} \div \frac{1}{2}$	13.	$16 \div \frac{4}{5}$	19.	$5\frac{3}{8} \div \frac{3}{4}$
8.	$3\frac{1}{3} \div \frac{1}{5}$	14.	$1\frac{2}{3} \div \frac{2}{3}$	20.	$8\frac{4}{9} \div 2\frac{2}{3}$

Divide:

21. 18 by $\frac{2}{3}$	**28.** 63 by $\frac{7}{8}$	**35.** 288 by $\frac{24}{5}$
22. 25 by $\frac{5}{8}$	**29.** 72 by $\frac{8}{9}$	**36.** 400 by $\frac{16}{9}$
23. 28 by $\frac{7}{9}$	**30.** 84 by $\frac{21}{25}$	**37.** 285 by $\frac{15}{9}$
24. 21 by $\frac{3}{7}$	**31.** 90 by $\frac{15}{18}$	**38.** 546 by $\frac{21}{35}$
25. 36 by $\frac{9}{10}$	**32.** 108 by $\frac{12}{13}$	**39.** 425 by $\frac{25}{12}$
26. 42 by $\frac{6}{11}$	**33.** 84 by $\frac{7}{12}$	**40.** 378 by $\frac{14}{15}$
27. 54 by $\frac{9}{7}$	**34.** 96 by $\frac{8}{9}$	**41.** 324 by $\frac{9}{14}$

42. Divide 36 by $3\frac{3}{5}$.

SUGGESTION. — Change the divisor to an improper fraction.

43. 27 by $2\frac{1}{4}$	**49.** 84 by $4\frac{1}{5}$	**55.** 780 by $7\frac{4}{5}$
44. 33 by $3\frac{2}{3}$	**50.** 75 by $2\frac{7}{9}$	**56.** 355 by $8\frac{7}{9}$
45. 44 by $4\frac{2}{5}$	**51.** 90 by $3\frac{3}{4}$	**57.** 295 by $6\frac{5}{9}$
46. 60 by $3\frac{3}{4}$	**52.** 92 by $2\frac{8}{10}$	**58.** 748 by $3\frac{2}{3}$
47. 76 by $4\frac{2}{4}$	**53.** 85 by $1\frac{2}{3}$	**59.** 549 by $8\frac{4}{7}$
48. 60 by $6\frac{2}{3}$	**54.** 245 by $5\frac{5}{6}$	**60.** 620 by $7\frac{3}{4}$

Find quotients:

61. $7\frac{1}{2} \div 1\frac{1}{2}$	**67.** $12\frac{1}{2} \div 6\frac{1}{4}$	**73.** $3\frac{1}{6} \div 9\frac{1}{2}$
62. $6\frac{2}{3} \div 1\frac{2}{3}$	**68.** $14\frac{2}{7} \div 2\frac{6}{7}$	**74.** $3\frac{2}{3} \div 3\frac{1}{7}$
63. $5\frac{1}{4} \div 1\frac{3}{4}$	**69.** $3\frac{1}{4} \div 2\frac{1}{2}$	**75.** $8\frac{2}{5} \div 6\frac{1}{7}$
64. $8\frac{2}{3} \div 2\frac{1}{3}$	**70.** $5\frac{2}{3} \div 4\frac{1}{6}$	**76.** $9\frac{1}{6} \div 3\frac{1}{2}$
65. $7\frac{1}{3} \div 1\frac{1}{6}$	**71.** $6\frac{1}{8} \div 7\frac{1}{4}$	**77.** $3\frac{9}{10} \div 2\frac{3}{5}$
66. $3\frac{1}{2} \div 2\frac{1}{2}$	**72.** $8\frac{1}{3} \div 9\frac{1}{6}$	**78.** $6\frac{1}{2} \div 4\frac{1}{6}$

Divide:

79. $11\frac{3}{5}$ by $3\frac{2}{5}$	**82.** $4\frac{8}{15}$ by $1\frac{5}{12}$	**85.** $10\frac{4}{5}$ by $2\frac{4}{7}$
80. $6\frac{3}{8}$ by $1\frac{3}{9}$	**83.** $7\frac{7}{12}$ by $1\frac{3}{4}$	**86.** $15\frac{5}{8}$ by $2\frac{7}{9}$
81. $10\frac{5}{9}$ by $2\frac{2}{11}$	**84.** $7\frac{2}{5}$ by $2\frac{3}{8}$	**87.** $12\frac{3}{4}$ by $5\frac{2}{3}$

88. How much are eggs per dozen when 72 cents are paid for $2\frac{1}{4}$ dozen?

89. A man's wages amounted to 46 dollars for $9\frac{1}{5}$ days' work. How much did he receive per day?

90. A piece of ribbon containing 10 yards is cut into badges each $\frac{1}{8}$ of a yard in length. How many badges can be cut from the piece?

91. A merchant sold 81 cents worth of ribbon. If he sold $6\frac{3}{4}$ yards, what was the price per yard?

92. At $\$1\frac{1}{2}$ apiece, how many pictures can be bought for $\$10\frac{1}{2}$?

93. At $\$1\frac{1}{4}$ each, how many straw hats can be bought for $\$14\frac{2}{5}$?

94. A man earns $\$16\frac{1}{2}$ in $5\frac{1}{2}$ days. How much is this per day?

95. At $\$2\frac{1}{2}$ per pair, how many pairs of shoes can be bought for $\$17\frac{1}{2}$?

Finding a number when a fractional part of it is given.

Oral Work

1. $\frac{2}{3}$ of a flock of sheep are 40. Find the number in the flock.

SOLUTION. — Since *two* thirds of the flock equal 40 sheep, *one* third of the flock equals $\frac{1}{2}$ of 40 sheep, or 20 sheep, and *three* thirds, or the flock, equal 3×20 sheep, or 60 sheep.

Find the number when:

2. $\frac{2}{3}$ of the number = 12	**7.** $\frac{4}{5}$ of the number = 12
3. $\frac{3}{4}$ of the number = 9	**8.** $\frac{9}{11}$ of the number = 36
4. $\frac{5}{6}$ of the number = 15	**9.** $\frac{12}{13}$ of the number = 60
5. $\frac{7}{8}$ of the number = 21	**10.** $\frac{5}{6}$ of the number = 45
6. $\frac{3}{7}$ of the number = 18	**11.** $\frac{7}{8}$ of the number = 28

12. There are 18 girls in a school. This number is $\frac{2}{3}$ of all the pupils in the school. How many pupils are there in the school?

13. James deposited $18 in the savings bank, which was $\frac{3}{4}$ of what he earned during the month. How much did he earn in the month?

14. May spelled correctly 27 words, which were $\frac{9}{10}$ of all the words given. How many words were given?

15. A farmer sold 42 lambs, which were $\frac{6}{7}$ of his flock. How many lambs had he at first?

16. John has attended school 40 days, which are $\frac{2}{9}$ of the number of days in the term. Find the number of days in the term.

17. Mr. Tanner pays $12 each quarter for his telephone. At the same rate, how much does he pay in a year?

Written Work

1. $500 is $\frac{5}{6}$ of a teacher's salary. Find her salary.

5 of the six parts of her salary = $500.

$$1 \text{ part} = \tfrac{1}{5} \text{ of } \$500, \text{ or } \$100.$$
$$6 \text{ parts, or her salary, } = 6 \times \$100 = \$600.$$

Or, Invert the divisor and multiply. Thus, $\dfrac{6}{5} \times \$\overset{100}{\cancel{500}} = \$600$.

Find the number when:

2. $\frac{8}{15}$ of the number = 56 **5.** $\frac{12}{18}$ of the number = 240 ft.

3. $\frac{12}{17}$ of the number = 108 **6.** $\frac{9}{20}$ of the number = 378 bu.

4. $\frac{5}{16}$ of the number = 275 **7.** $\frac{11}{12}$ of the number = 550 lb.

8. Mr. Arnold bought a horse and a cow. He paid $50 for the cow, which was $\frac{2}{7}$ of what the horse cost. How much did the horse cost?

READING AND WRITING DECIMALS

One tenth may be written .1 as well as $\frac{1}{10}$; $\frac{7}{10}$ may be written .7.

One hundredth may be written .01 as well as $\frac{1}{100}$; $\frac{7}{100}$ may be written .07.

One thousandth may be written .001 as well as $\frac{1}{1000}$; $\frac{7}{1000}$ may be written .007; $\frac{25}{1000}$ may be written .025; and $\frac{125}{1000}$ may be written .125.

Numbers like .1, .7, .01, .07, .25, .001, .025, and .125 are called **decimal fractions,** or **decimals.** The period is called the **decimal point.**

The first place to the right of the decimal point is called **tenths;** the second place, **hundredths;** and the third place, **thousandths.**

1. Write: Nine tenths; nine hundredths; twenty-nine hundredths; nine thousandths; twenty-five thousandths; two hundred twenty-nine thousandths.

2. Read .9, .09, .29, .025, .229, 2.9, 2.09, 2.29, 2.009, 2.229.

NOTE. — Always read the word *and* between the whole number and the decimal.

Read the following:

3. .5	.75	.6	.09	1.5	20.05
4. .05	.075	.66	.99	2.25	9.005
5. .005	.175	.666	.009	2.375	100.001

Write as decimals :

	a	b	c	d	e	f	g	h
6.	$\frac{3}{10}$	$\frac{8}{100}$	$\frac{83}{100}$	$\frac{9}{1000}$	$\frac{19}{1000}$	$\frac{195}{1000}$	$2\frac{1}{10}$	$5\frac{75}{100}$
7.	$\frac{7}{10}$	$\frac{5}{100}$	$\frac{75}{100}$	$\frac{8}{1000}$	$\frac{18}{1000}$	$\frac{125}{1000}$	$3\frac{5}{100}$	$6\frac{25}{1000}$
8.	$\frac{9}{10}$	$\frac{8}{100}$	$\frac{68}{100}$	$\frac{5}{1000}$	$\frac{15}{1000}$	$\frac{875}{1000}$	$6\frac{1}{1000}$	$19\frac{275}{1000}$

PARTS OF $1

1. Show by *addition* how many times 5¢ is contained in $1. Show in the same way how many times 10¢ is contained in $1; 20¢; 50¢; 25¢; $6\frac{1}{4}$¢; $12\frac{1}{2}$¢; $16\frac{2}{3}$¢; $33\frac{1}{3}$¢.

2. Is each of the integers or mixed numbers contained in $1 an integral number of times?

3. Show by addition that 5, 10, 25, 50, $6\frac{1}{4}$, $16\frac{2}{3}$, $12\frac{1}{2}$, $8\frac{1}{3}$, $33\frac{1}{3}$, are each contained in the abstract number 100 an integral number of times.

4. 5¢ is what part of a dollar? $12\frac{1}{2}$¢? $6\frac{1}{4}$¢? $33\frac{1}{3}$¢? 25¢? 50¢? $16\frac{2}{3}$¢? $8\frac{1}{3}$¢? 20¢?

5. 5 is what part of 100? $12\frac{1}{2}$? $33\frac{1}{3}$? 20? 25? 50? $16\frac{2}{3}$? $6\frac{1}{4}$?

6. Learn the following table :

FRACTIONAL PARTS OF $1:   FRACTIONAL PARTS OF 100:

50 ¢ = $$\frac{1}{2}$	$12\frac{1}{2}$ ¢ = $$\frac{1}{8}$	50 $= \frac{1}{2}$ of 100	$12\frac{1}{2} = \frac{1}{8}$ of 100
$33\frac{1}{3}$ ¢ = $$\frac{1}{3}$	10 ¢ = $$\frac{1}{10}$	$33\frac{1}{3} = \frac{1}{3}$ of 100	10 $= \frac{1}{10}$ of 100
25 ¢ = $$\frac{1}{4}$	$8\frac{1}{3}$ ¢ = $$\frac{1}{12}$	25 $= \frac{1}{4}$ of 100	$8\frac{1}{3} = \frac{1}{12}$ of 100
20 ¢ = $$\frac{1}{5}$	$6\frac{1}{4}$ ¢ = $$\frac{1}{16}$	20 $= \frac{1}{5}$ of 100	$6\frac{1}{4} = \frac{1}{16}$ of 100
$16\frac{2}{3}$ ¢ = $$\frac{1}{6}$	5 ¢ = $$\frac{1}{20}$	$16\frac{2}{3} = \frac{1}{6}$ of 100	5 $= \frac{1}{20}$ of 100

7. Tell at sight what fractional part of a dollar each of the following is :

8.	$6\frac{1}{4}$ ¢	**10.**	$8\frac{1}{3}$ ¢	**12.**	$12\frac{1}{2}$ ¢	**14.**	20 ¢	**16.**	50 ¢
9.	25 ¢	**11.**	$33\frac{1}{3}$ ¢	**13.**	$16\frac{2}{3}$ ¢	**15.**	10 ¢	**17.**	5 ¢

What part of 100 is :

18.	50	**20.**	$6\frac{1}{4}$	**22.**	20	**24.**	$16\frac{2}{3}$	**26.**	5
19.	25	**21.**	$8\frac{1}{3}$	**23.**	$12\frac{1}{2}$	**25.**	$33\frac{1}{3}$	**27.**	$12\frac{1}{2}$

28. If gingham costs $12\frac{1}{2}$¢ a yard, find the cost of 48 yards.

SOLUTION. — $12\frac{1}{2}$¢ = \$$\frac{1}{8}$, cost 1 yd. ; $48 \times$ \$$\frac{1}{8}$ = \$$\frac{48}{8}$, or \$6, cost 48 yd.

	NUMBER OF YARDS, POUNDS, ETC., PURCHASED	PRICE PER YARD, POUND, ETC.	TOTAL COST
29.	42 lb. coffee	$33\frac{1}{3}$¢	\$14
30.	32 lb. cheese	$12\frac{1}{2}$¢	?
31.	18 doz. eggs	$16\frac{2}{3}$¢	?
32.	48 baskets berries	$8\frac{1}{3}$¢	?
33.	80 qt. cranberries	$6\frac{1}{4}$¢	?
34.	160 baskets peaches . . .	20¢	?
35.	40 ball bats	25¢	?
36.	50 writing tablets	2 for 25¢	?
37.	60 lb. candy	25¢	?

38. Find the cost of 45 neckties at 75¢ each.

SOLUTION. — 45 neckties at 75¢ (\$$\frac{3}{4}$) cost \$33.75.
The sign @ means at so much a unit.
Thus, 12 lb. @ 10¢ means 12 lb. at 10¢ a pound.

Find the cost of :

39. 35 caps @ 50¢

40. 48 toys @ $6\frac{1}{4}$¢

41. 50 neckties @ 75¢

42. 72 yd. gingham @ $12\frac{1}{2}$¢

43. 40 baskets berries @ $8\frac{1}{3}$¢

44. 50 penknives @ $62\frac{1}{2}$¢

45. 80 boys' shirts @ 50¢

46. 42 yd. cloth @ $33\frac{1}{3}$¢

47. 42 baskets peaches @ $16\frac{2}{3}$¢

48. 90 balls @ $6\frac{1}{4}$¢

49. 40 oranges @ 2¢

50. 50 grapefruits @ $6\frac{1}{4}$¢

51. 20 lemons @ $3\frac{1}{3}$¢

52. 48 gas mantles @ $8\frac{1}{3}$¢

53. 32 books @ $37\frac{1}{2}$¢

54. 40 collars @ 2 for 25¢

55. The unit price of each article purchased is as follows. Give the price as a fractional part of \$1:

$8\frac{1}{3}$¢, $6\frac{1}{4}$¢, 20¢, $16\frac{2}{3}$¢, 25¢, 50¢, 10¢, 5¢, $12\frac{1}{2}$¢, $33\frac{1}{3}$¢.

56. Find the fractional part each number is of 100: 20, 10, 5, 25, 50, $6\frac{1}{4}$, $16\frac{2}{3}$, $8\frac{1}{3}$, $12\frac{1}{2}$, $33\frac{1}{3}$.

57. Find by a short method the number of baskets of strawberries purchased at $8\frac{1}{3}$¢ each, if \$2.00 worth are bought.

SOLUTION. — As $8\frac{1}{3}$¢ is \$$\frac{1}{12}$, at $8\frac{1}{3}$¢ each, you can buy 12 baskets for \$1. For \$2.00 you can buy 2 × 12 baskets, or 24 baskets.

Make and solve problems with the following conditions:

COST PRICE	AMOUNT OF PURCHASE	NUMBER PURCHASED
58. 10 ¢ each	\$ 3.00	?
59. $6\frac{1}{4}$ ¢ each	4.00	?
60. $12\frac{1}{2}$ ¢ each	5.00	?
61. $8\frac{1}{3}$ ¢ each	6.00	?
62. 20 ¢ each	7.00	?
63. 25 ¢ each	8.00	?
64. 50 ¢ each	20.00	?
65. $16\frac{2}{3}$ ¢ each	25.00	?
66. $6\frac{1}{4}$ ¢ each	10.00	?

Divide by the short method:

67. \$2 by 10¢	**71.** \$8 by 5¢	**75.** \$12 by $12\frac{1}{2}$¢
68. \$8 by 25¢	**72.** \$12 by 10¢	**76.** \$20 by $16\frac{2}{3}$¢
69. \$5 by $6\frac{1}{4}$¢	**73.** \$10 by $33\frac{1}{3}$¢	**77.** \$40 by 20¢
70. \$3 by $8\frac{1}{3}$¢	**74.** \$12 by 50¢	**78.** \$5 by $6\frac{1}{4}$¢

79. Make problems asking for the cost of a number of articles at $6\frac{1}{4}$¢, $16\frac{2}{3}$¢, 20¢, 25¢, 50¢, $8\frac{1}{3}$¢. and $12\frac{1}{2}$¢, purchased for a given amount of money.

MEASURES. DRAWING TO SCALE

DENOMINATE NUMBERS

Oral Work

A denominate number is a concrete number whose unit is a measure established by custom or law; as, 5 yards, or 8 bushels, in which 1 yard and 1 bushel are the units of measure.

1. Give orally the table for measuring *liquids;* then write this table with proper abbreviations.

2. Write the names of the articles found in a grocery store that are sold by *liquid* measure.

3. Give orally the table used for measuring dry and bulky articles found in a grocery store; then write this table with proper abbreviations.

4. Write the names of the articles found in a grocery store sold by *dry* measure.

5. Give orally the table· used for measuring *coal, sand, lime, cement, hay,* etc.; then write the table with proper abbreviations.

6. Give the names and prices of different articles sold by the *hundredweight* or by the *ton.*

7. Give orally the table for measuring *time;* then write it with proper abbreviations.

8. Name the uses that are made of a *foot* ruler and a *yard* stick. Are these used for measuring short or long distances? What units of measure are used for measuring long distances?

9. Give orally the table used for measuring *distance;* then write it with proper abbreviations.

10. Write the names of the different measures, and write each of the following under its proper measure : milk, oil, vinegar, spices, oats, hay, molasses, sugar, rice, cloth, coal, potatoes, the length of the blackboard, the width of a page of your book.

Written Work

To the Teacher. — Show the pupil that changing to a lower denomination in denominate numbers, as, 1 pk. to 8 qt., is just the same in principle as changing $\frac{1}{4}$ to $\frac{4}{16}$.

11. Change $1\frac{1}{2}$ pecks to quarts.

Solution. — 1 pk. = 8 qt.

$1\frac{1}{2}$ pk. = $1\frac{1}{2}$ × 8 qt., or 12 qt.

Copy, and fill the blanks :

12. $3\frac{1}{2}$ pk.	=—— qt.		**27.** $1\frac{1}{2}$ min.	=—— sec.
13. 1 bu.	=—— pk.		**28.** $1\frac{3}{4}$ gal.	=—— qt.
14. $1\frac{3}{4}$ bu.	=—— pk.		**29.** $3\frac{1}{4}$ pk.	=—— qt.
15. 16 qt.	=—— pt.		**30.** 18 sq. yd.	=—— sq. ft.
16. $3\frac{1}{2}$ qt.	=—— pt.		**31.** $8\frac{1}{4}$ yd.	=—— ft.
17. $8\frac{1}{4}$ ft.	=—— in.		**32.** $2\frac{1}{4}$ sq. ft.	=—— sq. in.
18. 6 yd.	=—— ft.		**33.** $\frac{7}{8}$ pk.	=—— qt.
19. 3 gal.	=—— qt.		**34.** $\frac{3}{4}$ pk.	=—— pt.
20. $5\frac{1}{2}$ gal.	=—— qt.		**35.** 5 lb.	=—— oz.
21. $6\frac{1}{2}$ lb.	=—— oz.		**36.** $3\frac{1}{2}$ yd.	=—— in.
22. $6\frac{3}{4}$ lb.	=—— oz.		**37.** $1\frac{3}{4}$ bu.	=—— pt.
23. $1\frac{3}{4}$ bu.	=—— qt.		**38.** $\frac{2}{3}$ yd.	=—— ft.
24. $1\frac{1}{4}$ da.	=—— hr.		**39.** $1\frac{7}{8}$ rd.	=—— ft.
25. $\frac{3}{4}$ pk.	=—— qt.		**40.** $1\frac{1}{4}$ mi.	=—— rd.
26. $1\frac{3}{4}$ T.	=—— lb.		**41.** $2\frac{1}{4}$ cu. ft.	=—— cu. in.

42. Change 36 feet to yards.

SOLUTION.—3 ft. = 1 yd. 36 ft. = (36 ÷ 3) yd., or 12 yd.

Change :

43. 92 ft. to yards.

44. 36 pt. to gallons.

45. 24 pk. to bushels.

46. 128 oz. to pounds.

47. 32 qt. to gallons.

48. 48 hr. to days.

49. 55 yd. to rods.

50. 129 ft. to yards.

51. 328 qt. to pecks.

52. 4000 lb. to tons.

53. 320 oz. to pounds.

54. 54 in. to feet.

DRAWING TO SCALE

1. This oblong represents a room 40 ft. long and 30 ft. wide, drawn on a scale of 40 feet to 1 inch. That is, the length, 40 ft., is represented by 1 in. and the width, 30 ft., by $\frac{3}{4}$ in. Represent the same room on a scale of 10 feet to 1 inch ; of 20 feet to 1 inch.

30 ft.

40 ft.
SCALE 40 ft. to 1 in.

2. Mary lives 50 rods from school. Draw a line to represent this distance, on a scale of 10 rods to 1 inch.

3. The following lines are drawn on a scale of 10 feet to 1 inch. Measure the lines with your ruler and find the distances represented.

a. —————————— —————————————

b. ————————————————————————————

c. ————————————————————

d. ————————————————

e. ——————————————

4. An oblong measures $3\frac{1}{2}$ in. by $4\frac{1}{2}$ in. If the scale is 10 feet to 1 inch, what is its length and width ?

On the scale indicated by the heading, find what lengths or distances are represented by the following:

1 in. = 10 ft.	**10 yd. = 1 in.**	**10 rd. = 1 in.**	**10 mi. = 1 in.**
5. 4 in.	10. 3 in.	15. 1 in.	20. $1\frac{1}{2}$ in.
6. $5\frac{3}{4}$ in.	11. 4 in.	16. $1\frac{1}{2}$ in.	21. $2\frac{3}{4}$ in.
7. $2\frac{3}{4}$ in.	12. $4\frac{1}{2}$ in.	17. $2\frac{1}{2}$ in.	22. $3\frac{1}{4}$ in.
8. $5\frac{1}{2}$ in.	13. $3\frac{1}{2}$ in.	18. 5 in.	23. $6\frac{1}{2}$ in.
9. $6\frac{1}{4}$ in.	14. $\frac{1}{2}$ in.	19. $5\frac{1}{2}$ in.	24. $2\frac{1}{4}$ in.

Draw oblongs on a suitable scale to represent the following dimensions:

25. 6 ft. by 9 ft.
26. 12 ft. by 15 ft.
27. 20 ft. by 30 ft.
28. 40 rd. by 80 rd.
29. 60 rd. by 110 rd.
30. 75 rd. by 120 rd.
31. 20 yd. by 45 yd.
32. 35 yd. by 55 yd.
33. 65 yd. by 95 yd.
34. 14 ft. by 28 ft.

TO THE TEACHER. — Explain to pupils that all maps, plans, etc., must be drawn on a suitable scale.

PROBLEMS IN DENOMINATE NUMBERS

1. James sold 8 qt. and 1 pt. of milk on Tuesday. How many pints did he sell?

8 qt. = 8 × 2 pt.	= 16 pt.
1 pt.	= 1 pt.
8 qt. 1 pt.	= 17 pt.

Since 1 qt. = 2 pt., 8 qt. = 8 × 2 pt., or 16 pt. Therefore, 8 qt. 1 pt. = 17 pt.

2. Henry picked 1 bu. 3 qt. of berries, and sold them for 10¢ a quart. How much did he receive for them?

3. Mary picked 1 bu. and 1 pk. of cranberries, and sold them at 8¢ a quart. How much did she receive for them?

4. William works 3 hr. and 20 min. each day at 10¢ an hour. How much does he earn in 6 days?

5. Nell hemstitched 1¾ doz. handkerchiefs at 3⊄ apiece. How much did she receive for her work?

6. George and Donald bought 1 bu. and 3 pk. of potatoes at 60⊄ a bushel, and planted them in the spring. In the fall they sold 6½ bu. at 20⊄ a half-peck. How much did they realize from the sale of these potatoes?

7. Mary helps her sister an hour and 40 minutes each day for 10⊄ an hour. How much does she earn in 6 days?

8. Clay and Willie picked chestnuts on Saturday. Willie picked 7 qt. and Clay picked 5 qt. If they sold them at 5⊄ a pint, how much did they get for them?

9. Mr. Smith feeds his horse 4 qt. of oats three times a day. How much are the oats worth in a month of thirty days at 40⊄ a bushel?

10. If a man earns 20⊄ an hour, how much will he earn at that rate in 3 hr. and 45 min.?

11. Play that you are picking fruit, such as cherries, plums, apples, etc., and selling it at so much per quart, peck, etc.

12. How much is gained in buying a bushel of apples at $1.00, and selling them at 20⊄ a half-peck?

13. How much is gained in buying a bushel of onions at $1.50 per bushel and selling them at 10⊄ a quart?

14. James sold from his school garden ½ a bushel of peas at 15⊄ a pint, ¾ of a bushel of string beans at 40⊄ a bushel, and a quarter-peck of potatoes for 10⊄. How much did he receive in all for these vegetables?

15. At 4⊄ a pint, find the cost of 6 gal. 1 pt. of milk.

16. At 8⊄ a quart, find the cost of 4 bu. 2 pk. of beans.

17. At 4⊄ an ounce, find the cost of 3 lb. 3 oz. of ginger.

18. At 24⊄ a pound, find the cost of 29 lb. of butter.

PRACTICAL PROBLEMS

To THE TEACHER. — Have pupils *first* interpret these problems; *second*, estimate the result; *third*, give the shortest method of solution.

1. If 2 pounds of butter cost 60 ¢, how much will 5 pounds cost?

HINT. — 5 pounds cost 2½ times the cost of 2 pounds.

$$2\tfrac{1}{2} \times 60 \text{¢} = ?$$

2. If 5 tablets cost 50 ¢, how much will 10 tablets cost?

3. If 2 pairs of gloves cost $1.50, how much will 4 pairs cost? 8 pairs? 6 pairs?

4. If 2 pencils cost 5 ¢, how much will 10 pencils cost? 6 pencils? 12 pencils?

5. If 3 pounds of butter cost $1.20, how much will 6 pounds cost?

HINT. — 6 pounds cost 2 times the cost of 3 pounds.

$$2 \times \$1.20 = ?$$

6. If 6 hats cost $9.00, how much will 12 hats cost? 24 hats?

7. If a train runs 52 miles in 2 hours, how far will it run in 22 hours? in 14 hours? in 12 hours?

8. If 4 sets of books cost $75, how much will 9 sets cost? 16 sets? 8 sets?

HINT. — 9 sets = 2¼ times 4 sets.

9. If 2 cans of corn cost 25 ¢, how much will 10 cans cost? 14 cans? 20 cans?

10. If 3 bolts of ribbon cost 10 ¢, how many bolts can be bought for 25 ¢? for 45 ¢?

HINT. — 25 ¢ is how many times 10 ¢?

11. If 10 hanks of yarn can be bought for $2, how many can be bought for $8? for $12? for $18?

12. If 5 quarts of strawberries cost 35 ¢, how much will 15 quarts cost? 25 quarts? 9 quarts?

HINT. — 9 quarts are 1 less than 10 quarts.

13. If 5 yards of serge cost $6.50, how much will 9 yards cost? 15 yards? 20 yards? 11 yards?

14. 3½ lb. beefsteak cost $1.05. At that rate find the cost of 28 lb.

15. A man's wages for 3¾ hours are 76¢. At that rate find his wages for 38 hours.

16. ¼ of a man's profits for a year is $219.37. Find ⅔ of his profits.

17. ⅝ of a ton of coal cost $3.60. At that rate find the cost of 1¼ tons.

18. 1¼ lb. veal cost 42¢. At that rate find the cost of 10½ lb.

19. 3½ lb. round steak cost 52½¢. Find the cost of 28 lb. at the same rate.

20. John's profits the first 3 months of the year are $379.85. Find his profits at that rate for ¾ of a year.

21. If 2 desks cost $15, how much will 9 desks cost? 11 desks? 15 desks?

22. If 3 yards of maline can be bought for $1.50, how many yards can be bought for $4.50? $6? $12?

23. When 8 yards of velvet cost $24, how much will ¾ of a yard cost?

24. When milk is selling at 5¢ a quart, how much will 3 gallons cost?

FIFTH GRADE — SECOND HALF

REVIEW

ADDITION

Oral Work

Add:

	a	b	c	d	e	f	g	h	i	j	k
1.	4	4	0	2	1	3	3	7	9	0	8
	3	2	1	5	6	5	4	2	1	8	7
	4	5	8	7	2	0	5	7	4	3	8
	5	4	7	2	8	2	6	8	5	4	8
	4	6	8	8	7	4	6	5	4	8	9

To the Teacher. — Let pupils add by groups orally. Thus, in 1 a, 9, 20. Time each pupil until speed and accuracy is attained. Give single column work of 5 numbers; then of 6, 7, 8, and 9 numbers until you are certain that the pupil has acquired a mastery of the 45 combinations in addition, and that he can see at a glance the groups making 5, 10, 15, or 20.

Written Work

Add, observing groups that make 5, 10, 15, 20, etc. Test by adding upwards:

	1.		2.		3.		4.
	14223		54724		60000		70927
	37245		57632		59725		52051
	45936		18206		34761		61847
	50649		32391		9892		42536
	3261		1001		47634		9910
	702		955		1078		49068
	30485		1417		21046		5099

5.		6.		7.		8.	
	33313		14542		21424		90009
	91737		43036		97835		56721
	41952		33355		42067		97856
	10090		44492		72109		23081
	79973		40001		19055		59039
	5681		10088		64562		17840
	22200		80063		8778		60029
	79942		56702		60021		60072

Explain what must be observed in setting down these sums to be added. Then add and test:

9.	$ 4.40	10.	$ 5.10	11.	$ 9.10	12.	$ 5.19
	82.02		26.98		15.10		29.10
	43.		34.25		27.75		88.25
	12.05		76.80		105.05		56.10
	498.04		155.05		73.10		34.50

13.	$ 8.07	14.	$ 3.75	15.	$ 6.50	16.	$ 8.20
	92.09		49.10		35.52		29.25
	21.10		34.50		1.		.01
	34.02		62.98		23.06		75.50
	82.75		1.65		17.10		63.50
	59.01		25.60		181.19		43.92

17.	$ 4.29	18.	$ 5.60	19.	$ 4.98	20.	$ 6.05
	23.72		29.01		14.09		12.99
	81.70		38.19		26.50		90.68
	36.01		42.05		95.27		.69.10
	49.25		74.92		77.76		76.23
	31.07		81.74		58.50		19.08
	64.01		25.40		23.17		29.00
	52.90		19.10		55.55		38.50

21.	$ 2.67	22.	$ 5.32	23.	$ 8.10	24.	$ 5.78
	45.80		98.48		91.00		81.00
	28.10		164.00		82.01		32.40
	91.67		23.69		76.49		172.10
	43.90		11.08		21.11		635.42
	97.00		25.27		135.01		21.00
	8.01		34.00		21.10		72.10

25.	$ 24.01	26.	$ 1.45	27.	$ 26.00	28.	$ 2.62
	8.77		108.20		145.10		61.78
	92.40		211.70		22.00		139.99
	72.00		934.01		9.10		43.00
	80.10		321.99		38.60		241.00
	55.00		534.11		42.50		22.90
	63.44		56.00		41.00		33.10

SUBTRACTION

Oral Work

1. Explain the terms used in subtraction.

2. From 74 take 53. Thus, $74 - 50 = 24$; $24 - 3 = 21$.

Find differences:

	a	b	c	d	e
3.	$95 - 39$	$63 - 47$	$76 - 54$	$15 - 5$	$60 - 45$
4.	$58 - 29$	$90 - 64$	$50 - 25$	$35 - 16$	$87 - 42$
5.	$71 - 48$	$81 - 47$	$78 - 39$	$56 - 42$	$67 - 45$
6.	$54 - 25$	$78 - 48$	$67 - 48$	$99 - 68$	$25 - 15$
7.	$71 - 69$	$88 - 29$	$58 - 39$	$23 - 15$	$89 - 52$
8.	$39 - 26$	$80 - 65$	$45 - 28$	$54 - 37$	$90 - 54$
9.	$58 - 34$	$45 - 26$	$60 - 40$	$77 - 62$	$85 - 43$

Subtract:

	a	b	c	d	e	f	g	h	i	j	k
10.	23	34	37	87	34	42	85	49	60	98	87
	7	8	9	8	9	7	8	8	6	9	8
11.	31	72	83	91	61	92	83	42	17	51	43
	9	7	8	8	5	7	5	6	9	8	6
12.	48	52	65	71	78	71	56	84	53	56	83
	9	7	8	7	9	8	9	7	9	4	7
13.	35	44	73	57	58	93	81	74	87	73	92
	8	9	8	8	7	7	9	5	0	9	7

Written Work

Subtract and test.   Time your work, and try to beat your record:

	a	b	c	d	e	f
1.	304	740	985	508	703	745
	289	387	507	390	598	690
2.	9708	4503	5207	3052	4163	8157
	3201	4292	1908	2954	2491	3299
3.	4140	4159	1908	4507	4562	3875
	2977	2786	1454	2972	3989	2651
4.	4062	5098	6703	5192	5099	3029
	3903	3673	4487	4066	4897	2544
5.	5007	6050	6505	4808	5806	.4501
	3762	2985	2436	2897	4263	2987

Subtract:

	a	b	c	d	e
6.	40092	70900	60251	70098	40932
	29819	54892	39807	58919	38766
7.	503459	623559	560734	600127	501134
	423602	454683	384842	576774	489289

	a	b	c	d
8.	$250.10	$892.15	$3.726	$22.60
	7.42	64.01	.998	1.60
9.	$328.67	$801.98	$27.021	$5241.09
	115.01	572.89	19.011	3174.77
10.	$3245.90	$9108.60	$285.99	$314.291
	2921.01	7299.01	192.10	189.189
11.	$2965.10	$1985.20	$7818.05	$421.987
	1824.01	465.99	659.16	128.243

	a	b	c	d	e
12.	$2009.70	$600.90	$709.05	$809.70	$700.00
	389.99	398.98	29.87	597.08	334.09
13.	$4008.60	$2690.00	$7785.00	$2531.11	$6100.01
	487.98	348.98	649.65	1348.38	78.49
14.	$9398.01	$8500.00	$2008.40	$3445.20	$2670.00
	7246.99	6356.29	1268.98	1398.77	1598.54
15.	$1400.00	$8780.20	$1980.00	$5008.00	$4467.01
	1250.75	2563.52	845.24	198.99	2247.87
16	$1240.11	$6475.14	$9550.01	$7498.00	$4498.11
	1150.00	4243.87	2480.19	4367.11	1177.98

Subtract:

	a	b	c	d	e
	a	*b*	*c*	*d*	*e*
17.	$2500.01	$6010.55	$7140.09	$8563.01	$3156.26
	1145.62	4198.80	5290.98	6199.99	2148.13
18.	$4190.40	$5007.00	$6145.01	$5501.56	$8675.19
	2989.56	1578.65	2498.45	3350.76	2515.98

MULTIPLICATION

Oral Work

1. Give products rapidly:

9 × 12	6 × 9	5 × 12	6 × 8	9 × 8
8 × 7	8 × 12	6 × 12	9 × 7	6 × 7
7 × 12	9 × 9	12 × 9	11 × 12	12 × 12

2. Multiply quickly by 10, by 100, by 1000: 4; 8; 12; 15; 18; 25; 30; 35; 40; 50; 75.

3. State how the addition of one naught, two naughts, three naughts, etc., to the right of a number affects its value.

Give products:

4. 40 × 20	**6.** 30 × 15	**8.** 18 × 30	**10.** 60 × 40
5. 20 × 20	**7.** 50 × 20	**9.** 80 × 70	**11.** 70 × 20

Written Work

To the Teacher. — This page is designed for a test in speed and accuracy. Many of these problems can be solved by *short methods*. Thus, 11 × 1860 = 10 × 1860 + 1 × 1860, or 18600 + 1860.

Find products:

1. 27 × 35	**3.** 92 × 36	**5.** 67 × 49	**7.** 74 × 39
2. 39 × 47	**4.** 37 × 42	**6.** 53 × 62	**8.** 27 × 69

Find products:

9. 64×82
10. 96×82
11. 99×34
12. 76×45
13. 87×56
14. 98×67
15. 56×14
16. 64×44
17. 74×18
18. 98×99
19. 75×65
20. 90×45
21. 84×67
22. 98×42
23. 34×23
24. 56×76
25. 30×102
26. 45×490
27. 68×304
28. 50×206
29. 54×600
30. 72×503
31. 40×725
32. 68×507
33. 75×462

34. 99×508
35. 99×598
36. 24×462
37. 34×533
38. 45×654
39. 76×871
40. 56×981
41. 53×666
42. 25×500
43. 19×989
44. 59×762
45. 59×63
46. 65×98
47. 45×80
48. 99×999
49. 12×1400
50. 15×2500
51. 16×1500
52. 20×1339
53. 35×3000
54. 22×25
55. 25×24
56. 32×25
57. 19×45
58. 18×35

59. 17×25
60. 17×1203
61. 19×5406
62. 21×3054
63. 307×6022
64. 305×4001
65. 607×7908
66. 656×4807
67. 789×3587
68. 999×9999
69. 100×5555
70. 401×4672
71. 399×5673
72. 968×3404
73. 957×1505
74. 968×9566
75. 490×5707
76. 960×7658
77. 850×4309
78. 869×2507
79. 960×5003
80. 101×8002
81. 110×4501
82. 948×7620
83. 859×6573

DIVISION

Oral Work

Divide the following numbers first by 6; then by 5; by 4;
by 7; 8; 9; 10; 11; 12; 20:

432	624	709	8140	9102
354	248	900	9210	7054
816	526	806	7000	6033

Written Work

Find quotients and test:

1. 9500 ÷ 20	19. 333333 ÷ 51	37. 77445 ÷ 860
2. 5060 ÷ 45	20. 459875 ÷ 65	38. 249681 ÷ 120
3. 5965 ÷ 56	21. 648650 ÷ 76	39. 386406 ÷ 200
4. 9543 ÷ 34	22. 185679 ÷ 18	40. 10505 ÷ 500
5. 4516 ÷ 87	23. 93567 ÷ 40	41. 60840 ÷ 600
6. 9842 ÷ 79	24. 50617 ÷ 10	42. 72400 ÷ 800
7. 9819 ÷ 95	25. 445119 ÷ 15	43. 97009 ÷ 694
8. 444444 ÷ 19	26. 590170 ÷ 25	44. 78007 ÷ 892
9. 459606 ÷ 72	27. 7688 ÷ 31	45. 52475 ÷ 623
10. 45675 ÷ 93	28. 567800 ÷ 24	46. 70601 ÷ 961
11. 486501 ÷ 41	29. 50875 ÷ 25	47. 70075 ÷ 560
12. 96543 ÷ 62	30. 65700 ÷ 40	48. 42882 ÷ 745
13. 548672 ÷ 33	31. 99660 ÷ 33	49. 48618 ÷ 125
14. 56400 ÷ 54	32. 36550 ÷ 10	50. 26099 ÷ 900
15. 10456 ÷ 30	33. 15789 ÷ 15	51. 14014 ÷ 145
16. $\frac{468502}{20}$	34. 484800 ÷ 48	52. 300909 ÷ 308
17. $\frac{859185}{7}$	35. 25560 ÷ 18	53. 56680 ÷ 706
18. $\frac{104858}{66}$	36. 4488 ÷ 22	54. 367047 ÷ 734

PROBLEMS

Oral Work

1. How many eggs are there in 6 dozen?

SOLUTION. — Since there are 12 eggs in one dozen, in 6 dozen there are 6 × 12 eggs, or 72 eggs.

2. How many trees are there in an orchard if there are 11 rows and 10 trees in each row?

3. James raised, on an average, 7 bushels of potatoes from each of 10 rows. How many bushels did he raise?

What is the cost of:

4. 10 quarts of cherries at 8 ¢ per quart?

5. 9 quarts of milk at 7 ¢ per quart?

6. 8 bushels of apples at $2 per bushel?

7. A twelve-pound cheese at 12 ¢ per pound?

8. 3 pecks of apples at 25 ¢ per peck?

9. How far does a boy ride on his bicycle in 4 hours at the rate of 9 miles per hour?

10. How many miles are there in 4 streets, if the streets average 12 miles?

11. There are 32 quarts in a bushel. Find the number of quarts in 13 bushels.

12. How far does an automobile run in 4 hours, if it averages 14 miles per hour?

13. Find the cost of posting 18 letters at 2 ¢ each.

14. Find the cost of a 10-pound turkey at 26 ¢ per pound.

15. A lady purchased 2 dozen oranges at 40 ¢ per dozen. How much did they cost her?

16. It takes John 15 minutes to walk to school. How many minutes will it take him to walk to school 60 times?

17. Frank used 12 tablets at 10¢ each. How much did they cost?

18. Find the cost of 3 dozen oranges at 20¢ per dozen.

19. At 12 cents a quart for berries, how much will 8½ quarts cost?

20. At 30¢ a peck, how much will ½ peck of beans cost?

21. When peaches are 50 cents a basket, how much will 7 baskets cost?

22. A ton of coal costs $4¾. How much will 9 tons cost?

23. A boy rides his wheel 6 miles in one hour. How far will he ride in 7⅔ hours?

24. A man leaves home at 6:30 A.M. and returns at 5:45 P.M. How long is he away from home?

25. A train leaves the station at 11:10 A.M. It requires 25 minutes to reach the station. At what time must I leave home in order to catch this train?

26. Harry leaves for school at 8:30 A.M., and reaches school at 3 minutes before 9 o'clock. How long is he on the way?

27. The morning session of school begins at 9 A.M. and closes at 11:30 A.M. The afternoon session begins at 1 P.M. and closes at 3:45 P.M. How long in hours and minutes are both sessions?

28. A dealer buys 150 bales of hay, averaging 90 pounds to the bale. How many tons and pounds over does he buy?

29. At 3 cents an ounce, how much will 1 pound of mustard cost?

30. Find the cost of 100 lb. of nails at 5½ cents a pound.

Written Work

1. 2 tons of rolled oats were packed in pound packages. How many packages were there?

2. How many ounces are there in a ton?

3. A load of hay weighed 3000 pounds. How many tons did it weigh? What was its value at $14 a ton?

4. Mr. Hosack feeds his horse 6 quarts of oats 3 times a day. How many bushels of oats does he feed the horse during November, December, and January?

5. How many 8-ounce packages of soda can be put up from 1 ton 300 pounds of soda?

6. A field is 80 rods long and 320 feet wide. How many yards is it around the field?

7. There are 5280 feet in a mile. How many feet is it from Albany to New York, a distance of 143 miles?

8. How many feet of picture molding will be required for a room 18 ft. long and 12 ft. wide? At 6¢ a foot, how much will it cost?

9. Each of 8 boxes holds 5 pounds 4 ounces of meal. How much do all the boxes hold?

10. What is the cost of 8 barrels of vinegar, averaging 41 gallons 3 quarts per barrel, at 18¢ a gallon?

11. A building is 46 feet 3 inches wide, and twice as long as wide. Find the distance around the building.

12. Mr. Bell picked 510 quarts, 380 quarts, 467 quarts of berries. How many bushels and quarts did he pick?

13. Let each pupil write and solve 5 *two-step* problems about farms, gardens, purchases in stores, etc.

COMMON FRACTIONS

REDUCTION OF FRACTIONS

Oral Work

1. Change to eighths : $\frac{1}{4}$, $\frac{1}{2}$, $\frac{3}{4}$, $\frac{2}{4}$.

2. Change to twelfths : $\frac{2}{3}$, $\frac{3}{4}$, $\frac{2}{4}$, $\frac{1}{6}$, $\frac{2}{6}$, $\frac{1}{3}$, $\frac{5}{6}$, $\frac{3}{6}$.

3. Change to sixteenths : $\frac{2}{8}$, $\frac{3}{4}$, $\frac{1}{2}$, $\frac{7}{8}$, $\frac{5}{8}$, $\frac{3}{8}$.

4. Change $\frac{8}{12}$ to thirds, $\frac{6}{10}$ to halves, $\frac{8}{16}$ to fourths.

Change to lowest terms :

5. $\frac{4}{6}$, $\frac{6}{12}$, $\frac{4}{8}$, $\frac{9}{12}$, $\frac{4}{16}$, $\frac{8}{24}$.

6. $\frac{2}{8}$, $\frac{10}{12}$, $\frac{8}{16}$, $\frac{11}{24}$, $\frac{10}{20}$.

7. $\frac{8}{32}$, $\frac{4}{24}$, $\frac{9}{36}$, $\frac{12}{45}$, $\frac{11}{22}$, $\frac{12}{18}$, $\frac{24}{16}$.

8. $\frac{4}{12}$, $\frac{8}{12}$, $\frac{10}{60}$, $\frac{8}{12}$, $\frac{18}{48}$, $\frac{24}{48}$, $\frac{15}{18}$, $\frac{16}{20}$.

9. $\frac{15}{24}$, $\frac{16}{32}$, $\frac{20}{36}$, $\frac{14}{28}$, $\frac{40}{28}$, $\frac{18}{27}$, $\frac{6}{8}$.

10. $\frac{18}{36}$, $\frac{18}{26}$, $\frac{14}{24}$, $\frac{9}{45}$, $\frac{7}{63}$, $\frac{6}{16}$, $\frac{9}{81}$, $\frac{15}{36}$.

Read as improper fractions :

11. $3\frac{3}{4}$, $4\frac{5}{8}$, $5\frac{7}{12}$, $5\frac{3}{8}$, $4\frac{3}{8}$, $5\frac{1}{8}$, $9\frac{5}{16}$.

12. $14\frac{3}{12}$, $5\frac{1}{6}$, $6\frac{2}{8}$, $8\frac{1}{4}$, $5\frac{3}{4}$, $4\frac{3}{8}$, $5\frac{2}{3}$.

13. $2\frac{1}{4}$, $3\frac{1}{8}$, $6\frac{2}{3}$, $4\frac{2}{8}$, $6\frac{2}{3}$, $5\frac{7}{12}$, $4\frac{3}{12}$.

14. $8\frac{2}{3}$, $3\frac{2}{8}$, $4\frac{6}{10}$, $5\frac{1}{8}$, $4\frac{1}{4}$, $3\frac{6}{12}$.

You have learned that a fraction is a **part of a unit.** Thus, $\frac{1}{2}=\frac{1}{2}$ of 1, $\frac{1}{3}=\frac{1}{3}$ of 1, $\frac{2}{3}=\frac{1}{3}$ of 2 or $\frac{2}{3}$ of 1, $\frac{3}{4}=\frac{1}{4}$ of 3 or $\frac{3}{4}$ of 1.

A fraction may also be regarded as an **expression of division** in which the numerator is the dividend and the denominator, the divisor. Thus, $\frac{1}{2}=1\div 2$, $\frac{1}{3}=1\div 3$, $\frac{2}{3}=2\div 3$; $\frac{16}{5}=16\div 5=3\frac{1}{5}$; $\frac{16}{4}=16\div 4=4$.

Change to integers or mixed numbers by dividing the numerator by the denominator :

15. $\frac{15}{8}$, $\frac{12}{4}$, $\frac{9}{3}$, $\frac{12}{6}$, $\frac{24}{6}$, $\frac{10}{5}$, $\frac{48}{12}$, $\frac{15}{5}$, $\frac{12}{6}$.

16. $\frac{9}{2}$, $\frac{18}{8}$, $\frac{11}{4}$, $\frac{15}{6}$, $\frac{7}{2}$, $\frac{17}{5}$, $\frac{19}{6}$, $\frac{21}{8}$, $\frac{19}{8}$, $\frac{21}{4}$, $\frac{47}{8}$.

17. $\frac{32}{8}$, $\frac{31}{6}$, $\frac{42}{7}$, $\frac{48}{12}$, $\frac{16}{8}$, $\frac{56}{9}$, $\frac{28}{7}$, $\frac{14}{6}$, $\frac{27}{8}$.

18. $\frac{64}{8}$, $\frac{25}{2}$, $\frac{21}{7}$, $\frac{45}{9}$, $\frac{72}{9}$, $\frac{14}{7}$, $\frac{84}{8}$, $\frac{18}{2}$, $\frac{15}{4}$.

19. $\frac{56}{8}$, $\frac{36}{9}$, $\frac{81}{9}$, $\frac{42}{4}$, $\frac{17}{4}$, $\frac{16}{3}$, $\frac{52}{8}$, $\frac{19}{12}$, $\frac{24}{12}$.

20. $\frac{64}{12}$, $\frac{37}{12}$, $\frac{14}{3}$, $\frac{48}{10}$, $\frac{50}{10}$, $\frac{64}{8}$, $\frac{29}{7}$, $\frac{45}{9}$.

The **ratio** of two numbers of the same kind is found by dividing the first by the second. Thus, the ratio of 4 to 8 is $4 \div 8$, or $\frac{4}{8}$, or $\frac{1}{2}$. The ratio of 8 to 4 is $8 \div 4$, or 2. The ratio of 3 ft. to 4 ft. is $3 \div 4$, or $\frac{3}{4}$. The ratio of 4 to 3 is $\frac{4}{3}$.

It is evident, therefore, that a fraction expresses a ratio.

Give the ratio of:

21. 2 to 3	**25.** 10 to 5	**29.** 5 to 10	**33.** 20 to 30
22. 4 to 5	**26.** 3 to 9	**30.** 15 to 20	**34.** 30 to 20
23. 6 to 8	**27.** 7 to 10	**31.** 8 to 12	**35.** 25 to 75
24. 1 to 7	**28.** 9 to 3	**32.** 5 to 15	**36.** 75 to 25

Change to like or similar fractions:

37. $\frac{1}{4}$ and $\frac{1}{6}$	**41.** $\frac{1}{6}$ and $\frac{1}{12}$	**45.** $\frac{1}{4}$ and $\frac{2}{7}$
38. $\frac{2}{3}$ and $\frac{5}{6}$	**42.** $\frac{2}{3}$ and $\frac{1}{4}$	**46.** $\frac{2}{3}$ and $\frac{1}{6}$
39. $\frac{3}{8}$ and $\frac{2}{3}$	**43.** $\frac{3}{4}$ and $\frac{1}{3}$	**47.** $\frac{3}{8}$ and $\frac{1}{4}$
40. $\frac{5}{8}$ and $\frac{3}{4}$	**44.** $\frac{2}{3}$ and $\frac{5}{8}$	**48.** $\frac{3}{4}$ and $\frac{1}{2}$

Change to fractions having their l. c. d.:

49. $\frac{3}{4}$, $\frac{1}{2}$, $\frac{3}{8}$	**51.** $\frac{1}{8}$, $\frac{3}{4}$, $\frac{8}{10}$	**53.** $\frac{3}{4}$, $\frac{5}{8}$, $\frac{3}{16}$	**55.** $\frac{3}{8}$, $\frac{1}{4}$, $\frac{7}{8}$
50. $\frac{3}{4}$, $\frac{1}{3}$, $\frac{2}{5}$	**52.** $\frac{2}{3}$, $\frac{3}{5}$, $\frac{3}{10}$	**54.** $\frac{1}{2}$, $\frac{3}{8}$, $\frac{3}{4}$	**56.** $\frac{3}{10}$, $\frac{4}{5}$, $\frac{3}{8}$

ADDITION AND SUBTRACTION OF FRACTIONS

Find the sums:

57. $\frac{3}{4}$ and $\frac{3}{16}$	**60.** $\frac{15}{16}$ and $\frac{5}{12}$	**63.** $\frac{7}{20}$ and $\frac{1}{4}$
58. $\frac{5}{8}$ and $\frac{1}{6}$	**61.** $\frac{9}{20}$ and $\frac{1}{4}$	**64.** $\frac{7}{12}$ and $\frac{7}{30}$
59. $\frac{11}{12}$ and $\frac{1}{8}$	**62.** $\frac{5}{8}$ and $\frac{1}{6}$	**65.** $\frac{3}{8}$ and $\frac{1}{6}$

66–74. Find the differences in each example from 57 to 65.

Find the sums or differences as indicated:

75. $\frac{3}{4} + \frac{1}{2}$ 79. $\frac{3}{4} - \frac{5}{8}$ 83. $\frac{3}{4} - \frac{2}{3}$ 87. $\frac{3}{4} - \frac{5}{12}$

76. $\frac{2}{5} + \frac{3}{10}$ 80. $\frac{7}{10} + \frac{1}{2}$ 84. $\frac{2}{5} + \frac{7}{10}$ 88. $\frac{5}{8} + \frac{2}{5}$

77. $\frac{1}{4} + \frac{2}{3}$ 81. $\frac{4}{5} - \frac{1}{2}$ 85. $\frac{5}{6} + \frac{4}{8}$ 89. $\frac{5}{6} - \frac{3}{4}$

78. $\frac{3}{4} + \frac{3}{8}$ 82. $\frac{4}{7} + \frac{4}{5}$ 86. $\frac{1}{2} - \frac{1}{6}$ 90. $\frac{1}{8} + \frac{2}{3}$

Written Work

Add:

1. $18\frac{3}{8}$ 3. $40\frac{3}{4}$ 5. $68\frac{2}{8}$ 7. $75\frac{5}{6}$
   $21\frac{5}{6}$ $12\frac{1}{8}$ $46\frac{3}{4}$ $60\frac{2}{3}$
   $14\frac{3}{8}$ $10\frac{1}{2}$ $34\frac{5}{8}$ $77\frac{5}{12}$

2. $45\frac{1}{2}$ 4. $16\frac{2}{5}$ 6. $77\frac{3}{10}$ 8. $89\frac{1}{2}$
   $48\frac{2}{3}$ $46\frac{5}{6}$ $65\frac{5}{8}$ $12\frac{3}{12}$
   $57\frac{5}{6}$ $75\frac{5}{6}$ $40\frac{3}{4}$ $76\frac{3}{4}$
   $62\frac{3}{4}$ $65\frac{3}{4}$ $97\frac{2}{3}$ $88\frac{5}{6}$

Subtract:

9. $24\frac{1}{2}$ 13. $59\frac{5}{6}$ 17. $25\frac{5}{8}$ 21. $95\frac{3}{7}$ 25. $87\frac{3}{8}$
   $15\frac{3}{4}$ $35\frac{5}{12}$ $15\frac{1}{6}$ $39\frac{3}{4}$ $45\frac{5}{12}$

10. $97\frac{1}{4}$ 14. $25\frac{1}{3}$ 18. $56\frac{7}{8}$ 22. $14\frac{1}{4}$ 26. $78\frac{1}{3}$
    $35\frac{1}{2}$ $10\frac{3}{4}$ $21\frac{2}{3}$ $9\frac{5}{6}$ $48\frac{1}{2}$

11. $58\frac{3}{4}$ 15. $73\frac{1}{4}$ 19. $61\frac{1}{3}$ 23. $98\frac{3}{7}$ 27. $99\frac{7}{10}$
    $35\frac{11}{16}$ $13\frac{5}{6}$ $53\frac{3}{8}$ $87\frac{1}{4}$ $63\frac{5}{12}$

12. $109\frac{5}{6}$ 16. $48\frac{2}{3}$ 20. $68\frac{3}{4}$ 24. $102\frac{1}{4}$ 28. $220\frac{7}{20}$
    $96\frac{5}{12}$ $25\frac{3}{4}$ $52\frac{3}{7}$ $87\frac{3}{5}$ $160\frac{5}{12}$

Give first the sum and then the difference of:

29. $\frac{3}{8}$ and $\frac{3}{10}$ 31. $1\frac{3}{8}$ and $\frac{7}{12}$ 33. $2\frac{1}{2}$ and $\frac{5}{12}$

30. $\frac{4}{9}$ and $\frac{3}{10}$ 32. $\frac{7}{8}$ and $\frac{7}{16}$ 34. $\frac{7}{8}$ and $\frac{3}{16}$

35. $\frac{3}{4}$ and $\frac{2}{3}$ **37.** $1\frac{5}{6}$ and $\frac{5}{12}$ **39.** $4\frac{1}{8}$ and $5\frac{1}{8}$

36. $\frac{5}{8}$ and $\frac{8}{9}$ **38.** $2\frac{5}{8}$ and $3\frac{1}{4}$ **40.** $\frac{9}{7}$ and $\frac{7}{8}$

Solve as indicated :

41. $2\frac{1}{2} + 1\frac{7}{8} - 3\frac{2}{3}$ **52.** $3\frac{9}{16} - \frac{3}{4} + 5\frac{5}{9}$

42. $5\frac{2}{3} - 1\frac{5}{9} + 4\frac{7}{12}$ **53.** $3\frac{11}{12} - 2\frac{7}{8} + 3\frac{5}{8}$

43. $5\frac{5}{6} + 2\frac{1}{2} - 5\frac{5}{16}$ **54.** $1\frac{9}{10} - \frac{3}{16} + \frac{7}{8}$

44. $7\frac{3}{4} + 2\frac{5}{16} - 7\frac{7}{8}$ **55.** $5\frac{1}{7} + 3\frac{11}{14} - 2\frac{1}{2}$

45. $10\frac{1}{10} + 3\frac{3}{4} - 4\frac{7}{12}$ **56.** $8\frac{3}{4} - 1\frac{9}{16} + 5\frac{7}{10}$

46. $5\frac{5}{8} - 4\frac{3}{4} + 7\frac{5}{9}$ **57.** $3\frac{5}{6} - 1\frac{1}{12} + 7\frac{2}{9}$

47. $6\frac{1}{6} + 4\frac{2}{3} - 1\frac{3}{16}$ **58.** $3\frac{4}{9} - \frac{7}{10} + 3\frac{3}{4}$

48. $8\frac{5}{6} + 2\frac{5}{16} - 7\frac{1}{8}$ **59.** $2\frac{1}{6} + 4\frac{7}{16} + 9\frac{7}{12}$

49. $3\frac{1}{3} + 5\frac{5}{8} - 6\frac{7}{10}$ **60.** $3\frac{9}{10} - 1\frac{11}{15} + 4\frac{8}{9}$

50. $2\frac{7}{12} + \frac{3}{10} - \frac{4}{5}$ **61.** $\frac{11}{12} + 3\frac{9}{10} - 3\frac{5}{8}$

51. $\frac{3}{4} + 2\frac{7}{8} - 3\frac{1}{12}$ **62.** $2\frac{5}{6} + \frac{3}{14} + 7\frac{7}{12}$

MULTIPLICATION OF FRACTIONS

Oral Work

Give products:

1. $8 \times \frac{3}{5}$ **4.** $7 \times \frac{8}{9}$ **7.** $12 \times \frac{5}{18}$

2. $6 \times \frac{5}{7}$ **5.** $8 \times \frac{4}{7}$ **8.** $8 \times \frac{4}{15}$

3. $11 \times \frac{2}{9}$ **6.** $10 \times \frac{5}{11}$ **9.** $9 \times \frac{11}{14}$

Find:

10. $\frac{3}{8}$ of 8   **13.** $\frac{4}{7}$ of 14   **16.** $\frac{2}{7}$ of 18   **19.** $\frac{3}{16}$ of 23

11. $\frac{5}{6}$ of 7   **14.** $\frac{3}{4}$ of 15   **17.** $\frac{3}{11}$ of 12   **20.** $\frac{2}{15}$ of 11

12. $\frac{7}{8}$ of 19   **15.** $\frac{4}{5}$ of 16   **18.** $\frac{5}{12}$ of 11   **21.** $\frac{3}{20}$ of 17

Find:

22. $\frac{3}{8}$ of $\frac{5}{9}$ **25.** $\frac{3}{8}$ of $\frac{7}{8}$ **28.** $\frac{5}{12}$ of $\frac{5}{7}$ **31.** $\frac{2}{7}$ of $\frac{4}{13}$

23. $\frac{3}{4}$ of $\frac{7}{8}$ **26.** $\frac{7}{8}$ of $\frac{3}{4}$ **29.** $\frac{6}{13}$ of $\frac{6}{11}$ **32.** $\frac{7}{20}$ of $\frac{7}{8}$

24. $\frac{5}{6}$ of $\frac{11}{12}$ **27.** $\frac{8}{9}$ of $\frac{10}{11}$ **30.** $\frac{2}{15}$ of $\frac{4}{5}$ **33.** $\frac{3}{16}$ of $\frac{5}{7}$

CANCELLATION

Written Work

1. Find $\frac{7}{8}$ of $\frac{8}{9}$.

(1) $\frac{7}{8}$ of $\frac{8}{9} = \frac{56}{72} = \frac{7}{9}$

(2) $\dfrac{7}{\cancel{8}} \times \dfrac{\cancel{8}}{9} = \dfrac{7}{9}$

In changing $\frac{56}{72}$ to $\frac{7}{9}$, both terms of the fraction are divided by 8. The work may be shortened by rejecting the factor 8 from both dividend and divisor, as indicated in second model. This reduces the answer to lowest terms by taking out common factors before multiplying.

Cancellation is the process of shortening operations by striking out equal factors from both dividend and divisor.

In the following operations, cancel whenever possible.

Multiply :

2. $16 \times \frac{3}{8}$

3. $24 \times \frac{7}{8}$

4. $27 \times \frac{2}{3}$

5. $45 \times \frac{4}{9}$

6. $18 \times \frac{10}{3}$

7. $12 \times \frac{11}{4}$

8. $13 \times \frac{18}{26}$

9. $14 \times \frac{11}{12}$

10. $12 \times \frac{17}{18}$

11. $11 \times \frac{7}{22}$

12. $\frac{8}{15}$ of 25

13. $\frac{7}{24}$ of 15

14. $\frac{13}{25}$ of 10

15. $\frac{13}{14}$ of 28

16. $\frac{7}{9}$ of 81

17. $\frac{8}{15}$ of 30

18. $\frac{5}{12}$ of 50

19. $\frac{7}{15}$ of 75

20. $\frac{8}{11}$ of 66

21. $\frac{7}{16}$ of 64

Find :

22. $\frac{1}{3}$ of $\frac{2}{4}$

23. $\frac{2}{3}$ of $\frac{3}{4}$

24. $\frac{1}{3}$ of $\frac{10}{11}$

25. $\frac{2}{5}$ of $\frac{10}{11}$

26. $\frac{4}{5}$ of $\frac{10}{11}$

27. $\frac{1}{4}$ of $\frac{11}{16}$

28. $\frac{3}{7}$ of $\frac{14}{15}$

29. $\frac{5}{6}$ of $\frac{14}{15}$

30. $\frac{1}{6}$ of $\frac{13}{18}$

31. $\frac{5}{6}$ of $\frac{14}{15}$

32. $\frac{2}{4}$ of $\frac{12}{13}$

33. $\frac{7}{12}$ of $\frac{12}{13}$

34. $\frac{11}{12}$ of $\frac{12}{13}$

35. $\frac{5}{11}$ of $\frac{22}{25}$

36. $\frac{7}{11}$ of $\frac{22}{25}$

Find :

37. $9 \times 2\frac{1}{4}$

38. $12 \times 3\frac{3}{4}$

39. $18 \times 5\frac{3}{8}$

40. $22 \times 4\frac{8}{10}$

41. $21 \times 2\frac{5}{14}$

42. $8\frac{1}{2} \times 7$

43. $2\frac{7}{15} \times 10$

44. $9\frac{5}{12} \times 16$

45. $8\frac{2}{7} \times 24$

46. $7\frac{13}{40} \times 32$

47. $27 \times 12\frac{5}{36}$

48. $19 \times 8\frac{1}{3}$

49. $26 \times 5\frac{7}{39}$

50. $36 \times 7\frac{1}{8}$

51. $42 \times 8\frac{1}{12}$

52. Find the product of $\frac{2}{3} \times 7\frac{1}{2} \times 3$.

$\frac{2}{3} \times 7\frac{1}{2} \times 3 =$

$\frac{2}{3} \times \frac{15}{2} \times \frac{3}{1} = 15$

Reduce the mixed number to an improper fraction. Cancel first the factor 2 from dividend and divisor; then the factor 3. The product is $\frac{45}{3}$, or 15.

Find the products, canceling when possible:

53. $5\frac{2}{5} \times \frac{5}{6} \times \frac{1}{2}$

54. $3\frac{1}{5} \times \frac{4}{5} \times \frac{3}{4}$

55. $7\frac{1}{5} \times \frac{4}{9} \times \frac{5}{9}$

56. $5 \times \frac{5}{8} \times \frac{1}{5}$

57. $\frac{7}{8} \times 4 \times \frac{3}{7}$

58. $4\frac{3}{4} \times \frac{3}{5} \times \frac{5}{9} \times \frac{1}{2}$

59. $4 \times \frac{3}{4} \times \frac{5}{9}$

60. $\frac{5}{12} \times \frac{2}{3} \times 2\frac{1}{4} \times \frac{3}{8} \times \frac{8}{16}$

61. $4\frac{3}{8} \times \frac{2}{7} \times \frac{3}{8} \times \frac{3}{8}$

62. $\frac{2}{9} \times \frac{3}{11} \times 7\frac{1}{8} \times \frac{1}{2}$

63. $5\frac{1}{4} \times \frac{3}{7} \times \frac{2}{5}$

64. $\frac{3}{10} \times \frac{5}{8} \times \frac{3}{4} \times \frac{3}{5}$

65. $\frac{7}{10} \times 5\frac{4}{5} \times \frac{5}{10} \times \frac{1}{2}$

66. $2\frac{7}{10} \times \frac{2}{5} \times \frac{5}{7} \times \frac{1}{2}$

67. $7\frac{1}{2} \times \frac{3}{5} \times \frac{8}{10}$

68. $\frac{4}{5} \times \frac{3}{10} \times \frac{1}{2}$

69. $\frac{5}{16} \times \frac{4}{5} \times \frac{3}{16}$

70. $\frac{2}{3} \times 4\frac{3}{4} \times \frac{2}{5}$

71. $5\frac{1}{8} \times 4\frac{3}{4} \times 8\frac{1}{2} \times \frac{1}{2}$

72. $\frac{4}{5} \times \frac{3}{4} \times \frac{3}{8} \times \frac{3}{10}$

Find the answers in lowest terms:

73. $\frac{1}{4}$ of $\frac{3}{10}$ of $\frac{2}{4}$

74. $\frac{3}{4}$ of $\frac{5}{8}$ of $\frac{5}{8}$

75. $\frac{4}{5}$ of $\frac{3}{10}$ of $\frac{2}{3}$

76. $\frac{1}{6}$ of $\frac{2}{3}$ of $\frac{1}{2}$

77. $\frac{3}{4}$ of $3\frac{1}{4}$

78. $\frac{2}{3}$ of $2\frac{1}{2}$

79. $\frac{4}{5}$ of $\frac{5}{6}$

80. $\frac{5}{8}$ of $\frac{3}{10}$

81. $\frac{2}{3}$ of $\frac{1}{8}$ of $\frac{2}{5}$

82. $\frac{1}{2}$ of $\frac{1}{4}$ of $\frac{1}{10}$

83. $\frac{3}{10}$ of $\frac{3}{8}$ of $\frac{7}{10}$

84. $\frac{3}{8}$ of $\frac{2}{3}$ of $\frac{4}{5}$

85. $\frac{5}{8}$ of $2\frac{1}{8}$

86. $\frac{4}{5}$ of $\frac{3}{4}$

87. $\frac{1}{8}$ of $\frac{3}{4}$

88. $\frac{1}{6}$ of $\frac{3}{8}$

89. $\frac{1}{2}$ of $\frac{1}{4}$ of $\frac{1}{8}$

90. $\frac{5}{8}$ of $\frac{3}{4}$ of $\frac{1}{4}$

91. $4\frac{1}{2} \times \frac{3}{4} \times 5\frac{1}{2}$

92. $4\frac{3}{8} \times \frac{1}{4} \times \frac{3}{10}$

93. $4\frac{1}{3} \times 6\frac{1}{4}$

94. $25\frac{1}{2} \times 13\frac{1}{4}$

95. $\frac{3}{4}$ of $4\frac{1}{2}$

96. $\frac{9}{10}$ of $2\frac{1}{2}$

DIVISION OF FRACTIONS

Written Work

Find quotients, canceling when possible:

1.	$4 \div \frac{1}{3}$	21.	$8\frac{1}{6} \div 4\frac{1}{4}$	41.	$77 \div 2\frac{1}{3}$
2.	$1\frac{1}{5} \div \frac{3}{5}$	22.	$1\frac{1}{6} \div \frac{5}{6}$	42.	$103 \div 10\frac{3}{10}$
3.	$5 \div \frac{3}{4}$	23.	$8 \div \frac{7}{8}$	43.	$10 \div \frac{2}{5}$
4.	$\frac{7}{8} \div \frac{3}{8}$	24.	$7 \div \frac{4}{5}$	44.	$1\frac{1}{4} \div \frac{1}{4}$
5.	$\frac{11}{16} \div \frac{7}{16}$	25.	$\frac{3}{4} \div \frac{7}{8}$	45.	$12 \div \frac{3}{7}$
6.	$\frac{5}{6} \div \frac{7}{8}$	26.	$\frac{5}{9} \div \frac{8}{10}$	46.	$1\frac{3}{8} \div \frac{3}{4}$
7.	$\frac{7}{12} \div \frac{5}{12}$	27.	$\frac{3}{7} \div \frac{5}{14}$	47.	$5\frac{5}{6} \div 2\frac{5}{6}$
8.	$\frac{15}{16} \div 8$	28.	$\frac{12}{16} \div \frac{3}{4}$	48.	$3\frac{3}{4} \div 4$
9.	$\frac{6}{7} \div 5$	29.	$\frac{18}{20} \div 19$	49.	$5\frac{7}{8} \div 10$
10.	$\frac{11}{12} \div 6$	30.	$\frac{11}{14} \div 14$	50.	$6\frac{3}{11} \div 12$
11.	$\frac{28}{33} \div 7$	31.	$\frac{9}{11} \div 11$	51.	$5\frac{6}{13} \div 14$
12.	$\frac{15}{16} \div 10$	32.	$\frac{15}{16} \div 32$	52.	$\frac{3}{10} \div 27$
13.	$22\frac{3}{4} \div 13$	33.	$\frac{11}{14} \div 28$	53.	$\frac{12}{13} \div 36$
14.	$35\frac{1}{6} \div 16$	34.	$41\frac{7}{8} \div 22$	54.	$\frac{18}{20} \div 38$
15.	$17\frac{2}{3} \div 12$	35.	$37\frac{1}{3} \div 10$	55.	$51\frac{3}{7} \div 20$
16.	$2\frac{1}{2} \div 2\frac{1}{3}$	36.	$18\frac{1}{4} \div 32$	56.	$29\frac{2}{11} \div 9$
17.	$12\frac{1}{2} \div 16\frac{2}{3}$	37.	$18\frac{2}{7} \div 7\frac{1}{9}$	57.	$46\frac{2}{3} \div 15$
18.	$8\frac{2}{3} \div 4\frac{2}{4}$	38.	$19\frac{3}{8} \div 1\frac{3}{5}$	58.	$67\frac{3}{16} \div 7\frac{3}{4}$
19.	$5\frac{1}{6} \div 4\frac{1}{4}$	39.	$160 \div \frac{5}{8}$	59.	$88\frac{2}{21} \div 16\frac{2}{3}$
20.	$7\frac{3}{8} \div 6\frac{1}{2}$	40.	$16\frac{3}{7} \div 6\frac{9}{11}$	60.	$30\frac{3}{16} \div 5\frac{3}{4}$

COMPARISON — WHOLE NUMBERS AND FRACTIONS

Oral Work

1. $\frac{1}{2} = \frac{4}{8}$; $\frac{3}{4} = \frac{6}{8}$. Then how do $\frac{1}{2}$ and $\frac{3}{4}$ compare?

2. If a unit is first divided into halves and then each half into halves, into how many parts is the unit divided? Is $\frac{1}{4}$ of a unit larger or smaller than $\frac{1}{2}$ of a unit?

3. Divide a unit into halves, fourths, eighths, and sixteenths, and show how many sixteenths of a unit it takes to make $\frac{1}{4}$ of the unit; $\frac{1}{8}$ of the unit; $\frac{3}{4}$ of the unit; $\frac{1}{2}$ of the unit.

4. Draw equal squares to show that $\frac{1}{2} = \frac{4}{8}$, or $\frac{8}{16}$.

5. Compare $\frac{6}{2}$ and $\frac{4}{1}$; $\frac{4}{2}$ and $\frac{12}{6}$; $\frac{6}{3}$ and $\frac{3}{2}$.

6. How does $\frac{1}{4}$ of 20 compare with $\frac{1}{2}$ of 20? $\frac{7}{8}$ of 16 with $\frac{3}{4}$ of 16? $\frac{3}{5}$ of 50 with $\frac{1}{4}$ of 20?

7. A has 40 acres of land, and B 60 acres. How does A's farm compare in size with B's?

8. 8 is what part of 12, 16, 24, 32, 48, 72?

SUGGESTION. — Make 8 the numerator in each case and the other numbers the denominators and reduce the fractions to their lowest terms. Thus, $\frac{8}{12} = \frac{2}{3}$; $\frac{8}{16} = \frac{1}{2}$; $\frac{8}{24} = \frac{1}{3}$, etc.

9. What is the ratio of 16 to 2? to 8? to 24?

SOLUTION. — $\frac{16}{2} = 8$; $\frac{16}{8} = 2$; $\frac{16}{24} = \frac{2}{3}$.

10. If 6 quarts of milk cost 60 cents, how much will 10 quarts cost? 15 quarts? 20 quarts?

SUGGESTION. — $\frac{10}{6}$ of 60 = ? $\frac{15}{6}$ of 60 = ? $\frac{20}{6}$ of 60 = ?

11. Elizabeth buys $3\frac{1}{2}$ yards of ribbon for 35 cents. At the same rate, how much would she pay for $10\frac{1}{2}$ yards?

SUGGESTION. — How many times $3\frac{1}{2}$ is $10\frac{1}{2}$?

12. A woodsman cuts 15 cords of wood in 6 days. How many cords, at the same rate, could he cut in 48 days?

13. Find the ratio of 8 to 2; of 6 to 4; of 5 to 10.

14. Compare in the same way $\frac{1}{2}$ and $\frac{1}{4}$; $\frac{1}{4}$ and $\frac{1}{2}$; $\frac{1}{6}$ and $\frac{1}{12}$.

15. Compare 24 with 4, 6, 8, 48, 72, 16, 20.

16. If $\frac{1}{4}$ of a man's weekly wages is $\$2\frac{1}{4}$, how much is $\frac{1}{2}$ of his weekly wages?

SUGGESTION. — $\frac{1}{2}$ is how many times $\frac{1}{4}$?

17. $3\frac{1}{2}$ pounds of rice cost 35 cents. At that rate how much will 7 pounds cost?

18. If $\frac{1}{4}$ of my money is $\$10$, and George has 6 times my money, how much has George?

19. In New York $9\frac{1}{4}$ inches of rain fell in 3 months. At that rate how much will fall in a year?

Written Work

1. If a man pays $\$3675$ for 60 acres of land, at the same rate how much should he pay for 120 acres?

2. My telephone bill for one year is $\$12.85$ a month. At that rate how much should I pay in $2\frac{3}{4}$ years?

3. My coal bill for $5\frac{1}{2}$ tons is $\$33$. What is the bill of my neighbor who buys $27\frac{1}{2}$ tons at the same rate?

4. If 30 bushels of oats sell for $\$13\frac{1}{5}$, how much will 60 bushels sell for?

5. If a boy receives $\$7\frac{1}{2}$ for two weeks' work, how much should he receive for 12 weeks' work?

6. How much will a clerk earn in a year if he earns $\$180$ in 3 months?

7. If 4 tons of coal cost $\$25$, how much will 16 tons cost?

8. When 5 books cost $\$17\frac{1}{2}$, how much will 25 such books cost?

9. A man walked $11\frac{1}{4}$ miles in 3 hours. At the same rate, how far would he walk in 6 hours?

10. Mr. Adams sold his farm for \$13,608, which was $\frac{6}{5}$ of the cost. What was the cost of the farm?

11. How many ribbons $\frac{2}{3}$ of a yard long can be cut from $46\frac{2}{3}$ yards of ribbon?

12. Two boys bought a sled. One paid $\frac{2}{3}$ of the cost and the other $\frac{1}{3}$ of the cost. If the boy that paid $\frac{2}{3}$ of the cost paid 90¢, how much did the sled cost?

13. If any two factors of a given product are known, how may we obtain the third factor? Make problems to illustrate this.

14. The dividend is $163\frac{3}{4}$, and the quotient is $218\frac{1}{3}$. Find the divisor.

15. $\frac{11}{12}$ is the product of 6 and what other number?

16. $\frac{4}{5}$ of the length of a flagpole is 60 feet. What is the length of $\frac{2}{3}$ of the pole?

17. $\frac{2}{3}$ of $\frac{3}{4}$ of \$60 is $\frac{1}{2}$ of what number of dollars?

18. The length of a certain city square is $338\frac{1}{4}$ feet. Find its length in rods.

19. An automobile ran $63\frac{4}{15}$ miles in $2\frac{1}{2}$ hours. Find the average rate per hour.

20. James weighs $160\frac{3}{4}$ pounds, Sarah $108\frac{1}{16}$ pounds, John $135\frac{3}{4}$ pounds, Mary $121\frac{5}{8}$ pounds, and Henry $124\frac{5}{6}$ pounds. Find $\frac{2}{3}$ of their combined weight.

21. Paul averages $2\frac{1}{2}$ feet at a step. How many steps does he take in going 1 mile?

22. The lead lining in a tank weighs $3\frac{2}{3}$ pounds to the square foot. How many pounds will be necessary to line a tank containing $275\frac{1}{2}$ square feet of inside surface?

DECIMALS

DECIMAL PARTS OF A DOLLAR

 **1.** How many dimes equal a dollar? Then what part of a dollar is a dime?

2. How many cents equal a dollar? Then what part of a dollar is a cent?

3. Ten mills equal one cent. How many mills equal a dollar? Then what part of a dollar is a mill?

Mills are not coined, but are used for exactness in computations.

When we think of a dollar as dimes, it has 10 equal parts; when we think of a dollar as cents, it has 100 equal parts; when we think of a dollar as mills, it has 1000 equal parts. A mill is $\frac{1}{10}$ of a cent; a cent $\frac{1}{10}$ of a dime ; and a dime $\frac{1}{10}$ of a dollar.

This division of the dollar into tenths, hundredths, thousandths, etc., we call **decimal parts of the dollar.**

The **decimal point** is the point separating dollars and cents. Thus, in $2.75 the point separates 2 dollars from 75 cents.

4. What decimal part of a dollar are 5 dimes? 6 dimes? 8 dimes? 9 dimes?

5. What decimal part of a dollar are 5 cents? 8 cents? 9 cents? 10 cents?

The first place to the right of the decimal point is occupied by *dimes* or *tenths* of a dollar; the second place, by *cents* or *hundredths* of a dollar; the third place, by *mills* or *thousandths* of a dollar.

Dimes, cents, and mills can always be written as decimal parts of a dollar. Thus, 8 dimes, 5 cents = $.85; 2 mills = $ .002.

114

READING AND WRITING DECIMALS

One tenth may be written .1 as well as $\frac{1}{10}$; **one hundredth** may be written .01 as well as $\frac{1}{100}$; and **one thousandth** may be written .001 as well as $\frac{1}{1000}$.

1. Read: .8 ft., .5 lb., .7 pk., .5 ft., .7 mi.

The **decimal point** is a period placed before tenths.

A **decimal fraction** is any number of 10ths, 100ths, 1000ths, etc., of a unit. When expressed after a decimal point and without a written denominator it is usually called a **decimal**.

The first place to the right of the decimal point is called **tenths**; the second place, **hundredths**; and the third place, **thousandths**.

2. In 55.55, the 5 hundredths is what part of the 5 tenths? the 5 tenths is what part of the 5 units? the 5 units is what part of the 5 tens?

3. In $1.256, state what each figure represents.

4. Name the parts of a dollar, first as tenths, hundredths, and thousandths; then as cents and mills: $.65, $8.05, $2.005, $.50, $.75, $.80, $.705.

5. Write in figures: six dollars, five cents; ten dollars, fifty cents; three mills; five cents; five mills.

6. Read: $.05; $.25; $0.07; $6.05; $7.09; $4.72.

7. Read; then change to cents. Thus, $3.85 = 385¢. $2.05; $.70; $0.05; $7.09; $8.00; $.75; $3.50.

8. Change to dollars and cents: 55¢; 85¢; 870¢; 1002¢.

9. Write: eighty-five cents; nine dollars two cents; twenty-two dollars; nine hundred dollars and six cents.

10. Write: one dollar one cent; 10 dollars and one cent; eighty-seven cents five mills.

In any number, whether a whole number or a decimal, *the value of a figure in any place is $\frac{1}{10}$ of the value of the same figure standing one place to the left.*

11. What is the largest decimal division of a unit? the second largest? the third largest?

12. $.06 = \frac{?}{100} = \frac{?}{1000}$ **15.** $.9 = \frac{?}{10} = \frac{?}{100} = \frac{?}{1000}$

13. $.25 = \frac{?}{100} = \frac{?}{1000}$ **16.** $.025 = \frac{?}{1000} = \frac{?}{10000}$

14. $.05 = \frac{?}{100} = \frac{?}{1000}$ **17.** $.349 = \frac{?}{1000} = \frac{?}{10000}$

Observe that a decimal is always less than unit.

Hundreds	Tens	Ones	Dec. Point	Tenths	Hundredths	Thousandths
5	2	5	·	2	5	6

This number is read, five hundred twenty-five *and* two hundred fifty-six thousandths.

18. What do we call the decimal point when we read a number? What word, then, always joins the whole number and the decimal?

Observe that we express every number as units, or ones, and parts of a unit. Thus, 525.256 is 525 units and .256 of a unit.

As the first decimal division of a unit is tenths, we always begin to enumerate the decimal at tenths' place; thus:

tenths	hundredths	thousandths
.0	0	5

19. At what place do we begin to enumerate whole numbers?

20. Read the following: .25, .025, 25.005, 7.05, 321.1, 0.875, 1.008, 100.001, 0.001.

21. Write as decimals: $\frac{5}{10}$, $\frac{7}{100}$, $\frac{25}{1000}$, $\frac{1}{10}$, $\frac{15}{1000}$, $\frac{2}{1000}$, $26\frac{6}{1000}$, $100\frac{1}{1000}$, $1\frac{8}{100}$, $70\frac{105}{1000}$.

Write as decimals :

22. Two thousandths.

23. Two and two thousandths.

24. Five hundredths.

25. Two hundred and two thousandths.

26. Two hundred two thousandths.

27. Three and five tenths.

28. Seventy-five hundredths.

29. Five hundred and five thousandths.

30. Thirty-three thousandths.

31. Ninety-five thousandths.

32. Two hundred and five hundredths.

33. Six and nine tenths.

CHANGING DECIMALS TO COMMON FRACTIONS

Oral Work

1. Change .5 to a common fraction.

$$.5 = \tfrac{5}{10} = \tfrac{1}{2}.$$

2. Express .25, .45, .025, each in the form of a common fraction.

3. Express .25 as a common fraction in its lowest terms.

Written Work

1. Change .875 to a common fraction in its lowest terms.

$.875 = \tfrac{875}{1000} = \tfrac{7}{8}$ Expressed in the form of a common fraction $.875 = \tfrac{875}{1000}$. By dividing both the numerator and the denominator of $\tfrac{875}{1000}$ first by 25 and then by 5, we reduce it to its lowest terms, $\tfrac{7}{8}$.

To change a decimal to a common fraction, write the decimal, omitting the decimal point, place the decimal denominator beneath the numerator, and change the fraction to its lowest terms.

Change to common fractions in their lowest terms:

2. .15	**4.** .9	**6.** .75	**8.** .125
3. .825	**5.** .325	**7.** .025	**9.** .425

10. Memorize the following equivalents:

$\frac{1}{2}$ = .5 or .50 $\frac{1}{5}$ = .2 or .20 $\frac{4}{5}$ = .8 or .80
$\frac{1}{4}$ = .25 $\frac{2}{5}$ = .4 or .40 $\frac{1}{8}$ = .125
$\frac{3}{4}$ = .75 $\frac{3}{5}$ = .6 or .60 $\frac{3}{8}$ = .375

11. Change to tenths : $\frac{1}{5}$; $\frac{1}{2}$; $\frac{2}{5}$; $\frac{3}{5}$; $\frac{4}{5}$.

12. Express as decimal hundredths: $\frac{1}{4}$; $\frac{3}{4}$; $\frac{1}{5}$; $\frac{3}{8}$.

Change to common fractions in their lowest terms:

13. .45	**19.** .20	**25.** .40	**31.** .075
14. .625	**20.** .60	**26.** .48	**32.** .0025
15. .75	**21.** .725	**27.** .150	**33.** .12
16. .65	**22.** .90	**28.** .50	**34.** .225
17. .375	**23.** .96	**29.** .875	**35.** .700
18. .80	**24.** .72	**30.** .08	**36.** .800

ADDITION OF DECIMALS

What kind of fractions can be added or subtracted?

In *adding* or *subtracting* decimals, like units must always be written under one another. Thus to add .8 + .85 + .096 write them as follows:

$$
\begin{array}{r}
.8 \\
.85 \\
\underline{.096}
\end{array}
$$

1. In how many of these decimals are there tenths to be added? hundredths? thousandths?

2. Why must tenths be written *under* tenths, hundredths *under* hundredths, etc. ?

A **mixed** decimal is a whole number and a decimal united. Thus, $4 + .05$, or 4.05, is a mixed decimal.

Written Work

1. Add $45.5 + 6.005 + 40$.

$$\begin{array}{r} 45.5 \\ 6.005 \\ 40. \\ \hline 91.505 \end{array}$$

Keep the decimal points and units of the same order in a column, and add as in whole numbers, placing the decimal point in the sum under the points above.

Test by adding downwards.

Write from dictation. Then add and test :

2. $.1 + .2 + .35 + .365 = ?$

3. $.02 + .05 + .095 + .056 = ?$

4. $.05 + .007 + .089 + .11 = ?$

5. $1.2 + 3.4 + 4.5 = ?$

6. $3.04 + 4.05 + 6.099 = ?$

7. $.005 + .007 + .009 + .0101 = ?$

8. $2.006 + 7.009 + 9.012 = ?$

9. $.001 + .001 + .0902 = ?$

10. $.5 + 2.5 + .003 + .60 = ?$

11. $.07 + 5.081 + .001 + .90 = ?$

12. $.7 + 1.07 + 1.007 = ?$

13. $.1 + 2 + .75 + 8.006 = ?$

14. $3 + .7 + 5.02 + 7.008 = ?$

15. $5 + 8 + .3 + .05 + .006 = ?$

16. $.7 + 89 + .60 + 8.75 = ?$

17. $.9 + .81 + .72 + 1.075 = ?$

18. $10 + 2.1 + 14.9 + 17.85 = ?$

19. $.9 + .85 + .005 + .25 + .895 = ?$

Write from dictation. Then add and test:

20.		22.		24.		26.	
	1.45		.424		.7		11.111
	3.7		8.2		.425		3.06
	10.01		6.16		18.54		.635
	2.005		19.009		7.011		.000

21.		23.		25.		27.	
	18.002		.040		89.400		.707
	2.056		48.010		75.800		101.101
	121.114		.708		761.612		96.086
	2.02		89.010		1245.000		27.409

28. Find the sum of 15.38, 9.17, 3.07, 20.35.

29. A boy picked on Monday, .75 of a bushel of berries; on Tuesday, .875 of a bushel; on Wednesday, 1.125 of a bushel. How many bushels did he pick in the three days?

30. Helen paid $.25 for a handkerchief, $2.75 for a pair of shoes, $.45 for lace, and $1.49 for a waist. How much did they all cost?

31. A train runs the first hour 19.625 miles; the second hour, 20.5 miles; the third hour, 20.75 miles; the fourth hour, 21.225 miles. How far does it run in the four hours?

32. Find the number of pounds in the following purchases: 1.25 lb. of cheese, 3.5 lb. of sugar, .5 lb. of cloves.

33. The distance from Harrington to Houston is 4.31 miles, thence to Ellendale 11.25 miles, thence to Georgetown 8.37 miles. How far is it from Harrington to Georgetown?

34. Find the sum of 24.36, 108.075, 20.009, 200.001, 654.03, 549.5, and 721.25.

SUBTRACTION OF DECIMALS

Oral Work

Perform the operations indicated:

1. $.5 - .3 = ?$
2. $.9 - .8 = ?$
3. $15.8 - 11.7 = ?$
4. $4.7 - 3.2 = ?$
5. $.008 - .002 = ?$
6. $.014 - .011 = ?$
7. $.08 + .09 - .12 + .04 + .02 = ?$
8. $.009 + .003 - .007 - .004 = ?$

Written Work

1. From 16.35 subtract 11.76.

$$\begin{array}{r} 16.35 \\ 11.76 \\ \hline 4.59 \end{array}$$

Keep the decimal points in a column and subtract as in whole numbers, placing the decimal point in the difference under the points above.

2. $\begin{array}{r} 7. \\ 1.21 \\ \hline \end{array}$

3. $\begin{array}{r} 16. \\ 3.046 \\ \hline \end{array}$

4. $\begin{array}{r} 1.101 \\ .796 \\ \hline \end{array}$

5. $\begin{array}{r} 265.36 \\ 84.468 \\ \hline \end{array}$

6. $\begin{array}{r} 25.2 \\ 9.18 \\ \hline \end{array}$

7. $\begin{array}{r} 151.003 \\ 78.076 \\ \hline \end{array}$

8. $\begin{array}{r} 954.1 \\ 258.375 \\ \hline \end{array}$

9. $\begin{array}{r} 86.291 \\ 17.456 \\ \hline \end{array}$

10. $\begin{array}{r} 144.001 \\ 12.256 \\ \hline \end{array}$

11. $\begin{array}{r} 300. \\ 261.385 \\ \hline \end{array}$

12. $\begin{array}{r} 86.59 \\ 53.594 \\ \hline \end{array}$

13. $\begin{array}{r} 728.3 \\ 619.333 \\ \hline \end{array}$

14. Warren had $7.50 and spent $3.75. How much had he remaining?

15. The distance between two towns is 9 miles. After I have walked 3.625 miles, how far have I yet to walk?

16. A man having 120 acres of land, sold to one man 28.75 acres, and to another, 35.5 acres. How many acres had he left?

17. If I pay $1.25 for car fare, $.65 for dinner, and $.90 for an umbrella, how much change have I left from $5?

18. The second floor of a house is 18.78 feet above the floor of the cellar, and the first floor is 7.92 feet above it. How far is it from the first floor to the second?

19. Four lots measure in width 123.08 ft. Three of them are respectively 25 ft., 32.72 ft., and 36.9 ft. wide. What is the width of the fourth?

20. A boy having $4.25 spent for skates $1.25, for a cap $.50, and for a hockey stick $.45. How much had he left?

21. A lady having 25.75 pounds of butter sold to one customer 3.25 pounds, to another 8.5 pounds, to another 7.25 pounds, and the remainder to a fourth customer. How many pounds did the fourth customer buy?

22. From a ham weighing 18.125 lb. a butcher sold 3.25 lb., 4.50 lb., 2.75 lb., and 2.5 lb. How many pounds had he left?

23. A fisherman brought home four trout weighing respectively 1.25 pounds, .875 pounds, 1.375 pounds, and 1.125 pounds. How much less than 5 pounds did they all weigh?

24. A farmer cut 40 tons of hay in 1913. He sold 6.85 tons to one man, and 5.50 tons to another. He fed the rest to his stock. How many tons did he feed to his stock?

25. A lady bought 4.75 yards of woolen cloth, 11.625 yards of cotton cloth, and 6.875 yards of silk. How many yards less than 30 yards did she buy?

26. A man having $20 spent $4.75 for board, $2.80 for a room, $.88 for laundry, $1.75 for a pair of gloves, and $3.50 for a pair of shoes. How much had he left?

27. A merchant purchased the following: coffee $15.25, sugar $18.35, cakes $11.65, fruit $27.75, and canned corn $8.45. How much less than $120 was the amount of his bill?

MULTIPLICATION OF DECIMALS

Multiplying a decimal by an integer or an integer by a decimal.

Oral Work

1. How much is $5 \times .3$? $.3 \times 5$?

SOLUTION. — $5 \times .3 = 5 \times \frac{3}{10} = \frac{15}{10} = 1.5$.
$.3 \times 5 = \frac{3}{10} \times 5 = \frac{15}{10} = 1.5$.

2. Find $4 \times .03$; $.03 \times 4$.

3. Find $6 \times .003$; $.003 \times 6$.

4. Find 6×1.3; 1.3×6.

SOLUTION. — $6 \times 1.3 = 6 \times \frac{13}{10} = \frac{78}{10}$, or 7.8.
$1.3 \times 6 = \frac{13}{10} \times 6 = \frac{78}{10}$, or 7.8.

In each problem above, how many decimal places are there in the multiplicand or in the multiplier? how many in the product?

In multiplying a decimal by an integer, the product contains the same number of decimal places as the multiplicand.

In multiplying an integer by a decimal, the product contains the same number of decimal places as the multiplier.

Written Work

1. Multiply 5.75 by 6.

5.75
6
——
34.50

6×5 hundredths = 30 hundredths, or 3 tenths and no hundredths. Write naught in hundredths' place and carry the three tenths. 6×7 tenths = 42 tenths; 42 tenths + 3 tenths = 45 tenths, or 4 units and 5 tenths. Write 5 in tenths' place and carry the 4 units. Write the decimal point. 6×5 units = 30 units; 30 units + 4 units = 34 units.

2. Multiply 575 by .026.

575
.026
‾‾‾‾‾
3450   Multiply as in integers. As there are 3 decimal places in
1150   the multiplier, point off 3 places from the right in the
‾‾‾‾‾   product.
14.950

3. Multiply 623 by 1.35.

623
1.35
‾‾‾‾‾
3115   As there are 2 decimal places in the multiplier, point off
1869   2 places from the right in the product.
623
‾‾‾‾‾
841.05

Find products :

4.	$8 \times .015$	**15.**	$.9 \times 117$	**26.**	$.066 \times 3455$
5.	$9 \times .005$	**16.**	$.02 \times 112$	**27.**	$.467 \times 2639$
6.	$18 \times .17$	**17.**	$.64 \times 236$	**28.**	$.095 \times 7148$
7.	$25 \times .207$	**18.**	$.004 \times 149$	**29.**	$.081 \times 3236$
8.	44×5.6	**19.**	3.04×415	**30.**	234.17×1099
9.	65×7.5	**20.**	10.34×308	**31.**	4.022×1402
10.	73×8.4	**21.**	$.005 \times 718$	**32.**	$.05 \times 2472$
11.	117×9.3	**22.**	$.024 \times 122$	**33.**	$.5625 \times 3122$
12.	208×6.8	**23.**	$.015 \times 215$	**34.**	$.003 \times 4144$
13.	306×5.8	**24.**	$.007 \times 283$	**35.**	1.02×596
14.	425×7.2	**25.**	2.042×212	**36.**	4.003×6407

37. How much will 7 arithmetics cost at $.82 apiece?

38. How many feet are there in .375 of a mile?

39. How many square inches are there in .75 of a square foot?

40. At $.35 each, how much will 24 chickens cost?

41. A rod is 16.5 feet. How many feet are there in 29 rods?

42. When a man earns $3.65 per day, how much does he earn in 26 days?

43. A pound of cream cheese costs $.115. How much do 126 pounds cost?

44. How many pounds are there in .875 of a ton?

45. If an automobile averages 17.75 miles an hour, how far will it travel in 14 hours?

46. Multiply each of the following by 10: .6, .8, .84, .86, .76, .65, .54, .005.

47. Multiply 500 by each of the following: .06, .04, .005, .42, .47, 42.3, 56.7, .478, 8.6, 9.8.

48. Multiply each of the following by 100: .6, .8, .84, .95, .86, .76, .06, .04, .005, 4.23, 56.7, .478, 8.6, 9.8.

49. Multiply 5000 by each of the following: .594, 5.94, 59.4, .007, .07, .7, 3.14, 2.5, .0025.

Find the cost of:

50. 24 lb. @ $.125		**56.** 64 bbl. @ $7.50
51. 27 yd. @ $.165		**57.** 16 ft. @ $18.75
52. 56 bu. @ $.375		**58.** 45 lb. @ $.052
53. .875 ft. @ $4		**59.** 66 gal. @ $.75
54. .375 yd. @ $2		**60.** 2 bu. @ $.375
55. .125 T. @ $4		**61.** 1.25 doz. @ $125

62. A girl sent 27 pieces to a laundry that charged her seventy-five cents a dozen for washing and ironing them. What was her bill?

63. A merchant bought 1200 gas fixtures at $.08⅓ each and sold them at $.10 each. How much did he gain?

64. If the rainfall in a certain state averages 4.62 inches per month, how much is the rainfall for the year?

65. A 24-story city building averages 14.75 ft. to a story. How high is the building?

66. In January we burned 36,000 feet of gas. At $1.00 per thousand, what was the bill?

67. A farmer paid $57.60 per acre for 36 acres of land. How much did he pay for the land?

DIVISION OF DECIMALS

Dividing by an integer.

Written Work

1. Divide .84 by 4 in this way: $\dfrac{4).84}{.21}$

2. Divide 6.648 by 6 in this way: $\dfrac{6)6.648}{1.108}$

3. Divide 24.600 by 8 in this way: $\dfrac{8)24.600}{3.075}$

Observe that in dividing a decimal or a mixed decimal by an integer, the dividend is simply separated or *partitioned* into equal parts. Thus, 6.9 ÷ 3 = ⅓ of 6.9, or 2.3.

Divide and test, placing a decimal point in the quotient before beginning to divide:

4. 6).66	**7.** 7).714	**10.** 7)7.847
5. 3).96	**8.** 5).535	**11.** 6)6.936
6. 8).808	**9.** 4).848	**12.** 8)8.896

13. Explain why adding *naughts* to the right of a decimal does not change its value. Thus, .8 = .80, .05 = .050.

It is sometimes necessary to add naughts to the right of the dividend to complete the division.

14. Divide .12 by 5. $5\overline{).12} = 5\overline{).120}$
$$\qquad\qquad\qquad\qquad\quad .024$$

Place **a decimal point directly above or below the decimal point in the dividend,** *before beginning to divide; then divide as in the division of integers.*

Find quotients and test:

15.	69.92 ÷ 23	**33.**	283.88 ÷ 47	**51.**	.018 ÷ 12
16.	29.54 ÷ 14	**34.**	6.497 ÷ 73	**52.**	.546 ÷ 21
17.	195.2 ÷ 32	**35.**	16.150 ÷ 34	**53.**	.003 ÷ 10
18.	401.4 ÷ 18	**36.**	55.660 ÷ 92	**54.**	.368 ÷ 16
19.	8.434 ÷ 34	**37.**	5.460 ÷ 84	**55.**	1.625 ÷ 25
20.	156.4 ÷ 46	**38.**	1.6272 ÷ 18	**56.**	24.36 ÷ 12
21.	1.014 ÷ 26	**39.**	1.25 ÷ 5	**57.**	172.8 ÷ 24
22.	5.084 ÷ 41	**40.**	.64 ÷ 16	**58.**	14.76 ÷ 41
23.	.945 ÷ 35	**41.**	.02 ÷ 40	**59.**	1.105 ÷ 65
24.	60.32 ÷ 52	**42.**	7.5 ÷ 60	**60.**	2.07 ÷ 46
25.	.968 ÷ 44	**43.**	4.9 ÷ 140	**61.**	31.2 ÷ 36
26.	.828 ÷ 23	**44.**	.01 ÷ 100	**62.**	2.31 ÷ 55
27.	5.18 ÷ 37	**45.**	.05 ÷ 500	**63.**	1.17 ÷ 65
28.	.0833 ÷ 49	**46.**	.03 ÷ 100	**64.**	16.5 ÷ 22
29.	1.566 ÷ 54	**47.**	.027 ÷ 18	**65.**	2.7355 ÷ 35
30.	2.144 ÷ 67	**48.**	4.44 ÷ 50	**66.**	31.288 ÷ 48
31.	8.437 ÷ 59	**49.**	125 ÷ 50	**67.**	137.95 ÷ 81
32.	233.32 ÷ 38	**50.**	9.66 ÷ 46	**68.**	106.32 ÷ 24

69. Divide 39.25 by 25.

```
     1.57
25)39.25
   25
   ----
   14.2
   12.5
   ----
   1.75
   1.75
```

70. Divide 12.648 by 24.

```
      .527
24)12.648
   12 0
   ----
     64
     48
    ---
    168
    168
```

Ex. 69. How many times is 25 contained in 39? in 14.2? in 1.75?

Divide as in the division of integers, placing the decimal point in the quotient immediately above the point in the dividend.

Ex. 70. Since 24 is larger than 12.648, the quotient must be a decimal.

71. Divide .1275 by 25.

```
     .0051
25).1275
   125
   ---
    25
    25
```

72. Divide 192.96 by 16.

```
       12.06
16)192.96
   16
   --
   32
   32
   --
     96
     96
```

In Ex. 71, two naughts must follow the decimal point in the quotient, since 25 is not contained in 1 or in 12.

In Ex. 72, one naught must follow the decimal point, since 16 is not contained in the third partial dividend, 9.

Find the quotients and test:

73. 4).3 **74.** 8).6 **75.** 6)6.27 **76.** 5)5.28

77. 1.6 ÷ 2 **85.** 1.024 ÷ 6 **93.** 3.108 ÷ 3

78. .9 ÷ 3 **86.** .102 ÷ 3 **94.** .08 ÷ 2

79. .12 ÷ 6 **87.** .039 ÷ 13 **95.** 1.125 ÷ 5

80. .005 ÷ 5 **88.** 1.44 ÷ 12 **96.** 4.16 ÷ 4

81. .008 ÷ 4 **89.** 3.015 ÷ 3 **97.** .35 ÷ 7

82. 2.7 ÷ 9 **90.** .063 ÷ 7 **98.** .077 ÷ 11

83. 1.2 ÷ 4 **91.** 9.04 ÷ 8 **99.** .022 ÷ 2

84. .24 ÷ 8 **92.** .72 ÷ 10 **100.** .036 ÷ 6

101. Mr. Johnston owns a piece of ground containing 1.565 A. He divides it into 4 lots. How much ground is there in each lot?

102. John divided $565.75 among his three sisters in equal shares. How much did each receive?

103. 4.9 miles of graded road will be built in Mercer County this year. Mr. Ames has a contract for ¼ of the distance. Find the distance Mr. Ames is to build.

104. $4695.98 is to be divided equally among 6 children. How much should each receive?

CHANGING COMMON FRACTIONS TO DECIMALS

Oral Work

1. Change ⅕ to tenths and express the result as a decimal. Thus, ⅕ = ²⁄₁₀, or .2.

2. Change ½ to tenths; to hundredths.

3. Change ¼ to hundredths.

4. Change ⅛ to thousandths.

Written Work

1. Divide 12 by 16.

$12 \div 16 = 12.00 \div 16.$

```
        .75
16)12.00
    11 2
    ─────
       80
       80
       ──
```

12 is equal to 12.00, which divided by 16 equals .75.

A decimal point must be placed after an integer before naughts are annexed.

Find the quotients:

2. 20 ÷ 75	**6.** 44 ÷ 99	**10.** 605 by 1210
3. 60 ÷ 150	**7.** 110 ÷ 220	**11.** 513 by 2052
4. 24 ÷ 228	**8.** 340 ÷ 1700	**12.** 208 by 1664
5. 30 ÷ 375	**9.** 510 ÷ 1020	**13.** 111 by 8888

Written Work

1. Change $\frac{3}{4}$ to a decimal.

$$\frac{3}{4} = 3 \div 4 = 4)\overline{3.00}$$
$$\phantom{\frac{3}{4} = 3 \div 4 = }0.75$$

Since a fraction may be regarded as an expression of division (p. 104), $\frac{3}{4} = 3 \div 4$. Annex naughts and divide as on p. 129. The result is 0.75.

Test: $0.75 = \frac{75}{100}$, or $\frac{3}{4}$.

NOTE. — A decimal point must be placed after an integer before naughts can be annexed.

2. Change $\frac{4}{9}$ to a decimal.

In changing $\frac{4}{9}$ to a decimal, thus, $9)\overline{4.000} \atop 0.444\frac{4}{9}$, it is evident that the divisor is not contained in the dividend an integral number of times. The quotient may be indicated as above, or a + sign may take the place of the fraction to show an undivided remainder. Thus, $9)\overline{4.000} \atop 0.444+$.

A common fraction is changed to an equivalent decimal by placing a decimal point after ones' place in the numerator and dividing by the denominator.

Change to equivalent decimals, and test:

3, $\frac{1}{5}$	**6.** $\frac{7}{8}$	**9.** $\frac{7}{20}$	**12.** $\frac{11}{16}$	**15.** $\frac{5}{11}$	**18.** $\frac{6}{7}$
4. $\frac{4}{5}$	**7.** $\frac{5}{8}$	**10.** $\frac{6}{25}$	**13.** $\frac{13}{25}$	**16.** $\frac{7}{12}$	**19.** $\frac{5}{9}$
5. $\frac{3}{8}$	**8.** $\frac{3}{10}$	**11.** $\frac{9}{16}$	**14.** $\frac{11}{20}$	**17.** $\frac{11}{16}$	**20.** $\frac{5}{6}$

Write as mixed decimals:

21. $5\frac{1}{4}$	**26.** $7\frac{1}{4}$	**31.** $3\frac{3}{4}$	**36.** $2\frac{2}{5}$	**41.** $2\frac{3}{20}$
22. $4\frac{1}{2}$	**27.** $8\frac{1}{10}$	**32.** $2\frac{7}{8}$	**37.** $2\frac{3}{4}$	**42.** $2\frac{1}{8}$
23. $3\frac{1}{8}$	**28.** $6\frac{1}{4}$	**33.** $2\frac{1}{2}$	**38.** $1\frac{3}{8}$	**43.** $2\frac{1}{4}$
24. $\frac{1}{5}$	**29.** $3\frac{2}{5}$	**34.** $3\frac{1}{4}$	**39.** $1\frac{5}{8}$	**44.** $1\frac{7}{20}$
25. $3\frac{2}{3}$	**30.** $5\frac{7}{8}$	**35.** $8\frac{1}{9}$	**40.** $7\frac{5}{6}$	**45.** $8\frac{3}{16}$

MEASUREMENTS AND SCALE DRAWING

MEASURING AND ESTIMATING LENGTH

Oral Work

1. Observe your foot ruler. Notice that it is first divided into inches; then into $\frac{1}{2}$ inches; then into $\frac{1}{4}$ inches, $\frac{1}{8}$ inches, and $\frac{1}{16}$ inches. These are all the divisions of the inch that are used in ordinary business.

2. Measure the lengths and widths of your various books to the nearest fraction of an inch.

3. Measure your teacher's desk, and the length and width of your own desk, etc., and express each in feet and fractions of a foot.

131

4. Observe the yardstick. Notice that it is divided in the same manner as the foot ruler. How many feet equal 1 yard? 2 yards is what part of a foot? 1 foot is what part of a yard?

5. How many feet equal a rod? Measure a rod on the blackboard with a yardstick. How many rods equal a mile?

6. Measure the length and the width of your school ground in yards. Determine the length and width of your school ground in rods in two different ways.

7. How many rods equal 1½ miles?

8. Secure a board 1 rod in length and divide it into feet and fractions of a foot. With this board, measure 20 rods from the schoolhouse and set a post.

To THE TEACHER. — Have pupils pace off certain distances until they are quite accurate in estimating the length of their own steps. In the country secure a 50-foot tapeline and get the boys and girls to measure or estimate the distances they live from the schoolhouse.

In the city divide the pupils into groups and have them measure certain city blocks in both yards and rods. Have each pupil estimate by distance between the hands — a foot, a yard, ½ yard, an inch.

9. Estimate the length of your school grounds in feet; then measure the length and compare the result with your estimate.

10. Show that 320 rd. = 5280 ft.

11. Show that 1760 yd. = 5280 ft.

12. Show that 5280 ft. = 1760 yd.

13. Show that 5280 ft. = 320 rd.

14. Draw lines on the blackboard and estimate their lengths. Then measure the lines and compare the results with your estimates.

15. Measure your steps. Then estimate distances by pacing.

16. Paul steps 2 ft. 3 in. at a step. How many feet does he travel in 240 steps?

17. May steps 2 ft. 2 in. at a step. How far does she live from the schoolhouse if she paces the distance in 596 steps?

18. How many feet of fence are required for a garden in the form of an oblong 26 yards long and 12 yards wide?

19. James lives 180 rods from the schoolhouse. How many feet does he travel in going to and coming from school each day?

20. A boy travels 135 yards each day in carrying the mail. How many yards does he travel in 6 days? How much less than a mile does he travel?

21. The doctor orders Mr. Jones to walk 4 miles each day. If the walk to the mineral spring and back is $\frac{1}{4}$ mile, how many trips must he make each day to equal 4 miles?

SCALE DRAWING

1. May lives 15 miles from Newark, and Susan lives 10 miles from the same city. Letting 1 inch represent 5 miles, you can readily show their relative distance from Newark.

NEWARK		SUSAN	MAY
0		10	15

This is called a *picture* or graphic representation.

2. Henry, Frank, Martha, and Arline received the following averages for the school term: 96, 84, 90, 78. Represent their averages by a line drawn on a scale of 1 inch to 24.

3. John, Raymond, Nellie, and Webb picked, respectively, 24, 18, 16, and 20 quarts of berries in one day. Show by a line drawn on a scale of 1 inch to 8 quarts a comparison of the amounts picked by each one.

4. Four boys put in the savings bank, in one year, the following amounts: John, $48; Henry, $40; Edward, $60; and Ben, $50. Represent their savings by a line drawn on a scale of 1 inch to $24.

5. Find the scale of the map of New Jersey in the geography you are using, and test it by actual measurement to see whether it is correct.

6. Measure your school ground and draw a map of the surface on a suitable scale.

7. Draw, on a suitable scale, a map of your schoolroom floor.

8. Draw, on a suitable scale, a map of the different blackboards in your schoolroom.

9. Find the scale of the map of Pennsylvania in your geography, and test it by measurement to see whether it is correct.

10. Mr. Ronald's farm is 40 rods by 80 rods. Show the surface on a scale of 1 inch to 20 rods.

11. Measure the walls of your schoolroom, and draw a diagram of the surface on a suitable scale.

12. The scale is frequently in fractions of an inch or of a foot. Draw a line, on a scale of $\frac{1}{16}$ of an inch to 20 miles, to show a distance of 3200 miles, which is about as far as from New York to San Francisco.

13. Mr. and Mrs. Jones and their three children, Mary, Martha, and Jane, are respectively 50, 40, 20, 15, and 10 years of age. On a scale of one inch for 20 years, show graphically a comparison of their ages.

14. The maximum temperature in Trenton for one week in September, 1912, was 80, 75, 85, 70, 65, 90, and 70 degrees. Represent these temperatures by a graph of an inch to 20 degrees.

15. John, Henry, Mary, Susan, and Harry attend the school term the following number of days, respectively, 180, 160, 170, 120, and 160. On a scale of an inch to 40 days, show graphically their attendance.

16. Show graphically, on a suitable scale, a comparison of the length, height, and width of your schoolroom.

17. Helen, John, Frank, and Martha have put respectively in the savings bank for the year 1913, $40, $48, $30, and $20. Show graphically a comparison of their savings, representing $8 to an inch.

A_____B

C_____D

E_____F

G_____H

18. These lines are drawn on a scale of 1 in. to 10 miles. How many miles does AB represent? CD? EF? GH?

BILLS

NEWARK, N.J., *Dec.* 1, 1912.

Mr. *L. M. Thomas,*

       Broad St.

  Bought of C. H. MORRISON & CO.,

TERMS: *Cash.*

Dec.	1	2 bu. apples	@	$0.75	1	50		
		3 doz. eggs	@	.25		75		
		2 bbl. flour	@	6.50	13	00		
		Total,					15	25
		Paid						
		Dec. 1, 1912						
		C. H. M. & Co:						

Observe that this bill shows: (1) the *place* and the *date;* (2) who *bought* the goods; (3) who *sold* the goods; (4) the *name of the goods* sold and the *price* and the *amount* of each sale.

To **foot** a bill means to add the cost of all the separate articles.

The word **total** means the amount of all the sales.

Make bills of the following sales, using a schoolmate's name as purchaser, and your grocer as the one who sells the goods.

1. Mch. 1   2 lb. butter @ 25¢
   Mch. 10   6 lb. meat @ 15¢
   Mch. 10   3 bars soap @ 10¢

Make out bills as suggested on previous page:

2. Jan. 10   3 skeins yarn @ $0.08
   Jan. 10   4 papers needles @   .05
   Jan. 10   5 yd. ribbon @   .50

3. Jan. 5   5 bu. potatoes @ $0.75
   Jan. 10   3 boxes peaches @   1.50
   Jan. 15   12 doz. lemons @   .40

4. Mch. 1   4 spools thread @ $0.05
   Mch. 10   6 papers pins @   .10
   Mch. 25   5 cards hooks and eyes @   .02

5. Feb. 10   5 lb. roast beef @ $0.15
   Feb. 10   3 lb. pork chops @   .15
   Feb. 10   4 lb. lamb chops @   .18

6. May   2   7 silver forks @ $2.00
   May   5   3 sterling spoons @   1.75
   May   5   4 napkin rings @   3.25

7. June 5   4 nickel sponge racks @ $2.25
   June 6   5 hairbrushes @   2.00
   June 7   6 nickel towel rods @   .75

8. June 6   4 doz. linen writing paper @ $0.20
   June 11   4 doz. linen envelopes @   .15
   June 20   15 pens @   .05

9. July 8   12 pkgs. flaxseed @ $0.05
   July 8   3 oz. cologne @   .10
   July 8   5 lb. paint @   .20

10. July 15   2 music cabinets @ $15.00

       July 16   5 rocking chairs @   5.00

       July 24   3 medicine cabinets @   3.00

Suppose your classmates to own different stores, as, *meat shops*, *grocery stores*, *hardware stores*, *dry goods stores*, etc., and pretend that you are the customer. Make out proper bills for the following purchases, giving day and date of purchase. Receipt the bills.

11. Sept. 10, 1 doz. oranges @ 40¢; 3 doz. bananas @ 20¢. September 14, 4 baskets grapes @ 2 for 35¢. September 20, 6 qt. peanuts @ 5¢ a pint.

12. October 4, 4 collars @ 12½¢; 1 necktie @ 50¢; 1 necktie @ 25¢; 1 shirt @ $1.00; 1 shirt @ $1.50; 3 handkerchiefs @ 20¢.

13. September 10, 1 set of 6 chairs @ $13.00; 1 rocker @ $8.50. September 14, 1 bedroom suite @ $30.00; 1 mattress @ $6.00; 1 spring @ $4.50.

14. April 1, 2 hoes @ 30¢; 1 rake @ 40¢; 1 shovel @ 60¢. April 10, to repairing hose $1.90. This bill was paid May 1.

15. February 10, 3 lb. steak @ 20¢; 1 lb. pork chops @ 18¢. February 12, 4 lb. roast @ 15¢; 2 lb. country sausage @ 18¢. February 17' 3 lb. veal cutlets @ 24¢. Receipt this bill if paid February 20.

16. February 19, 1 chisel @ 40¢; 1 handsaw @ $1.10; 5 lb. nails @ 5¢. February 24, 3 boxes tacks @ 10¢; 1 shovel @ 80¢. Receipt this bill if paid March 1.

17. July 1 John buys 1 fishing tackle @ $1.90; 1 doz. hooks @ 25¢; 6 fishlines @ $1.50; 1 tent @ $10.00; 2 skillets @ 40¢. Make out the bill if paid August 1.

18. August 1 Mary buys at a dry goods store 6 yd. of dress goods @ 30¢; 2 yd. of ribbon @ 20¢; 1 doz. buttons @ 30¢; 2 spools of thread @ 10¢; 4 yd. of muslin @ 6¢. Make out the bill if paid September 1.

MARKETING PROBLEMS

Oral and Written

Mr. Adams posted the following prices in his store window Saturday morning:

Grapefruits, fancy,	2 for	$0.25	Green corn, per dozen . .	$0.22
" "	common, 4 for	. 0.25	Baked beans, per dozen cans	0.95
Potatoes, per bushel . . .		. 0.75	Celery, per bunch	. 0.10
Butter, Elgin, per pound .		. 0.33	Eggplants, per dozen . . .	. 0.75
Butter, dairy, per pound .		. 0.25	Watercress, per dozen bunches	0.40
Sugar, per 25 lb. bag . .		. 1.30	Blackberries, per basket . .	0.12½
Flour, per half sack . . .		. 0.68	Sugar corn, per dozen cans .	1.00

Find the cost of :

1. 7 bu. potatoes.
2. 15 lb. Elgin butter.
3. 30 bunches celery.
4. 25 doz. watercress.
5. 12 lb. dairy butter.
6. 75 grapefruits, fancy.

7. 8¼ doz. corn.
8. 4 common grapefruits.
9. 2½ sacks flour.
10. 2 doz. cans baked beans.
11. 7 doz. eggplants.
12. 3 baskets blackberries.

13. James bought 10 lb. of sugar at 5½ cents a pound; 4½ lb. of butter at 20 cents a pound; 6 lemons at 3 for 5 cents; and two 8-cent loaves of bread. He gave the grocer a two-dollar bill. How much change did he receive?

14. Make out a bill for the above articles, and receipt it.

Find the cost of:

15. 2⅛ doz. eggplants.
16. 2¼ lb. Elgin butter.
17. 1½ lb. dairy butter.
18. 2¾ doz. watercress.

19. 1⅜ doz. eggplants.
20. 3 bags sugar.
21. 9 bunches celery.
22. 6 fancy grapefruits.

Let each pupil make out and receipt a bill for goods.
Fruit dealers on Tuesday sell at the following prices:

Grapefruits, fancy, 2 for	.	$0.25	Peaches, fancy, per peck	.	$0.60
" " " 1 for	.	0.15	Pears, fancy, per bushel	.	1.75
" " good, 3 for	.	0.25	Pears, good, per peck	.	0.55
" " fair, 4 for	.	0.25	Apples, fancy, per bushel	.	2.00
Blackberries, per crate	. .	2.80	Apples, fancy, per peck	.	0.60
Blackberries, per basket	. .	0.10	Apples, good, per bushel	.	1.20
Raspberries, per crate	. .	3.80	Apples, good, per peck	.	0.35
Raspberries, per basket	.	0.15	Oranges, fancy, per dozen	.	0.45
Peaches, fancy, per bushel	.	2.00	Bananas, per dozen	. .	0.20
Cranberries, per bushel	. .	3.80	Cranberries, per quart	. .	0.12½

Using this market report, find the cost of :

23. 4 crates blackberries.
24. 1¼ bu. peaches, fancy.
25. 27 baskets raspberries.
26. 3 pk. pears, fancy.
27. 6 bu. apples, fancy.
28. 4 grapefruits, fair.
29. 6 grapefruits, fancy.
30. 3 grapefruits, good.
31. 2¾ doz. oranges, fancy.
32. 2½ doz. bananas.
33. 1¾ bu. pears, fancy.
34. 1½ bu. apples, good.
35. 49 baskets blackberries.
36. 3 bu. apples, good.
37. 7¾ doz. oranges, fancy.
38. 75.6 doz. bananas.
39. 28 qt. cranberries.
40. 1 bu. 3 pk. cranberries.

41. 3¾ bu. apples, good.
42. 3 pk. peaches, fancy.
43. 3¾ bu. peaches, fancy.
44. 7¾ doz. bananas.
45. 17 baskets raspberries.
46. 22 qt. cranberries.
47. 3 pk. pears, fancy.
48. 3 pk. apples, good.
49. 10 grapefruits, fancy.
50. 9 grapefruits, good.
51. 8 grapefruits, fair.
52. 9 crates raspberries.
53. 12 crates blackberries.
54. 17 baskets blackberries.
55. 19 baskets raspberries.
56. 11 qt. cranberries.
57. 1 pk. pears, fancy.
58. 3 pk. pears, good.

59. Make out bills of your local grocery, showing the sale of three different articles to your mother, and the receipt for the payment of them.

The morning newspaper shows the following quotations:

Apples, fancy, per bushel $2.25	Eggs, storage, per dozen 0.22
Apples, fair grades, per bushel 1.20	Butter, creamery, per pound 0.28
Peaches, good, per bushel . . 1.80	Butter, dairy, per pound 0.25
Peaches, fancy, per bushel . 2.50	Cheese, full cream, per pound 0.12½
Pears, best quality, per bushel 1.50	Cheese, American, per pound 0.15
Grapes, Niagara, per 10-lb. basket 0.25	Potatoes, per bushel . . . 0.65
Concords, per 10-lb. basket 0.28	Sweet potatoes, Va.
Eggs, strictly fresh,	per bushel 0.80
per dozen $0.40	Jersey, per bushel 1.25

Using this market report, find the cost of:

60. 8¾ bu. fancy peaches.

61. 4½ lb. creamery butter.

62. 5½ bu. potatoes.

63. 8 10-lb. baskets Concord grapes.

64. 7 bu. apples, fancy.

65. 9 cases storage eggs, 30 doz. each.

66. 8 10-lb. baskets Niagara grapes.

67. 8¾ bu. sweet potatoes, Va.

68. ¾ bu. peaches, good.

69. 9 full cream cheese, 15 lb. each.

70. 7 10-lb. baskets Concord grapes.

71. 9¾ bu. pears, best quality.

72. 4¾ bu. apples, fair grades.

73. 3¾ lb. creamery butter.

74. 3½ doz. eggs, storage.

75. 6¾ doz. eggs, strictly fresh.

76. 10½ lb. cheese, full cream.

77. 3¾ bu. sweet potatoes, Va.

78. 3½ bu. sweet potatoes, Jersey.

79. 1¾ bu. apples, fair.

80. 1½ bu. pears, best quality.

81. 6 10-lb. baskets Concord grapes.

. 1 bu. 3 pk. apples, fancy.

· 1 bu. 3 pk. apples, fair.

10¾ lb. cheese, American.

82.
83. 6⅔ doz. eggs, strictly fresh.

86. 8⅞ lb. creamery butter and 3¾ lb. dairy butter.

87. 8¼ bu. fancy peaches, and 42 bu. good quality.

88. 8 10-lb. baskets Concord grapes and 6 10-lb. baskets Niagara grapes.

TEST EXERCISES

To the Teacher. — Divide the class into groups, giving to each group the same number of problems, and time the work of each group. Also let pupils act as customers at local stores and make out and receipt bills.

Observe short methods when possible and find the cost of

1. 14 yd. @ 7 ¢	**5.** 71 yd. @ 12 ¢
2. 40 lb. @ 7 ¢	**6.** 2 gal. @ 7½ ¢
3. 53 qt. @ 8½ ¢	**7.** 6 lb. @ 10 ¢
4. 27 gal. @ 9 ¢	**8.** 2 qt. @ 6¼ ¢

9. 26 qt. @ 8 ¢
10. 12 bu. @ 84 ¢
11. $3\frac{3}{4}$ yd. @ 8 ¢
12. $22\frac{3}{4}$ yd. @ 10 ¢
13. $41\frac{3}{4}$ gal. @ 12 ¢
14. $22\frac{3}{8}$ yd. @ 8 ¢
15. $66\frac{3}{4}$ lb. @ 10 ¢
16. $15\frac{2}{3}$ yd. @ 5 ¢
17. 4 pk. @ 9 ¢
18. $90\frac{1}{2}$ gal. @ 16 ¢
19. $23\frac{3}{4}$ lb. @ 15 ¢
20. $34\frac{3}{4}$ yd. @ 50 ¢
21. $29\frac{3}{4}$ gal. @ $8\frac{1}{2}$ ¢
22. $41\frac{3}{4}$ lb. @ 12 ¢
23. $22\frac{2}{3}$ yd. @ 15 ¢
24. $24\frac{3}{4}$ gal. @ 3 ¢
25. $41\frac{3}{4}$ yd. @ 8 ¢
26. $59\frac{3}{4}$ yd. @ 12 ¢
27. 25 yd. @ $12\frac{3}{4}$ ¢
28. $24\frac{3}{4}$ gal. @ 16 ¢
29. $90\frac{3}{4}$ lb. @ 20 ¢
30. $22\frac{2}{3}$ yd. @ 4ε $\frac{2}{3}$ ¢
31. $12\frac{1}{4}$ gal. @ 20 ¢
32. 49 pt. @ 8 ¢
33. 18 qt. @ $16\frac{2}{3}$ ¢
34. 60 yd. @ $12\frac{1}{2}$ ¢
35. $28\frac{3}{4}$ yd. @ 50 ¢
36. $19\frac{3}{4}$ qt. @ 12 ¢
37. $16\frac{2}{3}$ yd. @ 18 ¢
38. 36 yd. @ $12\frac{1}{2}$ ¢
39. 18 gal. @ $16\frac{2}{3}$ ¢
40. 12 pairs @ 75 ¢
41. 14 lb. @ 8 ¢
42. 102 lb. @ $12\frac{1}{2}$ ¢
43. 120 lb. @ $11\frac{1}{2}$ ¢
44. $16\frac{3}{4}$ lb. @ 9 ¢
45. $22\frac{3}{4}$ lb. @ $12\frac{1}{2}$ ¢
46. $83\frac{2}{3}$ yd. @ 12 ¢
47. $70\frac{1}{4}$ yd. @ $10\frac{1}{2}$ ¢
48. $56\frac{3}{4}$ yd. @ $6\frac{2}{3}$ ¢
49. $24\frac{3}{4}$ yd. @ 8 ¢
50. 54 yd. @ 16 ¢
51. 82 yd. @ $11\frac{3}{4}$ ¢
52. 51 gal. @ $54\frac{1}{2}$ ¢
53. 80 gal. @ 25 ¢
54. 69 gal. @ $33\frac{1}{3}$ ¢
55. 72 gal. @ 25 ¢
56. 90 gal. @ 20 ¢
57. 36 gal. @ $16\frac{2}{3}$ ¢
58. 45 gal. @ $8\frac{1}{3}$ ¢
59. 56 yd. @ $12\frac{1}{2}$ ¢
60. 100 yd. @ $12\frac{1}{2}$ ¢

INTERPRETING PROBLEMS

To THE TEACHER. — Pupils should be encouraged to frame problems relating to their own experiences and applying all the principles of arithmetic they have thus far learned. The following suggestions will be helpful in such work.

a. In city and town schools, the teacher should talk with the pupils about the purchases which they make for themselves or their parents; the amounts they earn or save from their allowances; the market prices of produce at the local stores; etc.

b. In rural or country schools, the teacher should talk with the pupils about selling vegetables, fruits, grains, etc., at near-by towns; about purchases made by the pupils for the home; about earnings of boys and girls at various employments, such as picking berries, gathering fruits, packing fruits, etc.

c. In fruit-growing communities, a great variety of problems can be formulated out of the child's experience in actual work.

d. Where children keep a savings account in some local savings bank, encourage them to make problems illustrating their actual experiences in depositing and drawing money.

e. Each pupil may be encouraged to plan a vacation spent in visiting some friend, or in making a trip, framing problems to include the cost of traveling, various purchases made while traveling, money spent for sight-seeing, etc.

f. Emphasis should be placed on the development of the child's *judgment* in selecting the best and shortest method of operation, where several methods are possible.

(1) In multiplying by 11 the pupil should multiply by 10 and add the multiplicand.

Thus, $11 \times 9990 = 99900$
$$\underline{9990}$$
$$109890$$

(2) In the problem, 24×25, the child who knows that $25 \times 25 = 625$, should instantly see that 24×25 is 25 less; that is, $625 - 25$, or 600.

(3) In the problem, .75 of 480, the pupil should be encouraged to multiply $\frac{1}{4}$ of 480 by 3; but in the problem .75 of $3520.87, the child should multiply by .75, since 3520.87 cannot be exactly divided by 4.

(4) The aliquot parts of the dollar and of 100 should always be used by the pupil in short methods of solution.

(5) The pupil should be encouraged to make use of *cancellation*, whenever possible.

(6) Whenever possible, pupils should make a mental estimate of the answer and compare it with the written result.

1. A Paterson gardener sold 4500 qt. of strawberries at an average price of 8⅓¢ a quart; and 600 qt. of raspberries at 12½¢. After deducting $150 for cultivation and labor, find his profit.

Statement. — The problem calls for the profit on the sales.

Explanation. — The amount received for each lot is found by multiplying the average price by the quantity. The profit is found by deducting the cost of labor, etc., from the sum of the amounts received for strawberries and raspberries.

Mental Estimate. — $\frac{1}{12}$ of $4500 = about $370; ⅛ of $600 = $75. $370 + $75 − $150 = $295.

Solution. — 4500 × 8⅓¢ = $375, the cost of 4500 qt.
600 × 12½¢ = $ 75, the cost of 600 qt.
$450, the cost of all
$450 − $150 = $300, balance.

2. Find the cost of 90 T. of soft coal at $1.62½ per ton, plus $19.85 freight charges.

Statement. — The problem calls for total cost of coal and freight.

Explanation. — The cost of the coal equals the price of one ton multiplied by the number of tons. The total cost equals the cost of the coal plus the cost of the freight.

Mental Estimate. — 100 T. would cost $162.50 and 90 T. $16.25 less. Instead of subtracting $16.25 and then adding freight charge, $19.85, add the difference, $3.60, to $162.50, making about $166.

Solution. — 90 × $1.62½ = 90 × $\frac{13}{8}$ = $\frac{1170}{8}$, or $146.25, cost of 90 T. $19.85 added to $146.25 = $166.10, total cost.

3. A western farmer raised 80 acres of oats that averaged 55¾ bushels per acre. What was the value of the crop at $.37½ per bushel?

4. When 1 pencil costs 10¢, how many can you buy for 60¢?

5. At 12¢ per yard, how many yards of ribbon can be bought for 96¢?

6. Cherries are 8¢ a quart. How many quarts can be bought for 72¢?

7. When 3 tons of coal cost $18, how much will 7 tons cost?

Why is it better in this example first to find the cost of 1?

8. How much will 9 dozen lemons cost when 3 dozen sell for 45 cents?

Why is it better in this example first to find the relation of 3 dozen to 9 dozen?

9. (*a*) If 3 men earn $30 in a certain time, how much will 8 men earn in the same time at the same rate?

(*b*) If 3 men earn $30 in a certain time, how much will 9 men earn in the same time at the same rate?

What is the most economical method of working (*a*)? of working (*b*)? why?

10. When ½ a bushel of potatoes sells for 25 cents, how much will 3 bushels cost?

$$25¢, \text{ cost } \tfrac{1}{2} \text{ bu.}$$
$$2 \times 25¢ = 50¢, \text{ cost } 1 \text{ bu.}$$
$$3 \times 50¢ = \$1.50, \text{ cost } 3 \text{ bu.}$$

11. At 20 cents a peck, how much will 2 bushels of apples cost?

12. When milk is selling at 5¢ a quart, how much will 3 gallons cost?

13. If ¼ of a yard of velvet costs 60¢, how much will 2 yards cost?

14. Mr. Thomas raised 640 bushels of peaches, that were sold on an average at 75¢ per bushel. His baskets cost $25, and his labor $50. Find the price per bushel he realized, after payment of baskets and labor.

15. Find the gain on 50 bushels of peaches bought at 65¢ per bushel and retailed at 25¢ per peck.

16. Frank worked $2\frac{1}{8}$ hours, for five days each week, and 12 hours each Saturday. Find his earnings for 10 weeks at 12¢ an hour.

17. If a boy pays $2.50 a hundred for papers, and sells them for 5¢ apiece, how much does he gain on 300 papers?

18. In an orchard there are 144 trees; 18 are cherry trees, 36 are apple trees, and $\frac{1}{5}$ of the remainder are peach trees. How many are peach trees?

19. Find the cost of 24,000 railroad ties at $62\frac{1}{2}$¢ each.

20. When lead pencils are selling at $1\frac{1}{8}$ per gross (144), find the cost of 3550 gross.

21. Find the cost of sewing buttons on 48 suits, at $2\frac{3}{4}$¢ a suit.

22. A contractor averages $6\frac{7}{8}$ rd. a day in digging a sewer. How long is the sewer if it takes him 39 days to dig it?

23. A rural mail carrier travels $23\frac{7}{8}$ miles for each delivery. Find the number of miles traveled in 310 deliveries.

24. An ocean steamer burns on an average $201\frac{5}{16}$ tons of coal in a day. How much coal will it consume in a voyage of 7 days?

25. I bought 46 lb. of sugar @ 7¢, 95 lb. of coffee @ 23¢, and 73 doz. eggs @ 18¢. Find the cost of all.

26. James had $24.36 in a bank and drew $17.49 out. How much had he left in the bank?

27. At $.34 each how much will 189 books cost?

28. If 54 cows cost $2430, how much does each cost?

29. At 25¢ each, find the cost of 20 gross of boys' caps.

30. If a school requires an average of 497 pads of paper for one month, how many will be needed for 3 years of 10 months each?

31. If 48 barrels of cement cost $208.55, how much will 86 barrels cost?

32. How much will 96 quarts of cranberries cost, if 24 quarts cost $3.60?

33. A western farmer raised 6741 bu. of oats. He kept 349 bu. for feed, and sold the remainder at $.35 a bushel. How much did he receive?

34. A merchant sold 32 yd. of cloth at 12¢ a yard, and 192 yd. at 8¢ a yard. How much did he receive?

35.   PAID ADMISSIONS TO A FAIR

TICKETS	PRICE	TUES.	WED.	THURS.	FRI.	TOTAL RECEIPTS
Children	15¢	864	1865	1226	1285	
Adults	25¢	2864	3245	2764	3768	
One-horse vehicles	35¢	376	364	176	472	
Two-horse vehicles	50¢	212	216	144	224	

36. At 4¢ a pint, find the cost of 7 gal. 1 pt. of milk.

37. At 9¢ a quart, find the cost of 5 bu. 2 pk. of beans.

38. At 4¢ an ounce, find the cost of 10 lb. 3 oz. of ginger.

39. At 32¢ a pound, find the cost of 18 lb. of butter.

40. Find the cost of sending a 15-word telegram from New York to Denver, at 75¢ for the first ten words and 5¢ for each additional word.

41. How much will it cost to telegraph a night letter of 50 words, from New York to Denver, if the night rate for 50 words is the same as the day rate for 10 words?

42. A telephone message from New York to Boston costs $1.25 for the first three minutes and $.40 for each additional minute. Find the cost of talking 17 minutes.

43. A grocer bought 57 crates of berries at $4.75 a crate. If he sells them at $5.15 a crate, how much will he gain?

44. How much will a fruit dealer gain if he buys 174 boxes of oranges for $739.50, and sells them at $5 a box?

45. A merchant bought 136 barrels of apples at $1.35 a barrel; but 30 barrels were damaged. If he sold the remainder at $1.75 a barrel, did he gain or lose, and how much?

46. A man bought 36 gallons of milk for $7.20. He sold it for 35¢ a gallon. How much did he gain?

47. A merchant bought 800 bushels of potatoes at 45¢ a bushel. He sold one half of them at 60¢ a bushel, and the remainder at 40¢ a bushel. Did he gain or lose, and how much?

48. Find the perimeter or distance around a room 18 ft. by 15 ft.

49. At 7¢ a foot, how much will it cost to put a picture molding around it?

50. Mr. Adams worked 8 hours a day for 26 days at $.37 an hour. How much did he earn?

51. Find the entire cost of:

6 yd. silk @ 85 ¢.

5 bu. peaches @ $2.25.

24 boxes soap @ $4.75.

52. Make out a bill for :

9 T. hard coal @ $6.75.

12 T. soft coal @ $3.75.

15 cwt. sugar @ $5.

53. A merchant buys from a farmer:

25 bu. corn @ 45 ¢.

36 doz. eggs @ 23 ¢.

16 bu. apples @ 75 ¢.

The farmer buys from the merchant :

6 brooms @ 25 ¢.

22 yd. carpet @ 90 ¢.

6 chairs @ $2.50.

Which person owes the other, and how much ?

54. In a factory there are 56 men employed at $2.25 a day; 12 men at $3.75 a day; 25 boys at $.87 a day; 8 women at $1.75 a day. The other expenses are $267 a day. How much does it cost to keep the factory going a month of 26 days ?

SIXTH GRADE — FIRST HALF

REVIEW

ADDITION

Oral Work

Drill on the following combinations of numbers, emphasizing the right-hand figure in the sums. Thus, $3 + 8 = 11$; $13 + 8 = 21$; $23 + 8 = 31$, etc.

	a	b	c	d	e	f	g
1.	11+3	21+3	31+3	41+3	51+3	61+3	81+3
2.	13+5	28+5	33+5	43+5	53+5	68+5	88+5
3.	17+6	27+6	37+6	47+6	57+6	67+6	87+6
4.	9+4	19+4	29+4	39+4	49+4	59+4	79+4
5.	7+9	17+9	37+9	57+9	67+9	77+9	97+9
6.	6+8	16+8	56+8	76+8	86+8	46+8	96+8

Add, giving results at sight:

	a	b	c	d	e	f	g	h	i	j	k	l
7.	5	6	7	8	6	8	9	7	8	9	7	6
	5	4	3	2	5	3	2	4	4	3	5	6
8.	9	7	8	9	8	9	8	9	8	9	7	8
	4	7	6	5	7	6	8	7	9	9	9	5

9. 11	11	11	11	11	11	12	12	12	12	12	12
4	5	6	7	8	9	3	4	5	6	7	8

10. 13	13	13	13	13	14	14	14	14	15	15	15
4	5	6	7	8	3	4	5	6	3	4	5

11. Count by 3's from 1 to 100; from 2 to 101.

12. Count by 4's from 1 to 101; from 3 to 103.

13. Count by 5's from 1 to 101; from 2 to 102; from 3 to 103; from 4 to 104.

14. Count by 6's from 1 to 103; from 2 to 104; from 3 to 105.

Add at sight:

	a	*b*	*c*	*d*	*e*	*f*	*g*	*h*	*i*
15.	12	12	12	13	15	14	15	15	18
	9	11	13	14	12	15	16	17	19
16.	18	17	15	19	16	19	17	19	25
	14	16	18	14	18	16	18	17	35
17.	29	37	27	28	27	39	44	38	48
	8	9	7	6	6	9	6	8	5
18.	67	88	75	56	97	85	48	54	65
	8	4	9	6	4	7	9	9	9
19.	67	78	89	34	32	57	97	47	76
	9	5	7	7	8	8	6	9	8
20.	49	63	45	67	76	95	27	55	99
	7	7	7	6	4	8	9	4	8
21.	27	36	56	59	39	57	59	66	87
	9	6	7	9	7	8	9	6	9

Add rapidly, looking for combinations that make 10:

22.	8	9	4	8	9	5	7	8	9
	5	4	2	8	2	2	4	5	1
	2	4	4	4	6	3	5	1	5
	3	2	4	6	2	5	1	4	4

23.		7	6	8	3	5	9	8	7
		2	6	3	7	5	1	2	3
		8	4	7	6	8	6	5	4
	5	6	5	9	8	3	4	5	5

	a	b	
24.	$9 + 8 + 7 + 3 = ?$	$8 + 2 + 5 + 4 = ?$	$8 + 7 + 3 + 9 = ?$
25.	$5 + 8 + 3 + 9 = ?$	$5 + 4 + 1 + 7 = ?$	$5 + 6 + 9 + 8 = ?$
26.	$6 + 5 + 4 + 2 = ?$	$6 + 9 + 4 + 7 = ?$	$3 + 9 + 7 + 5 = ?$

Add:

27.	$\frac{3}{4}, \frac{3}{8}, \frac{1}{2}$	$\frac{1}{2}, \frac{2}{3}, \frac{1}{6}$	$\frac{1}{10}, \frac{1}{5}, \frac{2}{15}$
28.	$\frac{2}{3}, \frac{1}{6}, \frac{1}{12}$	$\frac{1}{4}, \frac{3}{8}, \frac{1}{16}$	$\frac{3}{8}, \frac{1}{4}, \frac{1}{16}$
29.	$1\frac{1}{2}, 3\frac{1}{8}, 4\frac{1}{6}$	$5\frac{1}{4}, 7\frac{1}{8}, 2\frac{1}{2}$	$3\frac{1}{5}, 4\frac{1}{10}, 5\frac{2}{5}$

Add:

Written Work

	a	b	c	d	e	f	g
1.	86	68	51	89	89	79	78
	64	57	98	66	78	79	54
	73	89	65	78	97	89	97
	67	66	55	65	67	74	63
2.	98	91	77	68	99	99	88
	79	89	88	79	79	79	79
	89	98	56	89	83	77	84
	49	42	77	96	57	48	86

	a	b	c	d	e	f	g
3.	53	97	69	87	49	43	89
	67	98	45	93	45	37	13
	48	62	85	38	23	35	47
	89	48	57	79	82	98	37
	93	89	92	35	58	76	22
4.	98	88	42	91	75	64	65
	54	73	38	29	81	76	48
	33	35	39	87	29	24	39
	58	22	87	58	29	36	35
	84	56	56	59	66	56	85
	67	94	86	63	78	84	73

Add, and test by adding from the top downward. Observe the groups that make 10:

5.	2679	8.	7458	11.	8775	14.	7978
	6883		8739		6328		2563
	9427		2372		8687		8545
	2656		8923		3532		8778
	5834		9965		6748		5869

6.	4979	9.	6469	12.	4298	15.	5679
	3253		4535		3642		6587
	3858		5375		5842		·4271
	4794		5287		6789		4765
	3476		9798		4321		5244

7.	4579	10.	4590	13.	8948	16.	2301
	6531		5789		2167		7998
	8679		4510		5871		1611
	4510		7986		1907		5689
	8741		1902		5789		1001

Add:

17. 4671	**21.** 9876	**25.** 7680	**29.** 8457
4339	1780	3421	6323
2145	5897	7890	1330
5698	1041	8410	5690
4510	6803	2691	8190
18. 43875	**22.** 87962	**26.** 69186	**30.** 79667
63217	83226	22535	84873
54424	65542	82423	13389
87463	39369	76855	98597
77887	38796	89797	68298
95974	96468	47659	74978
19. 49397	**23.** 76718	**27.** 25431	**31.** 45394
57911	32895	67654	86728
16224	44124	23925	48656
38875	34449	98879	28856
69548	79547	68977	47984
49569	89757	47576	79575
20. 34759	**24.** 48539	**28.** 39764	**32.** 67458
68543	64998	96587	95866
59867	86599	58979	47597
35996	37746	77858	88669
44878	88796	69487	85948
98678	95647	79684	67745

Add:

33.	**34.**	**35.**	**36.**	**37.**	**38.**
$97\frac{5}{8}$	$68\frac{1}{6}$	$41\frac{1}{6}$	$54\frac{7}{8}$	$29\frac{1}{10}$	$62\frac{4}{7}$
$53\frac{3}{4}$	$79\frac{1}{8}$	$29\frac{1}{3}$	$88\frac{2}{3}$	$57\frac{5}{12}$	$94\frac{3}{7}$
$42\frac{7}{8}$	$52\frac{1}{12}$	$33\frac{5}{9}$	$72\frac{3}{4}$	$83\frac{3}{8}$	$31\frac{1}{4}$

SUBTRACTION

Oral Work

1. Count backward by 3's beginning with 100; with 101; with 102.

2. Count backward by 4's beginning with 100; with 101; with 95.

3. Count backward by 5's beginning with 99; with 98; with 101.

4. Count backward by 6's beginning with 100; with 101; with 102; with 103; with 105.

5. Count backward by 7's beginning with 100; with 101; with 102; with 103; with 104; with 105; with 106.

6. Count backward by 8's beginning with 100; with 101; with 102; with 103; with 104; with 105; with 106; with 107.

7. Count backward by 9's beginning with 100; with 101; with 102; with 103; with 104; with 105; with 106; with 107; with 108.

Subtract rapidly :

	a	b	c	d	e	f	g	h	i
8.	10	11	9	8	13	4	5	6	7
	3	3	5	4	8	3	2	5	4
9.	8	11	12	10	7	9	8	6	13
	2	5	4	7	2	7	3	3	9
10.	8	17	11	12	5	15	7	9	8
	7	9	7	9	4	7	5	8	6

	a	b	c	d	e	f	g	h	i
11.	7	12	11	12	10	9	8	13	9
	6	6	4	7	8	2	5	5	6
12.	15	12	11	10	9	12	10	14	13
	9	8	8	4	3	2	9	8	9
13.	11	12	14	16	15	14	13	15	16
	6	3	9	9	6	7	7	9	8

	a	b	c	d	e	f	g
14.	31	31	31	31	31	31	31
	17	13	16	19	14	18	12
15.	31	30	30	30	30	30	30
	15	12	14	19	16	18	18
16.	32	32	32	32	32	32	32
	16	19	14	18	17	13	15
17.	33	33	33	33	33	33	34
	18	16	12	19	17	15	18
18.	34	34	34	34	34	35	35
	15	19	17	16	18	17	16
19.	$35\frac{3}{4}$	$36\frac{5}{8}$	$36\frac{2}{3}$	$36\frac{1}{4}$	$37\frac{8}{20}$	$41\frac{3}{4}$	$41\frac{1}{6}$
	$18\frac{1}{4}$	$19\frac{1}{8}$	$18\frac{5}{12}$	$17\frac{7}{8}$	$19\frac{1}{20}$	$19\frac{1}{6}$	$17\frac{1}{10}$
20.	$41\frac{5}{16}$	$41\frac{2}{3}$	$41\frac{3}{8}$	$41\frac{3}{5}$	$41\frac{4}{5}$	$43\frac{7}{16}$	$43\frac{3}{4}$
	$13\frac{1}{8}$	$15\frac{1}{12}$	$18\frac{8}{10}$	$14\frac{8}{10}$	$16\frac{3}{5}$	$18\frac{1}{4}$	$14\frac{5}{8}$

	a	b	c	d	e	f	g
21.	43	43	43	43	44	44	44
	19	16	15	17	18	15	17
22.	44	44	45	45	45	45	46
	19	16	18	16	19	17	19

Give differences at sight:

	a	b	c	d	e	f
23.	15 − 5	16 − 8	42 − 9	56 − 9	78 − 9	56 − 8
24.	33 − 7	54 − 6	11 − 4	13 − 8	56 − 7	45 − 9
25.	9 − 5	14 − 9	17 − 8	11 − 9	48 − 5	15 − 8
26.	25 − 9	23 − 7	61 − 7	32 − 5	44 − 8	77 − 9
27.	36 − 7	43 − 7	21 − 8	72 − 7	14 − 5	66 − 8
28.	66 − 9	41 − 8	10 − 2	45 − 8	26 − 9	22 − 6
29.	56 − 9	45 − 7	51 − 4	50 − 2	59 − 7	43 − 7

Written Work

Subtract and test:

1.	58067	5.	49302	9.	61053	13.	90062
	28389		19479		55468		73974

2.	60051	6.	73040	10.	50697	14.	43021
	48978		38748		47898		26573

3.	53104	7.	89300	11.	20199	15.	65998
	48789		59969		19899		37899

4.	70065	8.	50007	12.	38240	16.	74627
	69866		36198		28754		45738

17.	95632 85836	**25.**	70051 29875	**33.**	86003 58279	**41.**	50000 45098
18.	10000 9009	**26.**	10000 1001	**34.**	10000 6974	**42.**	10000 9909
19.	73496 57898	**27.**	80407 37558	**35.**	59601 39674	**43.**	35820 15828
20.	39857 19869	**28.**	89076 77877	**36.**	75804 36839	**44.**	50087 30988
21.	42697 29879	**29.**	79081 28597	**37.**	51091 39098	**45.**	61019 28989
22.	91081 59809	**30.**	61001 29897	**38.**	91091 59009	**46.**	51093 27987
23.	51093 49789	**31.**	91003 49895	**39.**	81018 59889	**47.**	51097 39899
24.	61095 28698	**32.**	71093 59896	**40.**	90091 81979	**48.**	73013 29897

Subtract:

49.	$57\frac{5}{6}$ $48\frac{5}{12}$	**52.**	$52\frac{5}{12}$ $48\frac{5}{6}$	**55.**	$31\frac{1}{4}$ $29\frac{5}{6}$	**58.**	$40\frac{5}{12}$ $29\frac{7}{10}$	**61.**	$60\frac{2}{3}$ $48\frac{3}{4}$
50.	$78\frac{3}{16}$ $59\frac{3}{8}$	**53.**	$50\frac{4}{11}$ $33\frac{3}{10}$	**56.**	$71\frac{5}{7}$ $44\frac{3}{4}$	**59.**	$82\frac{7}{8}$ $57\frac{7}{10}$	**62.**	$93\frac{5}{8}$ $68\frac{5}{8}$
51.	$92\frac{7}{20}$ $49\frac{5}{12}$	**54.**	$71\frac{1}{10}$ $64\frac{1}{3}$	**57.**	$82\frac{1}{20}$ $59\frac{1}{10}$	**60.**	$73\frac{5}{16}$ $37\frac{3}{4}$	**63.**	$56\frac{5}{9}$ $49\frac{5}{11}$

MULTIPLICATION AND DIVISION

Oral Work

1. Drill on these tables until pupils know them by heart :

1	2's	3's	4's	5's	6's	7's	8's	9's	10's	11's	12's	13's	14's	15's
2	4	6	8	10	12	14	16	18	20	22	24	26	28	30
3	6	9	12	15	18	21	24	27	30	33	36	39	42	45
4	8	12	16	20	24	28	32	36	40	44	48	52	56	60
5	10	15	20	25	30	35	40	45	50	55	60	65	70	75
6	12	18	24	30	36	42	48	54	60	66	72	78	84	90
7	14	21	28	35	42	49	56	63	70	77	84	91	98	105
8	16	24	32	40	48	56	64	72	80	88	96	104	112	120
9	18	27	36	45	54	63	72	81	90	99	108	117	126	135
10	20	30	40	50	60	70	80	90	100	110	120	130	140	150
11	22	33	44	55	66	77	88	99	110	121	132	143	154	165
12	24	36	48	60	72	84	96	108	120	132	144	156	168	180
13	26	39	52	65	78	91	104	117	130	143	156	169	182	195
14	28	42	56	70	84	98	112	126	140	154	168	182	196	210
15	30	45	60	75	90	105	120	135	150	165	180	195	210	225

The first row of figures at the top stands for the *different tables*. By multiplying each of the numbers in *the left-hand row* by each of the numbers in the *top row*, the tables can all be made. Thus, in the table of the twos, the products are directly below the number of the table, etc.

Give answers at sight:

	a	b	c	d
2.	9×11	11×12	24×24	9×15
3.	13×73	11×13	18×18	11×15
4.	$\frac{2}{3}$ of 6	$8 \times \frac{3}{8}$	$1\frac{2}{3} \times 1\frac{1}{2}$	$\frac{2}{8}$ of $\frac{4}{5}$
5.	$\frac{2}{3}$ of $\frac{5}{8}$	$\frac{3}{4} \times \frac{2}{8}$	$1\frac{3}{4} \times 1\frac{1}{3}$	$\frac{5}{7} \times \frac{7}{10}$

The square of 2 is represented by 2^2; the square of 3 by 3^2; the square of 4 by 4^2; etc.

6. Memorize the following squares of numbers :

$1^2 = 1$	$6^2 = 36$	$11^2 = 121$	$16^2 = 256$	$21^2 = 441$
$2^2 = 4$	$7^2 = 49$	$12^2 = 144$	$17^2 = 289$	$22^2 = 484$
$3^2 = 9$	$8^2 = 64$	$13^2 = 169$	$18^2 = 324$	$23^2 = 529$
$4^2 = 16$	$9^2 = 81$	$14^2 = 196$	$19^2 = 381$	$24^2 = 576$
$5^2 = 25$	$10^2 = 100$	$15^2 = 225$	$20^2 = 400$	$25^2 = 625$

Written Work

To THE TEACHER. — Time the pupils in their work and encourage them to beat their own records.

First multiply each number in the multiplicands of Ex. 1 by 1 *a*, that is by 5; then by 1 *b*, that is by 6; then by 1 *c*, 1 *d*, 1 *e*. Then proceed in the same way with the multiplicands and multipliers in Ex. 2–7.

	Multipliers					Multiplicands			
	a	b	c	d	e	a	b	c	d
1.	5	6	7	8	12	1140	6380	4975	$9004\frac{3}{4}$
2.	8	9	7	13	14	1240	8764	5910	$2613\frac{7}{8}$
3.	29	38	57	96	78	5781	8888	9768	$6891\frac{5}{6}$
4.	470	900	500	890	807	5438	7614	9898	$9110\frac{2}{3}$
5.	608	709	580	769	857	8888	6570	6801	$8080\frac{4}{5}$
6.	$\frac{3}{4}$	$2\frac{3}{4}$	$5\frac{1}{2}$	$7\frac{1}{6}$	$4\frac{1}{5}$	7960	5919	4759	$4859\frac{4}{7}$
7.	$\frac{3}{8}$	$2\frac{7}{8}$	$6\frac{1}{4}$	$3\frac{3}{5}$	$3\frac{3}{8}$	6910	6917	5819	$3649\frac{2}{3}$

Divide the dividends in Ex. 8, first by 8 a, that is by 9; then by 8 b, 8 c, 8 d. Proceed in the same way with Ex. 9–16.

	Divisors				Dividends		
	a	b	c	d	a	b	c
8.	9	8	7	5	64990	57216	72170
9.	70	86	98	53	36790	66019	96510
10.	24	45	29	56	85190	71001	83890
11.	50	85	95	67	76080	83000	74578
12.	98	64	28	45	98901	14896	54444
13.	450	805	780	150	89870	45000	$64500\frac{3}{4}$
14.	720	650	705	480	77891	76710	$41003\frac{7}{8}$
15.	$4\frac{1}{4}$	$5\frac{2}{5}$	$3\frac{3}{4}$	$5\frac{3}{8}$	38710	93468	$13330\frac{3}{10}$
16.	$2\frac{1}{8}$	$5\frac{4}{5}$	$5\frac{5}{8}$	$6\frac{3}{4}$	97801	86710	$37800\frac{2}{5}$

Short Methods — Oral Work

1. Memorize these tables of **aliquot parts of $1**.

$ .05 $= \frac{1}{20}$ of $ 1	$ .37\frac{1}{2} = \frac{3}{8}$ of $ 1
$ .06\frac{1}{4} = \frac{1}{16}$ of $ 1	$ .40 $= \frac{2}{5}$ of $ 1
$ .08\frac{1}{3} = \frac{1}{12}$ of $ 1	$ .50 $= \frac{1}{2}$ of $ 1
$ .10 $= \frac{1}{10}$ of $ 1	$ .62\frac{1}{2} = \frac{5}{8}$ of $ 1
$ .12\frac{1}{2} = \frac{1}{8}$ of $ 1	$ .66\frac{2}{3} = \frac{2}{3}$ of $ 1
$ .16\frac{2}{3} = \frac{1}{6}$ of $ 1	$ .75 $= \frac{3}{4}$ of $ 1
$ .20 $= \frac{1}{5}$ of $ 1	$ .80 $= \frac{4}{5}$ of $ 1
$ .25 $= \frac{1}{4}$ of $ 1	$ .83\frac{1}{3} = \frac{5}{6}$ of $ 1
$ .33\frac{1}{3} = \frac{1}{3}$ of $ 1	$ .87\frac{1}{2} = \frac{7}{8}$ of $ 1

2. Make and memorize a similar table of aliquot parts of 100. Thus, $5 = \frac{1}{20}$ of 100; $6\frac{1}{4} = \frac{1}{16}$ of 100, etc.

Give products at sight:

3. Find the cost of 12 yd. of gingham at $.16⅔ per yard.

SOLUTION. — $.16⅔ = $⅙, cost of 1 yd.

12 × $⅙ = $¹²⁄₆, or $2, cost of 12 yd.

Find the cost of:

4. 8 yd. lace @ $.12½

5. 12 yd. muslin @ $.16⅔

6. 16 yd. lace @ $.06¼

7. 15 yd. muslin @ $.33⅓

8. 18 lb. tea @ $.66⅔

9. 24 lb. rice @ $.12½

10. 90 boxes crackers @ $.10

11. 40 boxes crackers @ $.21

12. 80 lb. starch @ $.05

13. 60 lb. beef @ 25 ¢

14. 40 cans corn @ 10 ¢

15. 96 qt. berries @ $.12½

16. 54 qt. berries @ $.16⅔

17. 16 lb. steak @ $.37½

18. 70 lb. sausage @ $.20

19. 50 yd. cloth @ $.75

20. 75 doz. eggs @ $.50

21. 40 yd. silk @ $.62½

22. 150 bars chocolate @ $.20

23. 120 lb. tapioca @ $.12½

24. 150 lb. coffee @ $.33⅓

25. 48 lb. cakes @ $.12½

26. 60 lb. rice @ $.05

27. 15 lb. coffee @ $.25

How many yards can be bought for:

28. $2 @ $.05

29. $5 @ $.62½

30. $3 @ $.37½

31. $16 @ $.06¼

32. $4 @ $.16⅔

33. $7 @ $.87½

34. $8 @ $.80

35. $25 @ $.12½

36. $20 @ $.20

37. $50 @ $.83⅓

38. $30 @ $.33⅓

39. $24 @ $.08⅓

40. $40 @ $.40

41. $20 @ $.50

42. $50 @ $.25

43. $36 @ $.37½

44. $42 @ $.66⅔

45. $32 @ $.06¼

Read and give approximate result; then solve, using all possible short methods of solution:

46. I bought 75 cows at $37½ each, and sold them at $50 each. Find my gain.

47. I buy 15 lb. of coffee at $.40 per pound, 16 lb. of rice at $.12½ per pound, and 16 lb. of sugar at $.06¼ per pound. How much change should I receive from $15?

48. From a piece of cloth containing 40 yd. three dress patterns of 12¼ yd. each were sold. How much was the remainder worth at $1.25 per yard?

49. I bought 6 lb. of codfish, at 2 lb. for $.25, 12 lb. of starch at 3 lb. for $.25, and 6 lb. of coffee at 3 lb. for $1. Find the amount of my bill, and my change from $5.

50. Find the cost of 12 yd. of gingham at 16⅔¢ per yard and 33⅓ yd. muslin at 12¢ per yard.

51. A cheese weighs 20 lb. How much is it worth at 8⅓¢ per pound?

52. A huckster sells 24 bushels of tomatoes at 62½¢ per bushel, and 36 dozen sweet corn at 16⅔¢ per dozen. Find the amount of the sale.

COMMON FRACTIONS

To the Teacher. — In addition to the drill in Addition, Subtraction, Multiplication, and Division of Common Fractions found on pages 153, 155, 157, 159, 161, and 162–164, review thoroughly the work on pages 104–113.

DECIMALS

DECIMAL DIVISIONS OF A UNIT

Any unit may be divided into 10ths, 100ths, 1000ths, etc.

A **decimal fraction** is any number of tenths, hundredths, thousandths, etc., of a unit. When expressed with a decimal point, without a written denominator, it is usually called a **decimal**.

Thus, $\frac{5}{10}$ and .5, $\frac{5}{100}$ and .05, $\frac{5}{1000}$ and .005 are *decimal fractions*, but the term *decimal* is usually restricted to the forms .5, .05, .005, etc.

1. Express decimally: $\frac{5}{10}$, $\frac{25}{100}$, $\frac{85}{1000}$, $\frac{2}{10}$, $\frac{855}{1000}$.

2. In 5.55, the figure in tenths' place equals how many times the figure in hundredths' place? The figure in ones' place equals how many times the figure in tenths' place?

In any decimal or whole number, 10 units of any place = 1 unit of the next place to the left.

The **decimal point** is used to separate the units and parts of units. It is always placed at the right of ones' place and before tenths' place.

3. What is the *first* place to the right of a decimal point called? the *second* place? the *third* place?

4. What is the largest decimal division of any unit? the *second* largest? the *third* largest?

165

READING AND WRITING DECIMALS

Oral Work

1. Since the first decimal division of an integral unit is tenths, what is the first place to the right of the decimal point?

2. What is the second place called? the third place?

3. Five tenths is written .5; five hundreths is written .05; five thousandths is written .005; etc. Write 7 tenths, 6 hundredths, 8 thousandths.

4. Express decimally $\frac{5}{10}$, $\frac{75}{100}$, $\frac{6}{1000}$, $\frac{54}{1000}$, $\frac{6}{100}$, $\frac{73}{1000}$.

Every decimal contains as many decimal places as there are naughts in the denominator of the equivalent fraction.

Table of places and names of *integral* and *fractional* units:

Millions	Hundred-thousands	Ten-thousands	Thousands	Hundreds	Tens	Ones	Decimal point	Tenths	Hundreths	Thousandths	Ten-thousandths	Hundred-thousandths	Millionths
3	5	7	8	4	5	5	.	0	0	7	4	8	9

5. In .555 what figure stands for tenths? for hundredths? for thousandths? .555 is read 555 thousandths.

6. Is the decimal point named in reading a decimal? Observe that the decimal is read as an integer and that the last figure is given the required denomination.

Read:

7. .25	**9.** .005	**11.** .101
8. .05	**10.** .375	**12.** .0045

13. .4045   **15.** .60745   **17.** .0065

14. .0002   **16.** .678705   **18.** .60005

19. 50.0745 is read 50 *and* 745 ten-thousandths. How are both the integer and the decimal read? How is the decimal point read? What name is given to the last decimal place?

20. When is the decimal point read?

21. What determines the value of any figure in a decimal?

22. In writing .5, .45, .075, .0075 as common fractions, what figure in each decimal tells us the size of the denominator?

A **mixed decimal** is a whole number and a decimal; as, 4.625.

23. What number in a mixed decimal is always read first?

24. How do the number of places in any decimal compare with the number of naughts in the denominator when the decimal is expressed as a common fraction?

Read :

25. 45.075   **28.** 72.003745   **31.** .00875

26. 50.3007   **29.** 1001.1001   **32.** .008090

27. 290.25387   **30.** 794.3085   **33.** 2.004890

34. In 5 thousandths how many decimal places are there? What part of the decimal (5 thousandths) stands for the numerator of the fraction? What part of this decimal stands for the denominator?

35. Name the numerator and the denominator in the following decimals : .05, .0006, .000025, .045.

In **writing a decimal** *write the numerator, and point off from the right as many decimal places as there are naughts in the denominator.*

Written · Work

Write:

1. 84 hundredths.
2. 675 ten-thousandths.
3. 16 and 75 millionths.
4. 400 and 45 thousandths.
5. 6006 and 66 ten-thousandths.
6. 89 and 5 thousandths.
7. Seven hundred and forty-six ten-thousandths.
8. Nine hundred and 84 millionths.
9. 5095 millionths.
10. 8 and 17 ten-thousandths.
11. 125 millionths.
12. 896 and 301 hundred-thousandths.
13. One thousand and one thousandth.
14. 18051 and 957 thousandths.
15. 97 and 3 ten-thousandths.
16. 9864 millionths.
17. 2135 and 32 millionths.
18. One and one millionth.
19. One million and one tenth.
20. 90 thousand and 71 thousandths.
21. 1830 and 11684 hundred-thousandths.
22. 429 thousand and 46 ten-thousandths.
23. 7035 and 97 hundredths.
24. 67375 and 35 hundred-thousandths.
25. 5815 hundred-thousandths.
26. 375 and 69 thousandths.

ADDITION AND SUBTRACTION OF DECIMALS

Oral Work

1. Why must ¼ and ⅔ be changed to fractions having a common denominator, as twelfths, before they can be added or subtracted ?

2. What kind of fractional units, then, can be added or subtracted ?

3. Observe that decimals are only so many 10ths, 100ths, 1000ths, etc., of a unit, expressed by placing a period before the numerator and omitting the denominator. In adding or subtracting decimals, then, the decimal must be written so that like units are under each other.

Written Work

1. Add .75, .055 and .096.

```
.75        Test:  .7   + .0   + .0   = .7
.055             .05   + .05  + .09  = .19
.096             .000  + .005 + .006 = .011
────                                  ────
.901                                  .901
```

2. From 15 take 6.387.

15 may be written 15.000.

```
15.000        Do naughts annexed to an integer change
 6.387        its value? Do naughts annexed to a decimal
 ─────        change its value?
 8.613
```

Add :

3. 25,005, .75, .005, .72, 8.445, .875, .05, .0745, .6475.

4. Subtract 75.005 from 485.007.

5. Subtract .8075 from 23.

6. Subtract .5075 from 23.004.

7. Add 2.5, 11.25, 18.042, 27.0548.

8. Add 1.45, 3.06, 6.605, .09.

9. Add 4.24, 8.2, 6.006, 19.098.

10. Add 11.01, 3.7, 10.01, 2.005.

11. Add .7, .4285, 18.054, 8.0108.

12. Add .002, 22.5607, 1.114, 18.

13. Add 126, 2578, 9.009, .00101, 2.02, .0245.

14. Add .0402, 48.0148, .07089, .1607, 17.0017.

15. Add 89.4004, 75.8002, 761.0612, 1245.0005.

Find differences:

16. 5.32 − 3.245

17. 10.004 − 6.205

18. 125.04 − 86.008

19. 12 − 3.001

20. 350.25 − 180.175

21. 221.201 − 175.1254

22. 434.5196 − 178.3021

23. 90.909 − 9.9009

24. 18 − 11.006.

25. 245.045 − 138.1256

26. 100.101 − 95.095

27. 300.333 − 195.033

28. 20.93875 − 15.55

29. 100 − .9999

30. From .06 + .0875 + 49.03 take .025 + 2.0025 + 43.701.

31. From the sum of .2305 + .9105 take 1.

32. From the sum of 27.045 and .7001 take their difference.

33. Add as indicated and test by adding totals:

$$.075 + 6.875 + .901 + 10.101 = \text{———}$$
$$6.375 + .057 + .057 + 9.704 = \text{———}$$
$$.598 + 2.079 + .864 + 12.006 = \text{———}$$
$$.803 + .868 + 9.805 + .011 = \text{———}$$
$$9.603 + 8.789 + 7.504 + .023 = \text{———}$$

Total: +   +   +   = ———

34. How many yards equal 5 pieces of cloth containing 25.25 yd., 32.625 yd., 40.81 yd., 45.5 yd., and 48.75 yd. respectively?

35. A clerk's income for one year is $600. He spends $487.75. How much does he save?

36. The sum of two numbers is 118.6, and one of the numbers is 14.247. What is the other number?

37. What is the weight in tons of 4 loads of coal, which weigh respectively 1.5 T., 1.25 T., 1.75 T., and 1.9 T.?

38. If a tailor uses 4.375 yards of cloth in making a suit, how much remains of a piece of cloth containing 12.25 yards?

39. Find the cost of three tables at $9.50, $12.75, and $15.80.

40. How many feet are there in three distances measuring 2675.25 ft., 6785.875 ft., and 5674.5 ft.?

41.. A farmer bought 75.5 pounds of clover seed and sowed 54.25 pounds. How much had he remaining?

42. Mary's temperature on Monday was 99.1 degrees, and on Wednesday 102.7 degrees. Find the increase in temperature each day over normal.

NOTE. — The normal temperature of the body of a person in good health should be 98.6 degrees.

43. The weather bureau in Atlantic City, N.J., for three days shows the following rainfall: Monday, .97 in.; Tuesday, 1.37 in.; Wednesday, .37 in. Find the total rainfall for the three days.

44. The rainfall in Chicago for five days was as follows: Monday, .25 in.; Tuesday, 1.3 in.; Wednesday, nothing; Thursday, 1.07 in.; Friday, 1.09 in. Find the rainfall for the five days.

45. A man bought 3 pieces of land, which a surveyor measured in acres and thousandths of an acre, as follows: .037 acres; .879 acres; and 1.089 acres. How much land in all did he buy?

46. The French franc is equal to $.193; the German mark to $.238$\frac{5}{10}$; the English shilling to $.243$\frac{4}{10}$. How much greater is our quarter in value than each of the other pieces?

47. A man owns a triangular piece of land. The perimeter of the triangle is 371.79 ft. Two of the sides are as follows: 97.9 ft.; 121.39 ft. Find the other side.

48. A dairyman tested the milk of 5 cows. Spot showed 4.3 lb. of *butter fat* to every 100 lb. of milk; White Star showed 3.93 lb.; Glory showed 4.75 lb.; Beauty showed 4.2 lb.; and Princess showed 3.99 lb. Find the total number of pounds of butter fat in 500 lb. of milk, 100 lb. from each cow.

MULTIPLICATION OF DECIMALS

Oral Work

1. Multiply 1.5 by 4.

2. How many are $5 \times .3$? $5 \times .03$? $5 \times .003$?

3. When a decimal is multiplied by an integer, what do you observe about the number of decimal places in the product?

Multiplying by 10, 100, 1000, etc., by moving the decimal point.

1. Multiply 5.25 by 10; by 100; by 1000.

Study of Problem

$10 \times 5.25 = 52.50$

$100 \times 5.25 = 525.00$

$1000 \times 5.25 = 5250.00$

a. How may we multiply a decimal by 10? by 100? by 1000?

b. How is the value of a number affected by moving the decimal point *one place* to the right? *two places? three places?*

c. How, then, may a decimal be multiplied by 10? by 100?

Multiply first by 10 ; then by 100 ; then by 1000:

2. 42.07	**5.** 16.94	**8.** 222.461	**11.** .005
3. 113.55	**6.** 849.02	**9.** 333.059	**12.** 4.009
4. 264.03	**7.** 500.09	**10.** 29.004	**13.** 13.655

Multiplying a decimal by a decimal.

1. Multiply .1 by .01. $\frac{1}{10} \times \frac{1}{100} = \frac{1}{1000}$, or .001.

2. Multiply 1.5 by .5. $1.5 = \frac{15}{10}$; $\frac{15}{10} \times \frac{5}{10} = \frac{75}{100}$, or .75.

3. When a decimal is multiplied by a decimal, what do you observe about the number of decimal places in the product?

Written Work

1. Multiply .75 by .3.

.75
. 3
———
.225

Since there are two decimal places in the multiplicand and one in the multiplier, point off three decimal places in the product, making the product .225.

Test : .75 = $\frac{75}{100}$, and .3 = $\frac{3}{10}$. $\frac{75}{100} \times \frac{3}{10} = \frac{225}{1000}$, or .225, a decimal.

2. Multiply .25 by .13.

.25
.13
———
75
25
———
.0325

Study of Problem

a. What is the sum of the decimal places in the multiplier and multiplicand?

b. The product, then, must contain how many places?

Test : .25 = $\frac{25}{100}$; .13 = $\frac{13}{100}$; $\frac{25}{100} \times \frac{13}{100} = \frac{325}{10000}$ = .0325.

Multiply as in integers, pointing off as many decimal places in the product as there are decimal places in both factors.

Find products :

3. .8 × .27	**8.** 3.21 × 4.5	**13.** 1.45 × .48
4. .5 × .45	**9.** 7.24 × .08	**14.** 11.4 × .15
5. .15 × .256	**10.** .011 × .42	**15.** .025 × .124
6. 6.5 × 1.5	**11.** .57 × .15	**16.** 22.5 × 4.04
7. 5.7 × 9.4	**12.** 2.03 × .4	**17.** .75 × .624

Find products :

18.	.4 × 5.6	**47.**	.101 × .012	**76.**	.009 × .099
19.	.5 × 7.5	**48.**	.14 × 89.76	**77.**	.1101 × .101
20.	.3 × 8.4	**49.**	.112 × .092	**78.**	.3756 × .124
21.	.7 × 9.3	**50.**	.363 × .003	**79.**	.0456 × .032
22.	.8 × 6.8	**51.**	.90 × 5.78	**80.**	.0038 × .097
23.	.6 × 5.8	**52.**	.36 × 6.48	**81.**	2.5 × 1.75
24.	.5 × 7.2	**53.**	.325 × .125	**82.**	14.4 × 1.325
25.	.7 × 6.5	**54.**	.043 × .057	**83.**	10.85 × 2.975
26.	.8 × 5.7	**55.**	.016 × .235	**84.**	60.95 × 3.03
27.	.12 × .25	**56.**	.534 × .223	**85.**	15.02 × 5.001
28.	.22 × .14	**57.**	.261 × .175	**86.**	40.023 × .021
29.	.25 × .25	**58.**	.022 × .022	**87.**	35.007 × 4.8
30.	.32 × .43	**59.**	.632 × .085	**88.**	1.78 × 1.024
31.	.41 × .55	**60.**	.99 × 234.17	**89.**	8.132 × 2.4
32.	.145 × .625	**61.**	.402 × 4.022	**90.**	10.001 × 7.07
33.	.046 × .752	**62.**	.472 × .0504	**91.**	52.5 × 7.07
34.	.125 × .246	**63.**	.122 × .5625	**92.**	57.135 × 8.56
35.	.414 × .601	**64.**	.144 × .00321	**93.**	9.901 × 1.99
36.	.851 × .004	**65.**	.96 × 1.0208	**94.**	456.375 × 44.8
37.	.654 × 1.08	**66.**	.407 × 4.003	**95.**	12.063 × 14.204
38.	.506 × 24.6	**67.**	.04 × .078	**96.**	101.1 × 11.01
39.	.15 × 3.04	**68.**	.64 × .016	**97.**	44.006 × 6.044
40.	.8 × 10.34	**69.**	.012 × .024	**98.**	5.117 × 1.88
41.	.18 × .004	**70.**	.625 × .001	**99.**	16.004 × 16.64
42.	.122 × .024	**71.**	.872 × .096	**100.**	32.406 × 15.108
43.	.215 × .015	**72.**	.0004 × .004	**101.**	41.041 × 9.009
44.	.83 × .007	**73.**	.0505 × .55	**102.**	56.561 × 6.01
45.	.212 × 2.042	**74.**	.0216 × .027	**103.**	3.054 × 1.405
46.	.432 × .078	**75.**	.0244 × .014	**104.**	5.237 × 3.701

Written Work

1. Turnips are .9 water. Find the amount of water in 50.5 lb. of turnips.

2. A lot is 75.9 ft. long and 32.7 ft. wide. How many square feet are there in the lot?

3. A lot in Newark, N.J., is 79.84 ft. long, and 22.6 ft. wide. It is sold at $3.50 a square foot. How much is the lot sold for?

4. A man paves his street 908.6 ft. long and 29.30 ft. wide. How many square feet are there in the street paved?

5. A cubic foot of water weighs 62.5 lb. A cubic foot of ice weighs .92 as much. Find the weight of a cubic foot of ice.

6. A cubic foot of cork weighs .24 as much as a cubic foot of water. Find the weight of the cork.

7. A block of ice is 2 ft. long, 1.5 ft. wide, and 1 ft. thick. Find its weight.

8. A piece of sandstone is 1 ft. square at the end, and 4 ft. long. How many cubic feet are there in it? Sandstone is 2.9 times as heavy as water. Find the weight of the sandstone.

9. Milk is 1.03 times as heavy as water. If a gallon of water weighs 8.5 lb., how much does a gallon of milk weigh?

10. If a man walks 4.75 miles in an hour, how far will he walk in 11.25 hours?

11. How much will 1.365 acres of land cost at $875.50 per acre?

12. A farmer averages 35.875 bushels of wheat from 12.95 acres. How many bushels does he harvest?

13. Find the cost of 6.25 dozen eggs at $.375 a dozen.

14. A cubic foot of water weighs 62.5 lb. Coal is 1.3 times as heavy as water. How much does the coal in a wagon box weigh, if the box is 3.5 ft. wide, 8.5 ft. long, and 2.5 ft. deep?

15. John's mother has a cistern in the cellar which is 4 ft. long, 3 ft. wide, and 3 ft. deep. How many cubic feet of water are there in it? A cubic foot contains 7.5 gal. How many gallons are there in the cistern?

16. The boys at school form a circular race track 79.6 ft. across. Find the length of the race track, if the circumference is 3.1416 times the distance across.

DIVISION OF DECIMALS

Dividing a decimal or a mixed decimal by an integer.

Oral Work

1. Find $\frac{1}{8}$ of 48 hundredths; of 64 hundredths.

2. Find $\frac{1}{5}$ of .25; of .35; .45; .75.

3. Find $\frac{1}{6}$ of 6 and 36 hundredths; 12 and 24 hundredths.

4. Find $\frac{1}{6}$ of 12.36; 24.42; 48.06; 54.06.

Observe that in each problem a decimal or a mixed decimal when divided by an integer is simply separated or *partitioned.*

Give quotients at sight :

5. $\frac{1}{4}$ of .16	**9.** $\frac{1}{3}$ of 6.6	**13.** 6.42 ÷ 6	**17.** .006 ÷ 3
6. $\frac{1}{5}$ of .25	**10.** $\frac{1}{4}$ of 8.08	**14.** 12.04 ÷ 4	**18.** .024 ÷ 6
7. $\frac{1}{2}$ of .08	**11.** $\frac{1}{5}$ of 10.10	**15.** 15.05 ÷ 5	**19.** .008 ÷ 4
8. $\frac{1}{4}$ of .04	**12.** $\frac{1}{6}$ of 12.06	**16.** 24.18 ÷ 3	**20.** .105 ÷ 5

21. .64 ÷ 16	**25.** .003 ÷ 10	**29.** 14.76 ÷ 41
22. .02 ÷ 40	**26.** .368 ÷ 16	**30.** .105 ÷ 65
23. .75 ÷ 60	**27.** .2286 ÷ 127	**31.** 2.07 ÷ 46
24. .49 ÷ 140	**28.** .0124 ÷ 20	**32.** 31.2 ÷ 36

33. .01 + 100	41. .5058 + 18	49. 2.31 + 55
34. .05 + 500	42. .1728 + 24	50. 1.17 + 65
35. .03 + 100	43. .0001 + 1000	51. 16.5 + 22
36. .444 + 50	44. .0343 + 14	52. 2.7855 + 35
37. .125 + 50	45. .5184 + 96	53. 31.288 + 48
38. .966 + 46	46. 1.625 + 25	54. 137.95 + 31
39. .018 + 12	47. 24.86 + 12	55. 106.32 + 24
40. .546 + 21	48. 172.8 + 24	56. 17.172 + 53

Find the sum of the quotients:

57.	.02 + 2 =	58.	.035 + 14 =
	.034 + 17 =		.868 + 16 =
	.06 + 60 =		.18 + 12 =
	.024 + 15 =		.045 + 18 =
	.168 + 7 =		.25 + 125 =
	.0044 + 11 =		.48 + 24 =

59. If 8 yards of muslin are sold for $.80, what is the price per yard?

60. When 144 pens are sold for $.72, what is the price per pen?

61. If 10 sheets of paper are sold for $.05, what is the price per sheet?

62. If $1.00 is paid for 1000 cubic feet of gas, what is the price per cubic foot?

Dividing a decimal by a decimal.

Oral Work

1. In what short way may a decimal be multiplied by 10? by 100? by 1000?

2. In .5, .25, .025, move the decimal point one place to the right and read the result; two places to the right; three places to the right.

Explain why :

3. $.2\overline{)\,.24} = 2\overline{)\,2.4}$; $.04\overline{)\,.0164} = 4\overline{)\,1.64}$; $1.6\overline{)\,25.6} = 16\overline{)\,256}$

4. $.6\overline{)\,18} = 6\overline{)\,180}$; $.12\overline{)\,.144} = 12\overline{)\,14.4}$; $.08\overline{)\,.48} = 8\overline{)\,48}$

5. $.09\overline{)\,8.1} = 9\overline{)\,810}$; $.25\overline{)\,2.25} = 25\overline{)\,225}$; $.05\overline{)\,5} = 5\overline{)\,500}$

Multiplying both dividend and divisor by 10, by 100, or by 1000, etc., does not change the quotient.

Tell the number of places the decimal point must be moved to the right in both dividend and divisor in each of the following problems in order to make the divisor an integer; then give quotients:

	a	b	c	d	e
6.	$.5\overline{)\,.25}$	$.2\overline{)\,2.6}$	$4\overline{)\,4.4}$	$.6\overline{)\,66}$	$8\overline{)\,8.08}$
7.	$.03\overline{)\,.15}$	$.12\overline{)\,.96}$	$.07\overline{)\,.21}$	$.09\overline{)\,.81}$	$.05\overline{)\,.05}$

Written Work

1. Divide 5.68 by .8.

$$7.1$$
$$.8\overline{)\,5.68} = (a)\ 8\overline{)\,56.8}$$

$$7.1$$
$$\text{Or, } (b)\ .8\overline{)\,5.6_{\wedge}8}$$

In (a) both dividend and divisor are multiplied by 10 to make the divisor a whole number. This is done by moving the decimal point in the divisor and the dividend one place to the right.

In (b) the example is written in its original form and the changed position of the decimal point in the dividend is indicated by a caret, placed as many places to the right of the decimal point as there are decimal places in the divisor. As there is one decimal place in the divisor, place the caret one place to the right of the decimal point in the dividend. Then divide as in integers, placing the decimal point in the quotient immediately after all the numbers to the left of the caret have been used in the process of division.

NOTE. — The method illustrated in (b) is called the **Austrian Method.** It retains the identity of the problem and saves the time consumed in restating it.

2. Divide 96.8 by .004.

$$.004\overline{)96.800_\wedge} \quad \frac{24\ 200.}{}$$

As there are three decimal places in the divisor, place the caret three places to the right of the point in the dividend. Two naughts must be added to give three decimal places. Place the decimal point in the quotient immediately after all the numbers to the left of the caret have been used. The answer is a whole number.

3. Divide 1.2864 by .032.

$$.032\overline{)1.286_\wedge4} \quad \frac{40.2}{}$$
$$\underline{1\ 28}$$
$$64$$
$$\underline{64}$$

As there are three decimal places in the divisor place the caret three places to the right of the point in the dividend. Place the decimal point in the quotient immediately after all the numbers to the left of the caret have been used in the division.

Mark off by a caret the same number of decimal places from the right of the decimal point in the dividend as there are decimal places in the divisor. Divide as in integers, placing the decimal point in the quotient immediately after all the numbers to the left of the caret have been used in the process of division.

Divide and test :

4. 6 by .3	**12.** 70 by .0056	**20.** 1 by .001
5. 9 by .06	**13.** 154 by .28	**21.** 10 by .001
6. 21 by .7	**14.** 78 by .052	**22.** 17 by .68
7. 10 by .01	**15.** 190 by .076	**23.** 112 by .032
8. 25 by .125	**16.** 115 by 6.25	**24.** 324 by .27
9. 80 by .125	**17.** 18 by .375	**25.** 1904 by .119
10. 36 by .75	**18.** 4 by .016	**26.** 114 by .76
11. 128 by .032	**19.** 48 by .1875	**27.** 896 by .256

Find quotients and test results :

28. .04 ÷ .002	**31.** .004 ÷ .004	**34.** .728 ÷ .13
29. .8 ÷ .25	**32.** .1952 ÷ 16	**35.** .8136 ÷ .224
30. .125 ÷ .5	**33.** .00624 ÷ .8	**36.** .4725 ÷ .2

37. $.112 \div 7$	50. $.0247 \div .019$	63. $.4375 \div .125$
38. $.036 \div 4$	51. $.8799 \div .7$	64. $.17225 \div .325$
39. $.0001 \div .01$	52. $.08799 \div .007$	65. $.7665 \div .365$
40. $.0187 \div .011$	53. $.15158 \div .286$	66. $.2944 \div .512$
41. $.555 \div .37$	54. $.408375 \div .135$	67. $.421875 \div .125$
42. $.655 \div .131$	55. $1.5 \div .005$	68. $8.686 \div 86.86$
43. $.75 \div .125$	56. $10.8 \div .12$	69. $.01 \div .001$
44. $.3625 \div .125$	57. $1.32 \div 11$	70. $100 \div 1000$
45. $1.44 \div .036$	58. $31.75 \div .025$	71. $7.25 \div .025$
46. $.9 \div .015$	59. $.5475 \div 1.5$	72. $1.225 \div 3.5$
47. $.1 \div 1.25$	60. $1.728 \div 17.28$	73. $139.956 \div 3.21$
48. $10 \div 2.25$	61. $1.111 \div 11.11$	74. $86.784 \div 226$
49. $.3 \div .03$	62. $100.5 \div 1.005$	75. $46.695 \div 1.65$

REDUCTION OF DECIMALS

Changing decimals to common fractions.

Oral Work

Change to common fractions in lowest terms:

1. $.25$	3. $.04$	5. $.5$	7. $.45$
2. $.50$	4. $.02$	6. $.8$	8. $.75$

9. Give the steps in changing a decimal to its fractional equivalent.

A **complex decimal** is a decimal and a fraction united; as, $.16\frac{2}{3}$ which is read $16\frac{2}{3}$ hundredths.

Written Work

1. Change $.87\frac{1}{2}$ to a common fraction in its lowest terms.

$.87\frac{1}{2} = .875 = \frac{875}{1000}$, or $\frac{7}{8}$ Express the denominator of the decimal and reduce the resulting fraction to its lowest terms.

2. Change $.66\frac{2}{3}$ to a common fraction.

$.66\frac{2}{3} = \frac{200}{3} \div 100 = \frac{200}{300}$, or $\frac{2}{3}$

Since $66\frac{2}{3} = \frac{200}{3}$, $.66\frac{2}{3} = \frac{200}{3} \div 100$, or $\frac{200}{300}$. This reduced to its lowest terms equals $\frac{2}{3}$.

3. Change .075 to a fraction in its lowest terms.

$.075 = \frac{75}{1000}$, or $\frac{3}{40}$

Express .075 with its denominator 1000 and reduce the fraction to its lowest terms, $\frac{3}{40}$.

Change the following decimals to fractions in their lowest terms:

4. .35	9. $.16\frac{2}{3}$	14. $.83\frac{1}{3}$	19. $.03\frac{1}{3}$
5. .24	10. $.41\frac{2}{3}$	15. $.33\frac{1}{3}$	20. $.12\frac{1}{2}$
6. .25	11. $.62\frac{1}{2}$	16. $.06\frac{1}{4}$	21. $.14\frac{2}{7}$
7. .125	12. $.37\frac{1}{2}$	17. $.04\frac{1}{6}$	22. $.58\frac{1}{3}$
8. .375	13. $.87\frac{1}{2}$	18. $.08\frac{1}{3}$	23. $.88\frac{1}{3}$

Changing common fractions to decimals.

Written Work

1. Change $\frac{4}{5}$ to a decimal.

$\frac{4}{5} = 4 \div 5 \quad 5\overline{)4.0}$ quotient $.8$

Since $\frac{4}{5} = 4 \div 5$, annex a naught and divide by 5.

2. Change $\frac{3}{7}$ to a decimal.

$\frac{3}{7} = 3 \div 7 \quad (a)\ 7\overline{)3.000}$ quotient $.428\frac{4}{7}$

Or, $(b)\ 7\overline{)3.000}$ quotient $.428^{+}$

Since the division does not terminate, the quotient may be shown as a complex decimal, as in (*a*) or as an incomplete decimal, as in (*b*).

Change to decimals:

3. $\frac{5}{8}$	5. $\frac{3}{20}$	7. $\frac{7}{8}$	9. $\frac{9}{16}$	11. $\frac{7}{16}$
4. $\frac{3}{5}$	6. $\frac{5}{16}$	8. $\frac{11}{25}$	10. $\frac{19}{20}$	12. $\frac{17}{25}$

Change to complex or incomplete decimals of not more than three places:

13. $\frac{2}{3}$	15. $\frac{17}{27}$	17. $\frac{25}{27}$	19. $\frac{1}{60}$	21. $\frac{12}{17}$
14. $\frac{9}{31}$	16. $\frac{9}{19}$	18. $\frac{37}{49}$	20. $\frac{6}{13}$	22. $\frac{7}{24}$

REVIEW PROBLEMS

1. How many books, at $.50 each, can be bought for $37.50?

2. Find the value of .06 × .03 divided by 2.5.

Find the sum of the quotients:

3.	*a.* 4.65 + 1.55 = ?		**4.**	*a.* 31.906 + 1.06 = ?
	b. 12.5 + 6.25 = ?			*b.* 40.804 + 1.01 = ?
	c. 9.03 + 3.01 = ?			*c.* 3.861 + 3.51 = ?
	d. 11.75 + 2.35 = ?			*d.* 6.012 + 5.01 = ?
	e. 25.75 + 5.15 = ?			*e.* 36.542 + 33.22 = ?

5. If a man earns $1¾ in a day, how much will he earn in 34 days?

6. Change ½, ¾, ⅞, ₉/₁₆ to decimals.

7. Two men bought a store for $6000, one paying .375 of it, and the other .625 of it. How much did each pay?

8. Divide ¹⁴/₂₅ by .014.

9. My wheat crop was 272 bu. I sold ⅚ of it at $1.085 a bushel. How much did I get for it?

10. Multiply 1000 by .001, and divide the product by .01.

11. There are 31.5 gallons in a barrel. How many barrels are there in 2378.25 gallons?

12. A farm of 171¼ acres was sold in plots of 1.25 acres each, the price for each plot being $75.50. How much was received for the farm?

13. Change 12¾ to a mixed decimal.

14. A cubic foot of pure water weighs 62½ lb. Find the weight of 10.75 cu. ft.

15. Change .625 to a common fraction.

16. Multiply 3.18 by 5.5, and write the result in words.

17. How many miles will a freight train travel in $5\frac{5}{8}$ hours, if it travels $20\frac{3}{4}$ miles an hour?

18. Divide forty-two ten-thousandths by five hundred twenty-five thousandths.

Find the cost of:

19. 90 lb. meal @ $3\frac{1}{8}$ ¢. **23.** 120 yd. flannel @ 75 ¢.

20. 256 lb. lard @ $12\frac{1}{2}$ ¢. **24.** 288 yd. linen @ $37\frac{1}{2}$ ¢.

21. 186 doz. eggs @ $33\frac{1}{2}$ ¢. **25.** 68 lb. butter @ 25 ¢.

22. 128 yd. gingham @ $6\frac{1}{4}$ ¢. **26.** 315 yd. prints @ $33\frac{1}{8}$ ¢.

27. How many yards of muslin, at $12\frac{1}{2}$ cents a yard, can be bought for $6?

28. If a boy earns $1.25 a day, and a man earns $3.25 a day, how long will it take the boy to earn as much as the man can earn in 15 days?

29. What is the cost of 4879 feet of boards at $.02 a board foot?

30. At $12\frac{1}{2}$ ¢ a pound, how many pounds of meat can be bought for $9?

31. The product of two numbers is .0006. If one of the numbers is .04, what is the other?

32. A grocer bought 75 bbl. of flour at $4.75 a barrel. He sold 31 barrels at $5.37 a barrel and the remainder at $4.87 a barrel. Find his gain.

33. Explain the effect of removing the naught in the following: .015, .250, .018, .240, .016.

34. If the rainfall in a certain place is, on an average, .1 of an inch a day, in how many days does the rainfall amount to 3 inches?

35. How many layers of gold leaf will be required to form a tablet 5 inches thick, if each layer is .001 of an inch thick?

36. If one yard of flannel is worth $.625, how many yards can be bought for $575?

37. How many pencils can be bought for $324 at $.0075 each?

38. When pencils average $.025 apiece, how many can be purchased for $100?

39. At $33\frac{1}{8}$ ¢ a yard how many yards of linen can be bought for $10?

40. At $87\frac{1}{2}$ ¢ a dozen how many dozen eggs can be bought for $7?

PERCENTAGE

In *Common Fractions* we learned that a unit may be divided into *any number* of equal parts and any number of these parts may be taken.

Thus, ¼ of 60 means that 60 is divided into 4 equal parts and 3 of these parts are taken.

In *Decimal Fractions* we learned that a unit may be divided into 10, 100, 1000, etc., equal parts, and that any number of these parts may be taken.

Thus, .9 of 60 means that 60 is divided into 10 equal parts and that 9 of these equal parts are taken.

We now come to a subject that divides a unit into 100 equal parts only. We call this subject **Percentage**.

In common fractions we compute by *halves, thirds, fourths, sixths,* etc.; in decimal fractions we compute by *tenths, hundredths, thousandths,* etc.; but in percentage we compute by *hundredths* only.

Another name for *hundredths* is *per cent,* usually written " *%.*"

Hundredths may be written in three different ways; thus, $\frac{8}{100}$, .08, 8%; $\frac{25}{100}$, .25, 25%.

Percentage is simply an application of decimal fractions.

Oral and Written Work

1. Write the following numbers as per cents:

.02 .05 .03 .15 .20 .25 .40 .06 .75 .87 .56

2. Write the following as decimals:

5% 20% 7% 15% 25% 16% 18% 24% 50% 75%

3. Show by several examples that naughts added to the right of a decimal do not affect its value.

4. Write as decimals and as per cents:

$$\tfrac{1}{2} \quad \tfrac{3}{4} \quad \tfrac{1}{20} \quad \tfrac{1}{25} \quad \tfrac{1}{10} \quad \tfrac{2}{5} \quad \tfrac{4}{5} \quad \tfrac{9}{10} \quad \tfrac{3}{5} \quad \tfrac{2}{20} \quad \tfrac{7}{25}$$

5. What is the difference between .05 of $100 and 5 % of $100?

6. $5\% = \tfrac{?}{20}$; $10\% = \tfrac{?}{10}$.

7. 25 % of $100 may be found in two ways: (a) $25\% = \tfrac{1}{4}$; $\tfrac{1}{4}$ of $100 = $25. (b) $25\% = .25$; $.25 \times $100 = 25.

Learn the following:

$50\% = \tfrac{1}{2}$	$10\% = \tfrac{1}{10}$	$75\% = \tfrac{3}{4}$
$25\% = \tfrac{1}{4}$	$5\% = \tfrac{1}{20}$	$40\% = \tfrac{2}{5}$
$20\% = \tfrac{1}{5}$	$60\% = \tfrac{3}{5}$	$80\% = \tfrac{4}{5}$

Give results at sight:

8. 20 % of $50   **17.** $33\tfrac{1}{3}\%$ of 30 da.   **26.** 10 % of 80 ¢

9. 25 % of $60   **18.** 50 % of 60 min.   **27.** 1 % of $200

10. 10 % of $40   **19.** 75 % of 100 books   **28.** 3 % of $30

11. 50 % of $80   **20.** 40 % of 20 rods   **29.** 5 % of 40 qt.

12. 40 % of $75   **21.** 5 % of 40 weeks   **30.** 20 % of 20 pk.

13. 5 % of $40   **22.** 15 % of 100 lb.   **31.** 50 % of 80 pt.

14. 6 % of $6   **23.** 10 % of 70 bu.   **32.** 2 % of 100 bu.

15. 75 % of $20   **24.** 25 % of 24 hr.   **33.** 20 % of 75 hr.

16. 25 % of $72   **25.** 10 % of 800 bu.   **34.** 25 % of 32 pk.

Written Work

1. Find 28 % of 7500 bushel of oats.

7500 bu.

.28
_____

60000

15000
_____

2100.$\emptyset\emptyset$ bu.

Since per cent means hundredths, 28% of 7500 bushels equals .28 of 7500 bushels, or 2100 bushels.

Find:

2. 27 % of $395	**7.** 85 % of $90.60	**12.** 75 % of $605
3. 14 % of $478	**8.** 40 % of $20.50	**13.** 37 % of $2005
4. 24 % of $527	**9.** 10 % of $2004	**14.** 45 % of $6745
5. 6 % of $57.40	**10.** 5 % of $200.60	**15.** 80 % of $905
6. 5 % of $90.80	**11.** 7 % of $500.50	**16.** 98 % of $7008

17. Mr. Jordan bought a horse for $175 and sold it for 90 % of the cost. For how much did he sell the horse?

18. Raymond has $165 in the savings bank and Bertha has 80 % as much. How much more money has Raymond in the bank than Bertha?

19. The distance between two cities is 1080 miles. After 45 % of the distance has been traveled, how much of the distance remains to be traveled?

20. Mr. Watson earned $1580 in a year, and his son Henry 65 % as much. Find the amount Henry earned.

21. The salary of a school teacher last year was $40 a month, and this year her salary was increased 25 % of last year's salary. Find her present salary.

22. Paul lives 560 rd. from the schoolhouse and David 72 % as far. Find the number of rods David lives from the schoolhouse.

23. Mr. Adams borrows $365 for one year, and pays 6% for the use of the money for the time. How much money will pay the debt when due?

24. Mr. Brown has loaned $1200 to one man and $1600 to another man. How much does Mr. Brown get each year for the use of the money if each man pays him 5% of the amount borrowed?

25. Find 5% of 20; of 40; of 50; of 60; of 80.

26. A newsboy sells $18 worth of papers and gets 40% for selling. How much does he earn?

27. Mary has $24 in the savings bank, and deposits 25% as much as she has in the bank. Find the amount deposited.

28. A boy borrows $200 to go to school, and pays the lender 5% for the use of the money for one year. How much does he pay for its use?

29. A boy bought a pony and a cart. The pony cost $80, and the cart 60% as much as the pony. Find the total cost.

30. A man had 400 sheep. On Monday he sold 25% of them. On Tuesday he sold 25% of the remainder. How many sheep had he then?

31. In a spelling test of 30 words, James missed 20%. How many words did he spell correctly?

32. $\frac{1}{4} = \frac{?}{100} = ?$ %. Compare $\frac{1}{5}$ and .20; $\frac{1}{5}$ and 20%.

33. A man bought a house for $2800. He paid 78% of the amount in cash, and gave his note for the balance. For how much did he give his note?

DENOMINATE NUMBERS

PROBLEMS

1. Henry sold 40 qt. of berries one afternoon, at 20 ¢ per quart. He gave his mother 40 % of what he earned. How much did he give his mother?

2. Agnes helped her mother to put up jelly in half-pint glasses. If they made 25 glasses, how many pints did they make?

3. James delivers milk for his father and sells on Tuesday 8 gal. At 8 ¢ a quart, how much does he get for the milk sold that day?

4. Mr. Wyman averaged 300 gal. of milk from his dairy for two weeks in July, and Mr. Henry averaged 70 % as much milk as Mr. Wyman. How much did Mr. Henry's dairy produce?

5. Kate, Henry, and George picked 3 pk. of cherries, and sold them at 16 ¢ per quart. How much did they receive?

6. How many bushels of potatoes, at $1.40 a bushel, can be bought for $22?

7. A man bought a bushel of nuts for $3 and sold them at 96 ¢ a peck. What was his gain?

8. If you can fill one pie with 1 pt. of berries, how many pies can you fill with 3 qt.?

9. A retail grocer bought 10 bu. of apples at $1 a bushel, and sold them at 30 ¢ a peck. What was his gain?

189

10. Robert and Edward gathered 15 qt. of chestnuts, and sold them at 7 ¢ a pint. How much did they receive?

11. Doris made a strawberry shortcake for which she used 3 cups of flour (1½ pt.) at 4 ¢ a pint, 2 teaspoonfuls of baking powder (⅛ ¢), ¼ lb. butter at 38 ¢ a pound, ½ pt. of milk at 4 ¢ a pint, 1 qt. 1 pt. of strawberries at 15 ¢ a quart, and ½ pt. of cream at 20 ¢ a pint. Find the cost of the cake.

12. John Wanamaker of New York sent by parcel post to Mrs. James Walton of Paterson, N.J., 4 blankets costing $2.15 each and weighing 5 lb. each when packed. If the postage on 5 lb. is 9 ¢, find the total cost of the blankets.

13. The rate of postage on books is 1 ¢ for each 2 ounces. How much will it cost to mail a package of books weighing 3¾ lb.?

14. Park and Tilford's chocolates cost 80 ¢ a pound. How much must I pay for 5 half-pound boxes?

15. Our grocer found that 9 hams weighed 82⅞ lb. What was the average weight?

16. One ton of starch was packed into boxes, each containing 5 lb. How much was received, if each box was sold for 6¼ ¢?

17. A butcher sold 30¼ lb. of lard at $0.12 a pound, and purchased with the money flour at $0.03 a pound. How much flour did he buy?

18. At $720 a year, find the rent of a cottage for 3 months.

19. A motorman receives 20 ¢ an hour, and works 6 days of 8 hours each. Find the amount he receives per week.

20. A clerk pays $4.50 a week for board. How much does his board cost him for September, October, and November?

21. John spends 30 minutes each night in studying. How many minutes does he spend in 6 nights? How many hours is that? John values the 30 minutes study each night at 25 ¢. How much will this study be worth in 20 days?

22. John worked 6 hr. and 45 min. on Saturday at 20 ¢ an hour. How much did he earn during that time?

23. James, Henry, and John live respectively, 100 rd., 80 rd., and 60 rd. from the schoolhouse. On a scale of 20 rd. to an inch, show graphically their relative distances from the schoolhouse.

24. What change should you receive from a five-dollar bill if you buy 8 lb. of steak at 18 ¢ a pound, 3 cans of tomatoes at 96 ¢ a dozen, and 2 gal. of gasoline at 15 ¢ a gallon?

25. What is my January milk bill, if I use 5 pt. every day, at 8 ¢ a quart?

26. If ⅝ T. of coal costs $3.75, how much will 3½ T. cost?

27. How many pint cans can be filled from 26 gal. of tomato soup?

28. How much will 3½ bu. of plums cost at 9 ¢ a quart?

29. Find the amount of the following sales :

1 doz. boxes of cocoa at 15 ¢ a box
8 cans of tomatoes at $1 a dozen
11¼ lb. turkey at 22 ¢ a pound.

30. Walter picked 4¼ bu. of blackberries and sold them to a grocer for 6 ¢ a quart. How much did he receive?

31. The Adams Coal Company sold 8 loads of coal as follows: 2470 lb., 3680 lb., 1974 lb., 2985 lb., 1741 lb., 3164 lb., 3749 lb., and 4278 lb. Find the number of tons, hundredweight, and pounds sold.

32. 500 bu. of peaches were packed in baskets, each holding 2 pk. How many baskets were needed?

33. 20 cwt. of starch was packed into boxes, each containing 10 lb. How much was received, if each box was sold for 12½ ¢ ?

34. George has a can of milk containing 10 gal. If he sells 10 qt. to his first customer, 4 qt. to the second customer, 3 gal. to the third customer, and 2 gal. 1 pt. to the fourth, how much has he left in the can?

35. The distance from Newark to Pittsburgh is 431 mi.; from Pittsburgh to Chicago, 468 mi.; from Chicago to Denver, 1025 mi.; and from Denver to San Francisco, 1377 mi. Estimate, on a scale of ½ in. to 50 mi., the distance of each city from Newark.

36. Mr. Adams has a field 80 rd. long and 40 rd. wide. Draw a picture to show the surface of this field, on a scale of 1 inch to 20 rd.

37. A New Jersey farm is 160 rd. long and 40 rd. wide. Draw a rectangle, on a scale of 1 in. to 20 rd., to represent this farm.

38. Frank is 56 in. tall. Express his height in feet and fractions of a foot.

39. Alice made an apron for her sister's birthday present. She used 1 yd. of dotted swiss @ 25¢, 4 yd. of edging @ 12½ ¢, 3 yd. of beading @ 7¢, and 1¾ yd. of ribbon @ 8¢. How much did the apron cost her?

PRACTICAL MEASUREMENTS

MEASURES OF SURFACE

Observe that the straight lines *AB* and *CD* cannot meet, however far they may be extended. Such lines are called **parallel lines.**

Lines that meet, making a square corner, form a **right angle.**

1. Point out several parallel lines in articles in the classroom; several right angles.

A figure that has four straight sides and four right angles is called **a rectangle.**

A rectangle having its four sides equal is called a **square.**
Rectangles that are not squares are sometimes called **oblongs.**

> **144 sq. in. = 1 sq. ft.**
> **9 sq. ft. = 1 sq. yd.**

The number of square inches, square feet, square yards, or square miles in any surface is called its **area.**

2. How many dimensions has every rectangular surface?

3. Draw on the blackboard a line 4 feet long. From each end draw lines in the same direction 3 feet in length, making square corners with the 4-foot line. Connect by a straight line the ends of the 3-foot lines.

4. Find the area of the figure.

5. Show by a diagram the number of square feet in a square yard.

6. Draw a diagram on a scale of 1 inch to 3 feet to represent a rectangle 24 ft. long and 18 ft. wide.

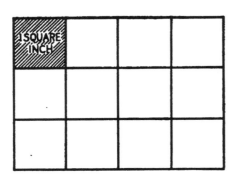

7. The surface to the left is drawn on a scale of $\frac{1}{2}$ inch to the inch. What is the length? the width? How many square inches of surface are there? Notice that the *unit of measure* is 1 sq. in.

Then $3 \times 4 \times 1$ *sq. in.* equals the number of square inches of surface.

SOLUTION. — 1 sq. in. = unit of measure.

$3 \times 4 \times 1$ sq. in. = 12 sq. in., area of surface.

The **area of a rectangle** *is a number of square units equal to the product of its two dimensions when expressed in like units.*

Thus, if the dimensions of a rectangle are 2 inches and 4 inches, the area is 8 square inches. If the dimensions are 2 feet and 4 feet, the area is 8 square feet.

Written Work

1. Find the area of a room 45 ft. long and 26 ft. wide.

Draw on a scale of $\frac{1}{4}$ of an inch to the foot and find surfaces of the following:

NOTE. — The sign ' represents feet and the sign ", inches.

2. A rug $9' \times 12'$.

3. A window $4' \times 8'$.

4. A room $10' \times 12'$.

5. A door $3\frac{1}{2}' \times 8'$.

6. A wall $12' \times 20'$.

7. A hall $9' \times 16'$.

8. A show window $16' \times 20'$.

9. A mat $8'' \times 12''$.

10. A 12-in. square.

11. An 8-in. square.

12. A 16-in. square.

13. A page $3\frac{1}{2}'' \times 6''$.

14. How many square inches equal a square foot? Show by a diagram, on a scale of $\frac{1}{12}$ of an inch to the inch, the number of square inches in a square foot.

15. How many square feet equal a square yard? Show by a diagram, on a scale of $\frac{1}{2}$ inch to the yard, the number of square feet in a square yard.

Find the area and the perimeter of the following. Draw rectangles to represent them on a scale of 1 inch to 1 yard :

16. A schoolroom 10 yd. long and 8 yd. wide.

17. A hall 15 yd. long and 3 yd. wide.

18. A sidewalk 12 yd. long and 2 yd. wide.

19. Matting for a room 5 yd. long and 4 yd. wide.

20. Measure the length and the width of your schoolroom floor, and find the number of square feet it contains. Also find the number of square feet in the walls and the ceiling.

21. Find the number of square feet of window light in your schoolroom.

22. Find the number of square inches in the surface of your desk ; of your teacher's desk ; of your different books.

23. Draw on the blackboard a square yard ; a square foot ; a square inch. These are called units of surface measure. What other units of surface measure are there?

24. Measure off on your school grounds or on some vacant lot a square rod. How many square feet should this contain? How many feet long should it be on each side? Observe that all these units of measure are squares. An acre contains 160 sq. rd., but it need not be a square.

25. Measure your school ground with a tapeline 50 ft. in length, divided into feet and tenths of a foot.

NOTE.—The division of *tenths* and *hundredths* of a foot are now in common use because they can be so readily changed from one denomination to another.

26. There are 10 lots on Grant Street, each 40 ft. in front, and 120 ft. in depth. Draw a plan to show these lots on a scale of 1 inch to 20 ft. and find the area of each lot.

27. Lots near the center of big cities are generally sold by so much per square foot. Find the value of a lot on Broadway, New York, 28 × 56 at $7.50 per square foot.

PRACTICAL PROBLEMS

1. Mr. Stokes, a real estate agent, purchased the lots shown in this plan at $70 per front foot on Main Street. Find the cost of the lots.

2. Mr. Rand bought lot A at $110 a front foot, and built on it a house for $6450. Find the cost of his property.

3. Mr. Remington purchased lots B and C, and put an iron fence around them at $1.10 per foot. Find the cost of the fence.

4. The concrete sidewalk on Main Street is 10 ft. in width. Find the cost of Mr. Remington's walk at 19¢ per square foot.

5. Mr. Stokes sells lots D, E, and F to L. F. Holtzman for $9500. Find his profit on these lots.

6. Find the scale on which this plan is made. What is the width of Main Street?

7. If a pane of glass is 10 inches by 12 inches, how many square inches does it contain?

8. How many square feet of glass are there in 32 such panes?

9. A garden is 73 feet by 50 feet. How many square feet does it contain?

10. The page of a book is $7\frac{1}{4}$ inches by 5 inches. How many square inches are there on the page?

11. How many square inches are there in a page of your book?

12. Measure the blackboard in your schoolroom and find how many square feet it contains.

NOTE. — Reduce inches to the fraction of a foot; as, 8 ft. 6 in. = $8\frac{1}{2}$ ft.

13. How many square inches are there in the surface of your schoolroom door?

14. At $1\frac{1}{8}$ per square yard, how much will it cost to cover a floor 12 feet by 15 feet with linoleum?

15. A plate glass window is 9 feet 8 inches wide and 12 feet 3 inches long. How much will such a window cost at $.36 per square foot?

16. Brussels carpet is $2\frac{1}{4}$ feet wide. How many square feet are there in a yard of it?

17. A room is 16 feet long and 14 feet wide. How much will it cost to paint the ceiling of this room at 12¢ per square yard?

18. Which is the larger, a surface 26 in. long and 5 in. wide or a surface 20 in. long and 12 in. wide?

19. At 15¢ per square foot, how much does a sidewalk 50 feet long and 5 feet wide cost?

20. At 12¢ per square foot, how much will it cost to cement the floor of a cellar 28 ft. 4 in. by 22 ft. 6 in.?

21. At $1.05 per square yard, how much will it cost to pave the street in front of a 50-foot lot, the street being 33 feet wide between the curbs?

22. John painted a signboard 4½ ft. wide and 20 ft. long, at 10¢ per square yard. How much did he get for painting it?

23. Measure your school ground and find the cost of a 4½ foot fence around it at $1.92 per lineal foot?

24. Find the cost of sodding the part of the school yard in front of your school at 6¢ per square foot.

MEASURES OF VOLUME

Oral Work

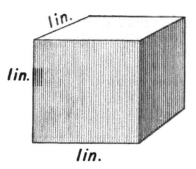

1. What is the length of the figure? the width? the height?

2. How many dimensions has it?

3. How many sides or faces has it?

4. Show that each side is a square.

5. How many square surfaces has it?

A solid with six equal square surfaces is a **cube**.

6. Look at the figure and tell how many edges it has. What is the length of each edge?

A cube whose edge is 1 inch is called a **cubic inch**.

A solid bounded by six rectangular surfaces is called a **rectangular solid**.

7. Draw on paper or on the blackboard a square foot.

8. Divide each side into 12 equal parts and connect them by straight lines.

9. How many square inches equal a square foot?

10. The base of a 1-inch cube has how many square inches?

11. 144 cubes 1-inch on an edge can be placed on a surface of 1 square foot, thus:

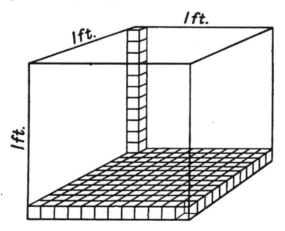

12. What is the height of 12 such layers of cubes? How many cubic inches are there in the first layer? in 12 layers?

13. How many cubic inches can be placed in the cube?

1728 cubic inches (cu. in.) = 1 cubic foot (cu. ft.)

14. How many feet on an edge is the cube in the figure below ?

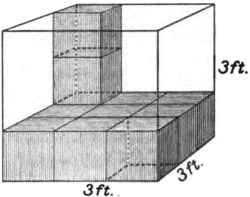

15. How many cubes 1 foot on an edge are there in the first layer ?

16. How many cubes 1 foot on an edge are there in the 3 layers ?

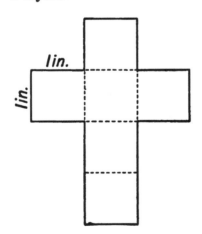

17. Measure the dimensions of your schoolroom and find the number of cubic feet of air in it.

18. A rectangular solid is 1 ft. square and 4 ft. long. Find the surface of its six faces.

19. Cut from cardboard a surface as shown in the drawing. Fold along the dotted lines into a box; find the surface of the six faces and the contents of the box in cubic inches.

20. Measure the walls and ceiling of your schoolroom and find the cost of plastering them at 20 cents per square yard.

21. Cut from cardboard a diagram to show a rectangular box 18 in. long, 12 in. wide, and 8 in. deep.

22. Find the number of cubic inches in the box.

23. Measure the surface of your schoolroom floor, and find the cost of oiling it at 10 cents per square yard.

24. What is the difference in cubic feet between 9 cubic feet and a cube 3 feet on an edge?

To THE TEACHER. — Get, if possible, 144 cubical blocks 1 inch on an edge.

25. Build a cube 3 blocks on an edge. How many cubic inches are there in the cube?

26. Build a cube 4 blocks on an edge. How many cubic inches are there in the cube?

27. Build a cube 2 blocks on an edge. How many cubic inches are there in the cube?

28. The cube 4 blocks on an edge is how many times the cube 2 blocks on an edge?

29. Build a rectangular solid as shown in the figure. How many cubic inches are there in the solid?

30. Show that the solid in problem 24 has 6 rectangular surfaces. Find the surface of each rectangle.

31. Build a cube 5 inches on an edge. How many more than 100 cubes are needed?

27 cu. ft. = 1 cubic yard (cu. yd.)
A cart load of earth = 1 cu. yd.

The contents, *or* volume, *of any body that has six rectangular surfaces is a number of cubical units equal to the product of its three dimensions, when expressed in like units.*

Written Work

1. A schoolroom is 30 ft. wide, 40 ft. long, and 16 ft. high. Find the number of cubic feet of air in it.

2. Find the number of cubic yards of air in the room.

3. A piece of timber is 1 ft. square at the end and 12 ft. long. How many cubic feet are there in it?

4. How many 1-in. cubes are necessary to make a rectangular solid 12 in. long, 8 in. wide, and 4 in. high?

5. A box is 4 ft. long, 2 ft. wide, and 2 ft. high. Find the number of square feet in its six surfaces.

6. Find the number of cubic inches in the box.

7. A bin for grain is 12 ft. long, 8 ft. wide, and 5 ft. deep. Find the number of cubic feet in it.

8. If there are 2150.42 cu. in. in a bushel of wheat, find the number of bushels of wheat the bin will hold.

9. A water tank is 8 ft. long, 4 ft. wide, and 3 ft. deep. If a cubic foot of water weighs $62\frac{1}{2}$ lb., find the weight of the water when the tank is full.

10. A stone wall is 40 ft. long, 4 ft. high, and 2 ft. thick. Find the number of cubic feet of stone in it.

11. Find the cost, at 30 cents a cart load (1 cubic yard), for excavating the ground for a cellar 30 ft. in length, 20 ft. in width, and 4 ft. in depth.

12. A laborer digs a ditch 100 ft. long, 18 in. wide, and $2\frac{1}{2}$ ft. deep in 1 day. Find the number of cart loads of earth removed, and the cost at 30 cents a load.

BILLS, RECEIPTS, AND CHECKS

BILLS

Oral and Written Work

The following is a common form of *bill*.

			NEWARK, N. J., *Oct. 10,* 1913.					

Mrs. James Brown,

42 Market St.

Bought of L. S. PLAUT & Co.,

707 BROAD ST.

TERMS: *Cash.*

Oct.	10	10 yd. Shirting @ $ .06½	$	65				
		10 " Crash @ .06¼		63				
		20 " Calico @ .07½	1	50		2	78	

Received payment,

L. S. Plaut & Co.

Per J. B.

Who sold the goods? Who purchased the goods? When and where was the purchase made? What words show that the bill has been paid?

Who received the money? When a clerk receives payment for a bill, he always writes the receipt of the firm, per his own name or initials. The receipted bill should be kept by the buyer to show that the bill has been paid.

Every bill should show: the *place* and *date* of purchase; the *names* of the buyer and the *seller;* the *quantity*, the *price*, and the *cost* of each item, and the *amount* of the bill.

1. Get blank bill heads from your grocer, your butcher, or your coal dealer, and fill in some of the actual purchases that your parents have made.

2. Let the teacher be represented as owning a grocery store, butcher shop, coal yard, or dry goods store, and let the pupils be the purchasers. Make out blank bill heads, giving date and day of purchase, amount and price of purchase, and receipt for payment.

3. Mr. Smith makes the following purchases at a hardware store : 1 saw @ 75 ¢ ; 3 gas heaters @ $4.90 ; 3½ doz. screws @ 12 ¢ ; 10½ lb. lawn seed @ 20 ¢ ; 8 joints stove pipe @ 30 ¢ ; 2 elbows @ 40 ¢. Fill in the name of your local hardware merchant, and receipt the bill, if cash is paid at the time of the purchase.

4. Walter Brown rents a vacant lot from Mrs. Smith and prepares to raise garden vegetables. He buys at the grocery store : 1½ bu. potatoes at 80 ¢ a bushel ; 3 qt. small onions at 15 ¢ a quart ; 1 package onions @ 20 ¢ ; 1 package lettuce seed @ 10 ¢ ; 1 package lima beans @ 10 ¢ ; 1 package peas @ 10 ¢. He pays cash. Make out and receipt the bill, using your local grocer's name.

5. Walter buys at the hardware store April 10, 1 rake @ 40 ¢ ; 1 hoe at 30 ¢ ; 1 Acme cultivator @ $3.20 ; a bushel basket @ 25 ¢. Make out and receipt the bill.

Make out receipted bills for the following sales :

6. 3½ lb. rice   @ $0.08.   7. 12 yd. muslin @ $0.09.
   10 lb. prunes @ 0.12½.   10 yd. lace   @ 0.12½.
   2 bags salt @ 0.10.   2 pair socks @ 0.35.

8. William Thomas bought of Acker, Merrall & Condit Company, Newark, N.J., Oct. 10, 1913, 15 lb. butter at 28 ¢ per pound ; 10 doz. eggs at 24 ¢ per dozen. Make out the receipted bill, representing yourself as clerk.

Another form of bill is commonly used when *services* have been rendered, as well as *material* furnished. For example:

			CAMDEN, N.J., June 1, 1913.					

Mr. *J. R. Burroughs,*

　　　　To R. W. Jones, Dr.

May	6	To 6 days' Labor	@ $ 1.50	$9	00			
	8	6 lb. Lawn Seed @	0.25	1	50			
	10	8 lb. Nails @	0.06		48	10	98	

Received Payment,
　　　June 12, 1913.
　　　　R. W. Jones.

The **creditor** is the person who sells the goods or does the work.

The **debtor** is the person who buys the goods or for whom the work is done.

In the bill on p. 203 Mrs. Brown is *debtor* to L. S. Plaut & Co., since she owes for the goods purchased, and L. S. Plaut & Co. are the creditors, since they furnished the goods. In the bill above Mr. Burroughs is the *debtor* for work received, and Mr. Jones is the creditor for work he has done.

9. T. S. Ball owes Dr. S. N. Pool, Bayonne, N.J., for services as follows: Jan. 1, 1914, to 1 call, $2; Jan. 12, 1914, to 1 call, $2; Jan. 14, 1914, office, $1; Jan. 16, 1914, to 1 call, $2. Make out and receipt the bill if paid Feb. 1, 1914.

10. Boydson & Co. owe Charles Frampton, Orange, N.J., for services as follows:

March 10, 1913, 6 hr. delivering goods @	$0.20
March 11, 1913, trip to country	2.00
March 12, 1913, " " "	2.00
March 13, 1913, " " "	2.00
March 14, 1913, repairs to wagon	3.75

Write the receipted bill of Boydson & Co., if paid April 1, 1913.

11. James Brown owes Stamm Bros. for labor and material as follows: June 1, 1913, 189 ft. lumber at 8 ¢ per foot; June 4, 1913, 50 lb. cement at 4 ¢ per pound; June 8, 1913, 15 days' labor at $4.50 per day.

Receipt this bill if paid July 1, 1913.

12. John and William agree to repair Mrs. Brown's lawn. They work four days each at $1 per day each. They furnish $2 worth of sodding, and 75 ¢ worth of lawn seed. Mrs. Brown asks them to make out a bill, and pays them April 15th. Write the bill and receipt it.

13. Claire Austin worked for 10 days for Mrs. Elliott Keller, repairing the lawn and a fence. He received $1.10 per day and furnished 150 ft. of lumber at 2 ¢ a foot; 1 gal. paint $1.60; 1 paint brush 90 ¢; and 4 lb. nails @ 6 ¢ a pound. Make out and receipt the bill if paid April 20.

RECEIPTS

Oral and Written Work

1. John Watson pays James Adams $35.50 for work for one month, and asks Mr. Adams for a receipt. Write the receipt to show that the money was paid by Mr. Watson and received by Mr. Adams.

```
$_____                    Passaic, N.J., June 1, 1913

Received from_____

_____Dollars

for_____

                       _____
```

2. What must every receipt show?

3. Write the receipt your grocer would give you in payment of $18.50 on account.

4. Henry Smith received $75 from James Brown for 3 months' rent. Make out a receipt for the amount.

5. Ralph Taylor pays H. W. Henderson $15 for a month's tuition. Write the receipt Ralph Taylor should receive.

6. Mr. John Wylands has a reservoir that supplies several families with water. Charles Hoffman owes him $6.75 water rent for January, February, and March, 1913. Write Mr. Wyland's receipt for payment April 1, 1913.

7. Write a receipt for $75 which John Page paid Edgar Peale for balance due on a buggy.

8. Make out and receipt the bill for the following articles bought by James Thomas from Joseph Horne & Co.:

3 shirts @ $1.75		2 neckties @ $.75
6 collars @ .20		4 pairs cuffs @ .20

9. Assuming that you are a collector for a newspaper, make out a receipt to a subscriber who has paid you $2.60 in full for account.

10. Elliott Keller sells $35 worth of coal to William Phipps. January 20, $20 is paid on the account. Write the receipt.

11. Mary Peck owes Dr. George Hallock $90 for medical service. The doctor gives her 25% discount for paying cash. Write the receipt.

CHECKS

Oral and Written Work

A **check** is an order on a bank where a person keeps a deposit, directing the bank to pay money.

STUB	CHECK
No. 875	*No. 875*
————	*Montclair, N. J., Jan. 10, 1914.*
$67⁰⁰	*The First National Bank.*
	PAY TO THE
Jan. 10, '14	*Order of......James Ward......$ 67⁰⁰*
To James Ward	*Sixty-seven $\frac{no}{100}$ ~~~~Dollars.*
For Labor	*W. J. Moore.*

1. Name the different things stated in this check.

2. Observe that this check is payable to the *order of* James Ward. He orders it paid by writing his name across the back of it. This is called **indorsing** the check.

3. Write the check your father would give your teacher in payment of $18 for September tuition.

4. Bring to school some canceled checks, if you can get them, and compare them with this check.

5. Edwin works for John D. Walker, and receives a check on a local bank for $20. Write the check.

6. Name a bank after your school, and write checks on it in payment of the bills on the previous pages.

SIXTH GRADE—SECOND HALF

To ·the Teacher.— Review thoroughly the work on integers, fractions, and decimals found in the preceding pages.

REVIEW OF COMMON FRACTIONS

Written Work

Note.— For development, model solutions, etc., review, if necessary, pp. 27 to 81.

Change to improper fractions:

1. $15\frac{3}{8}$	8. $60\frac{5}{12}$	15. $95\frac{7}{20}$	22. $65\frac{17}{25}$
2. $17\frac{3}{4}$	9. $45\frac{3}{20}$	16. $48\frac{13}{18}$	23. $28\frac{22}{25}$
3. $18\frac{5}{6}$	10. $56\frac{9}{14}$	17. $78\frac{8}{50}$	24. $47\frac{17}{25}$
4. $25\frac{5}{7}$	11. $75\frac{8}{15}$	18. $80\frac{11}{16}$	25. $75\frac{18}{19}$
5. $35\frac{7}{8}$	12. $40\frac{9}{16}$	19. $42\frac{14}{25}$	26. $37\frac{54}{81}$
6. $42\frac{7}{9}$	13. $85\frac{5}{18}$	20. $71\frac{9}{50}$	27. $46\frac{32}{75}$
7. $51\frac{2}{11}$	14. $92\frac{3}{20}$	21. $83\frac{12}{15}$	28. $53\frac{21}{25}$

Change to mixed numbers:

29. $\frac{18}{5}$	34. $\frac{29}{8}$	39. $\frac{875}{8}$	44. $\frac{75}{7}$ mi.
30. $\frac{19}{3}$	35. $\frac{45}{6}$	40. $\frac{99}{12}$ lb.	45. $\frac{98}{12}$ rd.
31. $\frac{25}{4}$	36. $\frac{75}{9}$	41. $\frac{148}{12}$ hr.	46. $\frac{82}{8}$ bu.
32. $\frac{87}{9}$	37. $\frac{84}{6}$	42. $\frac{124}{11}$ min.	47. $\frac{110}{9}$ in.
33. $\frac{46}{5}$	38. $\frac{65}{7}$ oz.	43. $\frac{57}{12}$	48. $\frac{116}{10}$ A.

Change to lowest terms:

49. $\frac{12}{24}$	52. $\frac{28}{60}$	55. $\frac{84}{108}$	58. $\frac{48}{100}$	61. $\frac{160}{180}$
50. $\frac{24}{36}$	53. $\frac{18}{40}$	56. $\frac{120}{144}$	59. $\frac{120}{360}$	62. $\frac{144}{216}$
51. $\frac{28}{42}$	54. $\frac{60}{96}$	57. $\frac{132}{192}$	60. $\frac{68}{144}$	63. $\frac{175}{375}$

Change to similar fractions having the l. c. d.:

64. $\frac{1}{2}, \frac{1}{3}$

65. $\frac{1}{4}, \frac{1}{6}$

66. $\frac{1}{5}, \frac{1}{4}$

67. $\frac{2}{3}, \frac{3}{4}$

68. $\frac{5}{6}, \frac{3}{8}$

69. $\frac{9}{10}, \frac{5}{6}$

70. $\frac{5}{8}, \frac{7}{12}$

71. $\frac{1}{2}, \frac{1}{3}, \frac{1}{4}$

72. $\frac{1}{6}, \frac{2}{3}, \frac{3}{4}$

73. $\frac{1}{2}, \frac{3}{8}, \frac{4}{5}$

74. $\frac{3}{4}, \frac{7}{9}, \frac{8}{9}$

75. $\frac{7}{8}, \frac{3}{4}, \frac{8}{10}$

76. $\frac{5}{12}, \frac{4}{5}, \frac{2}{3}$

77. $\frac{4}{7}, \frac{1}{3}, \frac{7}{9}$

78. $\frac{9}{16}, \frac{7}{10}$

79. $\frac{5}{9}, \frac{5}{12}$

80. $\frac{5}{8}, \frac{4}{5}, \frac{7}{12}$

81. $\frac{3}{11}, \frac{3}{4}, \frac{1}{2}$

82. $\frac{11}{16}, \frac{7}{24}, \frac{9}{32}$

83. $\frac{15}{25}, \frac{3}{4}, \frac{4}{7}$

84. $\frac{5}{13}, \frac{4}{13}, \frac{9}{26}$

85. $\frac{2}{15}, \frac{5}{12}, \frac{3}{4}$

86. $\frac{1}{12}, \frac{3}{4}, \frac{2}{3}$

87. $\frac{1}{20}, \frac{1}{2}, \frac{2}{3}$

88. $\frac{6}{7}, \frac{3}{8}, \frac{1}{8}$

89. $\frac{3}{8}, \frac{1}{7}, \frac{1}{4}$

90. $\frac{5}{6}, \frac{4}{7}, \frac{1}{8}$

Add:

91. $\frac{2}{4}, \frac{7}{8}$

92. $\frac{7}{9}, \frac{3}{8}$

93. $\frac{6}{7}, \frac{4}{8}$

94. $\frac{4}{8}, \frac{5}{8}$

95. $\frac{8}{10}, \frac{2}{3}$

96. $\frac{5}{6}, \frac{7}{9}$

97. $\frac{2}{12}, \frac{4}{5}$

98. $\frac{9}{10}, \frac{7}{8}$

99. $\frac{2}{3}, \frac{3}{4}, \frac{1}{2}$

100. $\frac{2}{3}, \frac{6}{7}, \frac{5}{6}$

101. $\frac{6}{7}, \frac{1}{8}, \frac{1}{4}$

102. $\frac{3}{4}, \frac{4}{5}, \frac{1}{2}$

103. $\frac{3}{8}, \frac{7}{8}, \frac{3}{4}$

104. $\frac{4}{5}, \frac{6}{7}, \frac{5}{8}$

105. $\frac{7}{8}, \frac{1}{8}, \frac{3}{4}$

106. $\frac{4}{5}, \frac{1}{2}, \frac{3}{8}$

107. $\frac{7}{8}, \frac{7}{18}, \frac{8}{9}$

108. $\frac{11}{12}, \frac{4}{5}, \frac{9}{10}$

109. $1\frac{5}{8} + \frac{3}{8} + 2\frac{7}{8} = ?$

110. $10\frac{8}{9} + 12\frac{2}{3} + 5\frac{8}{27} = ?$

111. $2\frac{5}{12} + 5\frac{3}{4} + 9\frac{4}{15} = ?$

112. $1\frac{3}{4} + 7\frac{1}{6} + 8\frac{7}{8} = ?$

113. $9\frac{3}{5} + 16\frac{7}{10} + 5\frac{11}{15} = ?$

114. $4\frac{7}{8} + \frac{11}{16} + \frac{13}{16} = ?$

115. $3\frac{1}{2} + 4\frac{3}{4} + 1\frac{7}{12} = ?$

116. $4\frac{3}{8} + 5\frac{7}{10} + 9\frac{5}{18} = ?$

117. $12\frac{2}{3} + \frac{3}{8} + \frac{7}{8} + 3\frac{9}{10} = ?$

118. $2\frac{4}{5} + 1\frac{1}{6} + 3\frac{1}{4} = ?$

Find the value of:

119. $\frac{6}{7} - \frac{4}{5}$

120. $3\frac{1}{8} - 1\frac{1}{2}$

121. $5 - 1\frac{3}{8}$

122. $5\frac{1}{4} - 2\frac{2}{3}$

123. $10\frac{3}{8} - 5\frac{3}{4}$

124. $20\frac{1}{12} - 13\frac{5}{6}$

125. $7\frac{5}{8} - 2\frac{1}{3}$

126. $50\frac{1}{4} - 30\frac{1}{2}$

127. $6\frac{5}{8} - 2\frac{3}{7}$

128. $31\frac{9}{16} - 20\frac{25}{28}$

129. $11\frac{3}{8} - 5\frac{2}{3}$

130. $15 - 9\frac{9}{13}$

131. $65\frac{4}{15} - 63\frac{9}{20}$

132. $44\frac{3}{8} - 12\frac{1}{2}$

133. $36\frac{11}{8} - 11\frac{4}{15}$

134. $27\frac{1}{4} - 19\frac{4}{11}$

135. $81\frac{1}{8} - 14\frac{9}{7}$

136. $\frac{3}{5} - \frac{2}{7}$

Find the value of :

137. $\frac{2}{3} + \frac{3}{4} - \frac{5}{6}$

138. $3\frac{1}{2} + 2\frac{1}{4} - 1\frac{5}{6}$

139. $5\frac{1}{8} + 4\frac{5}{6} - 4\frac{5}{12}$

140. $1\frac{5}{8} + 2\frac{1}{12} - 1\frac{5}{6}$

141. $9\frac{9}{15} - 4\frac{4}{5} + 3\frac{7}{10}$

142. $12\frac{3}{4} - 5\frac{1}{2} + 2\frac{2}{3}$

143. $\frac{3}{4} - \frac{1}{2} + \frac{7}{8} + \frac{17}{18}$

144. $20 + \frac{1}{16} + \frac{1}{12} - 2\frac{1}{4}$

145. $\frac{1}{3} + \frac{1}{4} + \frac{5}{6} - \frac{7}{12}$

146. $18\frac{11}{24} - 10\frac{3}{8} + 5\frac{1}{4} + 2\frac{5}{6}$

147. $15 - 3\frac{2}{3} + 4\frac{1}{5}$

148. $40 + 60 + 30\frac{11}{18}$

149. $19\frac{27}{32} - 11\frac{5}{8} + 1\frac{5}{16}$

150. $7\frac{7}{8} + 4\frac{1}{3} + 11\frac{5}{18}$

151. $3\frac{5}{12} + 9\frac{5}{8} - 7\frac{7}{24}$

152. $22\frac{2}{11} + 7\frac{1}{3} - 20\frac{19}{33}$

153. $8\frac{3}{4} + 22\frac{2}{5} + 43\frac{3}{8}$

154. $225\frac{5}{8} + 132\frac{5}{12} + 80\frac{17}{20}$

155. $12\frac{3}{4} + 19\frac{7}{8} - 27\frac{5}{6}$

156. $19\frac{1}{8} + 11\frac{3}{8} + 14\frac{1}{3}$

Find the products:

157. $16 \times \frac{3}{8}$

158. $24 \times \frac{7}{8}$

159. $27 \times \frac{2}{3}$

160. $45 \times \frac{4}{5}$

161. $18 \times \frac{2}{9}$

162. $12 \times \frac{11}{3}$

163. $13 \times \frac{12}{26}$

164. $14 \times \frac{11}{12}$

165. $12 \times \frac{17}{18}$

166. $11 \times \frac{7}{22}$

167. $25 \times \frac{8}{15}$

168. $15 \times \frac{11}{24}$

169. $10 \times \frac{12}{25}$

170. $21 \times \frac{25}{63}$

171. $36 \times \frac{85}{144}$

172. $16 \times \frac{109}{256}$

173. $27 \times \frac{56}{81}$

174. $33 \times \frac{78}{132}$

175. $45 \times \frac{56}{135}$

176. $36 \times \frac{21}{144}$

177. $9 \times 2\frac{1}{4}$

178. $12 \times 3\frac{2}{3}$

179. $18 \times 5\frac{3}{8}$

180. $22 \times 4\frac{8}{10}$

181. $21 \times 2\frac{5}{14}$

182. $7 \times 8\frac{1}{4}$

183. $10 \times 2\frac{7}{15}$

184. $16 \times 9\frac{5}{12}$

185. $24 \times 8\frac{3}{4}$

186. $36 \times 7\frac{1}{8}$

Find:

187. $\frac{2}{3}$ of $\frac{5}{6}$

188. $\frac{3}{8}$ of $\frac{5}{7}$

189. $\frac{3}{4}$ of $\frac{5}{6}$

190. $\frac{2}{3}$ of $\frac{7}{9}$

191. $\frac{5}{6}$ of $\frac{3}{8}$

192. $\frac{2}{5}$ of $\frac{8}{9}$

193. $\frac{25}{88}$ of $\frac{8}{15}$

194. $\frac{9}{10}$ of $\frac{5}{12}$

195. $\frac{11}{8}$ of $\frac{10}{17}$

196. $\frac{15}{16}$ of $\frac{3}{4}$

197. $\frac{16}{19}$ of $\frac{17}{24}$

198. $\frac{20}{7}$ of $\frac{49}{8}$

199. $\frac{15}{32}$ of $\frac{44}{8}$

200. $\frac{26}{45}$ of $\frac{55}{18}$

201. $\frac{2}{3}$ of $\frac{7}{8}$ of $\frac{3}{4}$

202. $\frac{9}{10}$ of $\frac{25}{33}$ of $\frac{18}{15}$

203. $\frac{7}{8}$ of $\frac{15}{32}$ of $\frac{24}{35}$

204. $\frac{25}{48}$ of $\frac{105}{32}$ of $\frac{34}{56}$

Find the value of:

205. $1\frac{1}{2} \times 1\frac{1}{3}$ 208. $1\frac{3}{8} \times 2\frac{5}{8}$ 211. $20\frac{1}{4} \times 5\frac{7}{8}$

206. $2\frac{1}{2} \times 3\frac{2}{3}$ 209. $6\frac{1}{3} \times 2\frac{1}{4}$ 212. $12\frac{1}{5} \times 1\frac{7}{8}$

207. $2\frac{3}{4} \times 3\frac{5}{8}$ 210. $16\frac{2}{3} \times 5\frac{1}{4}$ 213. $39\frac{3}{5} \times 66\frac{2}{3}$

Divide:

214. $\frac{4}{5}$ by 2   224. $\frac{24}{25}$ by 8   234. $\frac{99}{100}$ by 11

215. $\frac{8}{19}$ by 4   225. $\frac{21}{25}$ by 7   235. $\frac{96}{100}$ by 12

216. $\frac{12}{13}$ by 3   226. $\frac{32}{35}$ by 8   236. $\frac{54}{5}$ by 9

217. $\frac{15}{16}$ by 5   227. $\frac{45}{49}$ by 9   237. $\frac{20}{21}$ by 10

218. $\frac{13}{18}$ by 6   228. $\frac{108}{125}$ by 12   238. $\frac{121}{130}$ by 11

219. $\frac{16}{17}$ by 8   229. $\frac{88}{95}$ by 19   239. $3853\frac{3}{5}$ by 7

220. $\frac{6}{7}$ by 5   230. $\frac{20}{27}$ by 14   240. $4267\frac{3}{4}$ by 12

221. $\frac{15}{18}$ by 6   231. $\frac{18}{44}$ by 11   241. $2687\frac{7}{8}$ by 9

222. $\frac{24}{35}$ by 7   232. $\frac{17}{40}$ by 34   242. $5831\frac{2}{3}$ by 11

223. $\frac{15}{16}$ by 10   233. $\frac{21}{26}$ by 14   243. $1790\frac{7}{10}$ by 12

Find quotients:

244. $\frac{7}{8} \div \frac{3}{8}$ 258. $10 \div \frac{2}{5}$ 272. $2\frac{1}{2} \div 2\frac{1}{3}$

245. $\frac{7}{15} \div \frac{7}{15}$ 259. $1\frac{1}{4} \div \frac{1}{4}$ 273. $12\frac{1}{2} \div 16\frac{2}{3}$

246. $\frac{5}{6} \div \frac{7}{8}$ 260. $12 \div \frac{6}{7}$ 274. $8\frac{2}{3} \div 4\frac{2}{3}$

247. $\frac{7}{12} \div \frac{5}{12}$ 261. $13 \div \frac{5}{7}$ 275. $5\frac{1}{6} \div 4\frac{1}{4}$

248. $\frac{3}{4} \div \frac{7}{8}$ 262. $25 \div \frac{7}{8}$ 276. $7\frac{3}{8} \div 6\frac{1}{2}$

249. $\frac{5}{6} \div \frac{3}{10}$ 263. $121 \div \frac{11}{12}$ 277. $8\frac{4}{5} \div 4\frac{1}{4}$

250. $\frac{3}{7} \div \frac{5}{14}$ 264. $256 \div \frac{16}{21}$ 278. $18\frac{2}{3} \div 7\frac{1}{2}$

251. $\frac{4}{15} \div \frac{3}{4}$ 265. $42 \div \frac{5}{9}$ 279. $19\frac{3}{8} \div 1\frac{3}{20}$

252. $1\frac{2}{5} \div \frac{3}{4}$ 266. $55 \div \frac{6}{11}$ 280. $160 \div \frac{5}{8}$

253. $5\frac{4}{5} \div 2\frac{5}{6}$ 267. $\frac{11}{12} \div \frac{5}{6}$ 281. $16\frac{2}{3} \div 6\frac{9}{11}$

254. $3\frac{3}{4} \div 4$ 268. $\frac{13}{18} \div \frac{6}{11}$ 282. $77 \div 2\frac{1}{8}$

255. $5\frac{7}{8} \div 10$ 269. $\frac{7}{15} \div \frac{1}{6}$ 283. $103 \div 10\frac{3}{10}$

256. $4 \div \frac{1}{8}$ 270. $\frac{9}{16} \div \frac{1}{4}$ 284. $67\frac{3}{16} \div 7\frac{3}{4}$

257. $1\frac{1}{5} \div \frac{2}{3}$ 271. $\frac{9}{16} \div \frac{1}{8}$ 285. $88\frac{2}{21} \div 16\frac{2}{3}$

FRACTIONAL RELATION OF NUMBERS. RATIO

Finding what part one number is of another.

Oral Work

1. What part of 2 is 1? What part of 15 is 3? of 6 is 5? Express the answers also in the form of division. Thus, $\frac{1}{2} = 1 \div 2$; $\frac{3}{15} = 3 \div 15$; $\frac{5}{6} = 5 \div 6$.

2. What part of 16 is 2? 4? 8? 5?

3. What part of $\frac{3}{8}$ is $\frac{2}{8}$? $\frac{2}{8} \div \frac{3}{8} = 2 \div 3 = \frac{2}{3}$.

4. What part of $\frac{5}{8}$ is $\frac{4}{8}$? $\frac{4}{8} \div \frac{5}{8} = 4 \div 5 = \frac{4}{5}$.

5. How, then, can you find what part one number is of another?

To **find what part one number is of another,** *divide the number of which the part is to be found by the other number.*

On page 105 you learned that the quotient of two like numbers is called their **ratio.**

The ratio of 6 to 3 is 2; the ratio of 3 to 6 is $\frac{1}{2}$.

To **find the ratio of two numbers,** *divide the first by the second.*

What part of:	What is the ratio of:
6. 4 is 2?	**11.** 20 to 80?
7. 6 is 4?	**12.** 80 to 20?
8. 5 is 3?	**13.** 16 to 64?
9. 40 is 10?	**14.** 64 to 16?
10. $\frac{3}{4}$ is $\frac{3}{8}$?	**15.** $\frac{1}{4}$ to $\frac{1}{2}$?

Written Work

1. What part of 108 is 48?

$$48 \div 108 = \frac{48}{108}, \text{ or } \frac{4}{9}.$$

Divide 48 by 108, which gives the fraction $\frac{48}{108}$. This reduced to its lowest terms equals $\frac{4}{9}$.

2. What part of $4\frac{1}{2}$ is $\frac{3}{4}$?

$$4\frac{1}{2} = \frac{9}{2}, \quad \frac{3}{4} \div \frac{9}{2} = \frac{\cancel{3}}{\cancel{4}} \times \frac{\cancel{2}}{\cancel{9}}, \text{ or } \frac{1}{6}$$

Reduce the mixed number to the improper fraction $\frac{9}{2}$. Then divide $\frac{3}{4}$ by $\frac{9}{2}$. The result is $\frac{1}{6}$.

What part of : What is the ratio of :

3. 48 is 36?   **5.** 117 to 104?   **7.** $\frac{4}{5}$ to $\frac{2}{3}$?

4. 96 is 16?   **6.** 144 to 128?   **8.** $1\frac{1}{2}$ to $1\frac{1}{4}$?

9. If John had 60 cents and spent 20 cents, what part of his money did he spend?

10. The distance between two stations is 72 miles. What part of the distance would 18 miles be?

11. What part of a yard do I buy when I buy 18 inches?

12. John and Henry together received 80¢ for a piece of work at which John worked 2 hr. and Henry 3 hr. How should they divide the amount received?

SOLUTION. — Together they worked 5 hr.
John worked 2 hr. and earned $\frac{2}{5}$ of the whole amount.
Henry worked 3 hr. and earned $\frac{3}{5}$ of the whole amount.
$\frac{1}{5}$ of the whole amount is 16 ¢.
$2 \times 16 ¢ = 32 ¢$, $\frac{2}{5}$ of the whole amount, or John's earnings.
$3 \times 16 ¢ = 48 ¢$, $\frac{3}{5}$ of the whole amount, or Henry's earnings.

13. Mary and Lucy together picked 49 qt. of strawberries. Mary picked 4 qt. as often as Lucy picked 3 qt. How should they divide their berries?

14. Frank and Clay sell newspapers on Saturday. Frank works 6 hr. and Clay 9 hr. If they have $2.40 to divide, how much should each receive?

PROBLEMS IN FRACTIONS

1. The sum of two fractions is $\frac{17}{18}$. One fraction is $\frac{2}{3}$. What is the other?

2. The product of two fractions is $\frac{5}{8}$. One fraction is $\frac{11}{16}$. What is the other?

3. If $\frac{3}{4}$ of an acre of land costs $45, how much will 5 acres cost?

4. A boy, after spending $\frac{2}{3}$ of his money, had $1.60 remaining. How much money had he at first?

5. If butter is selling at 28 cents a pound, how much will $3\frac{1}{2}$ pounds cost?

6. A man owned $\frac{3}{4}$ of a farm and sold $\frac{1}{3}$ of his share for $1250. At that rate how much was the whole farm worth?

7. If $1\frac{1}{2}$ acres yield 120 bushels of potatoes, how many bushels will 3 acres yield at the same rate?

SUGGESTION. — 3 are how many times $1\frac{1}{2}$?

8. Compare $\frac{1}{2}$ and $\frac{1}{4}$; $\frac{1}{3}$ and $\frac{1}{6}$; $\frac{1}{4}$ and $\frac{1}{12}$.

9. At $16\frac{2}{3}$ cents a pound, how many pounds of meat can be bought for $1? $2? $4? $5?

10. What part of 60 is 10? 15? 20? 40?

11. I sold a bicycle for $15; this was $\frac{3}{4}$ of what I paid for it. How much did I pay for it?

12. If 20 dozen eggs are worth $5.60, how much are 50 dozen worth?

13. If a telegram of 10 words costs 25 cents, how much will a telegram of 30 words cost at 2¢ for each extra word?

14. Find the ratio of 6 to 18; of 8 to 48; of 36 to 12.

15. How does a city lot worth $6000 compare in value with another worth $5000?

16. If James's age is 16 years and his father is $2\frac{1}{2}$ times as old, what is the difference in their ages? James's age is what part of his father's?

17. How much will 9 lemons cost at the rate of 40 cents a dozen? How much will $1\frac{1}{4}$ dozen cost? 3 dozen?

18. A field is 40 rods long. The width is $\frac{3}{4}$ of its length. What is the distance around the field?

19. A traveler walked $25\frac{1}{2}$ miles the first day, $30\frac{1}{4}$ miles the second day, $29\frac{3}{4}$ miles the third day, $27\frac{5}{8}$ miles the fourth day, and $25\frac{1}{8}$ miles the fifth day. How far did he travel in the 5 days, and what was his average rate per day?

20. An errand boy receives $4 a week. In how many weeks can he earn enough to pay for a pair of shoes at $2\frac{1}{4}$, a suit at $10, a hat at $1\frac{1}{2}$, and 3 shirts at $\frac{3}{4}$ each?

21. Divisor $25\frac{1}{2}$; quotient 200. Find the dividend.

22. Multiply the sum of $4\frac{1}{2}$ and 3 by 8.

23. At $1\frac{1}{4}$ a yard, how many yards of Brussels carpet can be bought for $30?

24. A lady bought $12\frac{1}{2}$ yards of muslin at 8 cents a yard, $4\frac{1}{2}$ yards of ribbon at 4 cents a yard, 8 yards of dress goods at 75 cents a yard, and gave in payment a ten-dollar bill. How much change should she receive?

25. The sum of three numbers is 418; the first is $198\frac{3}{4}$, and the second, $206\frac{1}{2}$. What is the third?

26. Mr. Jones sold $\frac{7}{8}$ of his farm for $8890. At the same rate what was the value of the remainder?

27. Mr. White paid $2700 for a three-eighths interest in a store. How much was the entire store worth?

28. A teacher spends $1062.50 a year, which is $\frac{5}{8}$ of his salary. How much is his salary?

29. A dealer bought 960 bushels of wheat at 84 cents a bushel. He sold $\frac{1}{4}$ of it at 88 cents a bushel, $\frac{1}{3}$ of it at 87 cents a bushel, and the remainder at 89 cents a bushel. What was his entire gain?

30. What part of $\frac{5}{8}$ is $\frac{3}{4}$?

31. Thomas buys a necktie for $\$\frac{1}{2}$, a pair of gloves for $\$1\frac{2}{3}$, and 2 shirts at $\$\frac{3}{4}$ apiece. How much change should he receive from a five-dollar bill?

32. If a man's wages are $\$3\frac{1}{2}$ a day, and his daily expenses $\$1\frac{3}{4}$, how many weeks of 6 days each must he labor to save $\$105$?

33. Two men are 29 miles apart and travel toward each other, the one at the rate of $3\frac{1}{2}$ miles an hour and the other at $3\frac{3}{4}$ miles an hour. In how many hours will they meet?

34. Find the cost of $7\frac{1}{4}$ pounds of mackerel at 16 cents, $6\frac{1}{2}$ pounds of codfish at 12 cents, and 6 cans of salmon at the rate of 2 for a quarter.

35. What is the cost of 80 pounds of sugar at $6\frac{1}{4}\not\!c$, $48\frac{1}{2}$ pounds of tea at $36\not\!c$, and 124 pounds of coffee at $22\not\!c$?

36. What is the cost of $2\frac{1}{4}$ pounds of coffee at $28\not\!c$, $\frac{1}{2}$ pound of pepper at $24\not\!c$, $3\frac{3}{4}$ pounds of butter at $28\not\!c$, and $1\frac{1}{4}$ gallons of vinegar at $32\not\!c$?

37. Mrs. Baker had an estate valued at $\$2400$, of which she willed $\frac{1}{3}$ to her daughter, $\frac{1}{2}$ to her son, and the rest to her sister. How much did the son and the sister each receive?

38. A man spent $\$22\frac{1}{2}$; then earned $\$18.75$; and then had $\$50$. How much had he at first?

39. What is the cost of 3 barrels of sugar weighing 300 pounds, 310 pounds, 312 pounds, respectively, at $5\frac{1}{2}\not\!c$ a pound?

40. A farmer has 300 sheep. He sells $\frac{2}{3}$ of them at $3\frac{1}{2}$ each, and the remainder at $4 each. How much does he get for all?

41. If 6 dozen oranges cost $2.10, how much will 24 dozen cost?

Find the amount of the following bills:

42. 6$\frac{1}{2}$ yd. ribbon @ 10 ¢.
12$\frac{3}{4}$ yd. muslin @ 8 ¢.
11$\frac{1}{4}$ yd. silk @ 98 ¢.
10$\frac{1}{2}$ yd. lace @ 20 ¢.

43. 12$\frac{1}{4}$ lb. sugar @ 6 ¢.
2$\frac{1}{4}$ lb. butter @ 28 ¢.
4$\frac{3}{4}$ lb. lard @ 16 ¢.
3$\frac{1}{2}$ qt. beans @ 10 ¢.

44. A blackboard is 21 feet long and 3$\frac{1}{4}$ feet wide. The width is what part of the length?

45. I paid $4000 for 2$\frac{1}{2}$ acres of land, and after taking $\frac{2}{3}$ of an acre for streets, divided the remainder into lots of $\frac{1}{16}$ of an acre each. I sold the lots at $150 each. Find my gain.

46. I spent $\frac{1}{2}$, $\frac{1}{4}$, and $\frac{1}{5}$ of my money and had $20 left. How much had I at first?

47. Compare 42 and 28. If 28 men earn $56 in a day, how much will 42 men earn in the same time?

48. What is the distance around a lot 60$\frac{1}{2}$ feet front and 121 feet deep?

49. How many times can I fill a pail holding $\frac{1}{2}$ a gallon from a 12-gallon tank that is full?

50. $\frac{7}{8}$ of 16 is $\frac{7}{10}$ of what number?

51. Find the product of 104$\frac{3}{4}$ × 24.

52. Find the cost of the following articles: 12 yd. velvet @ $2.25; 12$\frac{1}{2}$ yd. suiting @ 50 ¢; 7$\frac{3}{4}$ yd. dimity @ 20 ¢.

53. A carpenter worked 4$\frac{1}{2}$ days one week, 5$\frac{1}{4}$ days the next week, and 5$\frac{3}{4}$ days the next week. How many days did he work in the three weeks?

REVIEW OF DECIMALS

Written Work

Change to decimals :

1. $\frac{3}{4}$	4. $\frac{5}{16}$	7. $\frac{4}{25}$	10. $\frac{4}{9}$	13. $\frac{8}{16}$
2. $\frac{7}{8}$	5. $\frac{3}{20}$	8. $\frac{4}{6}$	11. $\frac{10}{11}$	14. $\frac{7}{16}$
3. $\frac{3}{8}$	6. $\frac{8}{9}$	9. $\frac{6}{7}$	12. $\frac{5}{12}$	15. $\frac{7}{20}$

Change to common fractions :

16. .87$\frac{1}{2}$	20. .88$\frac{1}{8}$	24. .62$\frac{1}{2}$	28. .12$\frac{1}{2}$
17. .41$\frac{2}{3}$	21. .06$\frac{1}{4}$	25. .83$\frac{1}{3}$	29. .14$\frac{2}{7}$
18. .66$\frac{2}{3}$	22. .04$\frac{1}{6}$	26. .31$\frac{1}{4}$	30. .58$\frac{1}{3}$
19. .24$\frac{1}{6}$	23. .08$\frac{1}{3}$	27. .03$\frac{1}{3}$	31. .87$\frac{1}{2}$

Add and test :

32.	33.	34.	35.
.45	72.5	8.557	87.
8.07	5.07	0.039	6.5
62.093	89.09	16.909	11.9
40.987	20.98	41.005	00.5

36.	37.	38.
3.7	.001	.65
5.06	12.3	.001
8.023	15.0248	10.1
9.04	18.0149	25.004

Add :

39.	40.	41.
1.1	120.2601	36.15
4.01	230.31002	9.00999
1.0101	.05673	128.87
5.055	3.7	16.08753

Add:

42.	166.6	43.	.27	44.	185.057
	7.0425		.0616		127.0348
	28.318		.010912		216.253
	142.0101		1.940054		456.03456

Subtract and test:

45.	40.275	46.	9.0098	47.	219.75	48.	28.7
	39.009		6.7849		8.95		12.5

49.	.75	50.	.3216	51.	4.205	52.	15.
	.1825		.275		1.7856		5.007

53.	38.	54.	$45.67	55.	125½	56.	249¾
	18.276		12.09		87.432		178.625

57.	230.4897	58.	100.001	59.	1001.101	60.	105.55
	116.5988		99.9		900.909		79.067

61.	1.1	62.	5.05	63.	8.25	64.	1000.00
	.999		.6565		.0085		999.99

Find products:

65.	.25 × .22	74.	.1232 × .961	83.	.4986 × .086
66.	.17 × .28	75.	.2592 × 8	84.	.006 × 20
67.	.027 × .03	76.	65.65 × .65	85.	38.2 × .75
68.	27 × .12	77.	75.002 × 16.04	86.	.0045 × .05
69.	.35 × 42	78.	275 × .007	87.	9.876 × .786
70.	8.7 × 9.22	79.	.0018 × 720	88.	362.9 × .0076
71.	.085 × 50	80.	1500 × .004	89.	119.8 × 2.74
72.	.027 × 18	81.	124 × .064	90.	20.08 × .006
73.	1.005 × .011	82.	326 × .096	91.	.375 × 2.027

92. 26.8×34

93. 28.25×12

94. $324.6 \times 8\frac{1}{2}$

95. 39.10×18.4

96. $.0214 \times .016$

97. $12.134 \times .0025$

98. 15.684×8

99. $.1232 \times 345$

100. $627 \times .78$

101. $246 \times .3$

102. $29.4 \times .08$

103. 9.86×3.8

104. $39.75 \times .27$

105. 8.708×6.8

106. 368.9×8.5

107. $2009 \times .006$

108. 98.64×4.096

109. $.069 \times 8.92$

110. 6.34×2.34

111. $12.34 \times .004$

112. $6.08 \times .0001$

113. 1.002×1.004

114. $.05 \times .005$

115. 6.876×4.37

Find:

116. $\frac{1}{4}$ of .16

117. $\frac{1}{5}$ of .25

118. $\frac{1}{2}$ of .08

119. $\frac{1}{4}$ of .04

120. $\frac{1}{6}$ of 6.6

121. $\frac{1}{4}$ of 8.08

122. $\frac{1}{5}$ of 10.10

123. $\frac{1}{6}$ of 12.06

124. $6.42 \div 6$

125. $12.04 \div 4$

126. $15.05 \div 5$

127. $24.18 \div 3$

Find quotients:

128. $96.16 \div 8$

129. $849.6 \div 6$

130. $72.84 \div 12$

131. $22.5 \div 15$

132. $80.96 \div 16$

133. $2.5625 \div 25$

134. $.96064 \div 32$

135. $701.05 \div 35$

136. $2.268 \div 27$

137. $2.867 \div 47$

138. $36.54 \div 42$

139. $.666 \div 74$

140. $6.675 \div 89$

141. $4.05 \div .27$

142. $.252 \div .14$

143. $8.398 \div 3.8$

144. $2.173 \div 1.06$

145. $144 \div .12$

146. $.144 \div 12$

147. $31.36 \div .056$

148. $.41912 \div .338$

149. $3.125 \div .25$

150. $.3105 \div 15$

151. $.5 \div .625$

152. $6.705 \div .009$

153. $139.195 \div 14.35$

154. $46.5 \div .1875$

155. $.00522 \div .29$

156. .001705 + .31

157. 12.312 + 27

158. 2.25 + 15

159. 809.6 + 16

160. 256.25 + 25

161. 96.064 + 32

162. 7010.5 + 35

163. 61.472 + 68

164. 27.8142 + 307

165. 425.92 + 605

166. 901.57 + 97

167. .2322 + 86

168. 34.356 + 409

169. 45.76 + 650

170. 1000 + .001

171. .2375 + .095

172. 177.8028 + 72.87

173. 145.908 + 1.26

174. .0656 + .004

175. .1701 + 63

176. 85.75 + .0049

177. .025641 + .777

178. .0022 + 200

179. 222 + .002

180. .025 + .00025

181. .0003 + 1.5

182. $1.05 + $.005

183. .685 + 500

184. .01058 + 46

185. 125.625 + 1.005

PRACTICAL PROBLEMS

1. Find the weight of four silver bars weighing as follows: 15.75 pounds, .125 pounds, 14.3125 pounds, and 16.875 pounds.

2. Find the number of acres in four fields containing, respectively, 4.125 acres, .3125 acres, 8.8 acres, and 9.85 acres.

3. Find the sum of one hundred twenty-five and seven hundredths, eighty-nine and two hundred thirty-five thousandths, one hundred twenty-seven ten-thousandths, and sixteen and four tenths.

4. A farm cost $4225.50; stock, $745.25; buildings, $1825.75; and implements, $358.45. What was the total cost?

5. How many square feet are there in four floors measuring, respectively, 245.5 square feet, 278.75 square feet, 174.375 square feet, and 168.3125 square feet?

6. A contractor furnished 2,626,000 bricks at $7.75 a thousand, and a laborer for 65 days at $2.75 a day. What was the amount of his bill?

7. If there are 39.37 inches in a meter, how many inches are there in 12 meters? how many yards?

8. How much must be paid for 85 acres of land at $45 per acre?

9. Three brothers divided an estate worth $9600. The first received .125 of it, the second .375 of it, and the third the remainder. How much did each receive?

10. If a stone cutter earns $3.75 a day, how many days will it take him to earn $311.25?

11. If 4275 acres of land cost $1,731,375, what is the price per acre?

12. At $.22 a dozen, how many dozen eggs can be bought for $19.47?

13. If 16 stoves are sold for $292, what is the average price per stove?

14. Divide .18 by 20.

15. If the wheel of a bicycle is 9.25 feet around, how many times does it turn in going a mile?

16. The product of two numbers is .9375. One of them is .75. What is the other?

17. There are $30\frac{1}{4}$ square yards in one square rod. How many square rods are there in a plot containing 559.625 square yards?

18. A merchant, in closing out his stock of goods, sold .37$\frac{1}{2}$ of the stock the first month, .35 the second month, and the remainder, $5500 worth, the third month. What was the value of his stock of goods?

Find the cost of:

19. 60 bu. apples at 33$\frac{1}{3}$ ¢ per bushel.

20. 25 lb. butter at 25 ¢ per pound.

21. 960 yd. calico at 6$\frac{1}{4}$ ¢ per yard.

22. 50 lb. lard at 12$\frac{1}{2}$ ¢ per pound.

23. 80 lb. rice at 12$\frac{1}{2}$ ¢ per pound.

24. 120 yd. ribbon at 37$\frac{1}{2}$ ¢ per yard.

25. 500 books at 40 ¢ each.

26. 1200 doz. eggs at 25 ¢ per dozen.

27. 600 bu. oats at 33$\frac{1}{3}$ ¢ per bushel.

Find the quantity of each article if a grocer invested:

28. $160 in sugar at 6$\frac{1}{4}$ ¢ per pound.

29. $120 in sugar at 4 ¢ per pound.

30. $6.00 in rice at 10 ¢ per pound.

31. $100 in cloth at 50 ¢ per yard.

32. $13.00 in gingham at 5 ¢ per yard.

33. $3.00 in cheese cloth at 2 ¢ per yard.

SIMPLE ACCOUNTS

OUTLINE FOR SIXTH YEAR WORK IN KEEPING SIMPLE ACCOUNTS

To the Teacher. — 1. Each pupil should be provided with a neatly bound book, suitable for keeping accounts.

2. On the first page of the blank book should be entered the name and age of the pupil, and such other information as may be of value to the teacher and the parents.

3. On the second page of the blank book should be entered the receipts for September; and on the third page, the payments for September.

4. The pupils should be taught that "Balance cash on hand" should be written in red ink, and they should be directed to bring down the balance to the receipts for the next month.

5. They should be directed to enter in their books any earnings or payments or deposits made in savings banks.

6. As neatness and accuracy are the requisites of a good bookkeeper, pupils should be encouraged to make their books present the best possible appearance.

7. Earnings and payments should be entered in a notebook during the school vacation, to be transferred to the proper book at the beginning of the next school term.

A statement of money received and expended is a **cash account.**

There are *two* sides to every account. On the first side are entered all the receipts as well as cash on hand at the beginning; on the second side are entered all payments as well as " Balance, Cash on hand."

The " Balance, Cash on hand " added to the payment must equal the sum of the receipts.

The following is a simple account of a schoolboy for the month of September, 1913.

1913 Sept.		RECEIPTS			1913 Sept.		PAYMENTS		
	1	Cash on hand	10	65		2	Notebook and pencil		20
	4	Errand for Mrs. Long		10		4	Arithmetic		50
	6	Selling papers		25		5	Geography	1	00
	13	Worked one day	1	00		6	Copy-book, ink, and pens		18
	23	Mowing lawns		75		12	History	1	00
	27	Cleaning cellar		75		30	Balance, cash on hand	10	62
			13	50				13	50

Continue the *balance* of each month through the following months to September, 1914. " Balance cash on hand " should be written in red ink.

NOTE TO PARENTS. — Children should be encouraged to keep their own personal accounts.

1. *October.* Oct. 3, Bought 1 pair of shoes, $2.50. 1 hat, $1.50. Oct. 8, Repairs to bicycle, $.75. Oct. 15, Earned $1.50. Oct. 17, Worked for Mr. Black and received $.75. Oct. 25, Saturday outing, $.60.

2. *November.* Nov. 5, Bought a sled, $.95. Nov. 10, Bought a cap, $.75. Nov. 15, Shoveled snow off Mrs. Graham's walk, $.30. Nov. 17, Sawed kindling wood for Mr. Goff, $.50. Nov. 26, Bought a knife, $.25.

3. *December.* Dec. 3, Bought 1 pair of skates, $.75. Dec. 10, Received from Mr. Black for work in store, $1.00. Dec. 17, Expense for school supplies, $.17. Dec. 22, Received from Mrs. Williams for carrying in load of coal, $.30. Dec. 23, Bought Christmas presents, $3.75. Dec. 25, Christmas gift from Uncle James, $1.00. Dec. 29, Expense for having skates sharpened, 10¢.

4. *January,* 1914. Jan. 5, Received from Mrs. Jones for fixing doorbell, $.15. Jan. 8, Bought 1 pair mittens, $.50. Jan. 15, Delivered bills around town for Mr. Black, $.50. Jan. 26, Bought necktie, $.25. Jan. 30, Bought "History of French Revolution," $.75.

5. *February.* Feb. 6, Worked on Saturday for Mr. Black, $.75. Feb. 11, Shoveled snow from sidewalk for Mr. Hart, $.25. Feb. 16, Ran errands, $.40. Feb. 20, Helped unload car of feed, $1.00. Feb. 26, Copied 2 leases for Mr. Irwin, $.75. Feb. 28, Bought pair of gloves, $1.25.

6. *March.* March 2, Cleaned yard for Mrs. Williams, $.50. March 6, Bought 2 pairs of socks, $.30. March 11, Bought new umbrella for mother, $1.75. March 15, Repaired fence for Mr. Jones, $.25. March 27, Car fare, $.30. March 30, Sold my old bicycle for $5.00.

7. *April.* Apr. 1, Burned paper and refuse for Mr. Hart, $.25. Apr. 8, Made garden for Mrs. Black, $.50. Apr. 10, Whitewashed cellar for Mrs. Goff, $.35. Apr. 15, Wheeled load of coal for Mr. Brown, $.35. Apr. 25, Bought 4 collars and 2 pairs of cuffs, $.90.

8. *May.* May 4, Bought straw hat, $1.00. May 7, Mowed lawn for Mrs. Jones, $.25. May 13, Repaired Mr. Brown's sidewalk, $.40. May 29, Bought baseball, $.50. May 31, Received a reward of $5.00 for finding a pocketbook containing $50, which I returned to owner.

9. *June.* June 1, Made $.20 selling papers. June 6, Worked a day for Mr. Black, $.75. June 10, Delivered package, $.25. June 17, Bought ball bat, $.50. June 20, Wheeled a trunk for Mr. Hart, $.25. June 29, Bought 1 pair of baseball shoes, $1.00.

10. *July.* July 4, Fireworks, $.50. July 6, Received from Mr. Black salary for week, $5.00. July 11, Bought 2 shirts, $1.50. July 13, Received week's salary, $5.00. July 15, Bought outing suit, $6.50. July 20, Received my salary, $5.00. July 25, Expense for small articles, $.95. July 27, Received my week's salary, $5.00. July 30, Received for overtime, for month, $7.50.

11. *August.* Aug. 3, Received salary, $5.00. Aug. 8, Bought 1 pair of tan shoes, $2.50. Aug. 10, Received salary, $5.00. Aug. 15, Bought fishing tackle, etc., $3.75. Aug. 17, Received salary, $5.00. Aug. 31, Expenses for 2 weeks' vacation, $15.75.

Sept. 1, Balance, Cash on hand, ——.

Make out a statement at close of year, showing total receipts and disbursements, and proving final balance.

DENOMINATE NUMBERS

The standard or principal units of capacity, length, weight, and time are as follows:

Liquid — **gallon.**   Length or distance — **yard.**
Dry — **bushel.**   Avoirdupois — **pound** (16 oz.).
                                   Time — **day.**

All other measures are determined from the above unit measures. Thus, the ton is 2000 times 1 pound (16 oz.); the hour is $\frac{1}{24}$ of the day, the period of one revolution of the earth on its axis; the quart is $\frac{1}{4}$ of a gallon; the inch is $\frac{1}{36}$ of a yard, etc.

A **denominate number** is a concrete number whose unit is a measure established by custom or law; as, 10 feet, in which 1 foot is the unit of measure; 5 pounds, in which 1 pound is the unit of measure; 12 bushels in which 1 bushel is the unit of measure, etc.

A **simple denominate number** is a number of one denomination: as, 12 rods; 2 ounces; 15 days; 16 pounds; 25 gallons; 16 hundredweight, etc.

A **compound denominate number** is composed of two or more concrete numbers that express one quantity; as, 6 yards, 2 feet, 4 inches (yards, feet, and inches being used to express a single quantity), or 5 bushels, 3 pecks, 4 quarts (bushels, quarts, and pecks being used to express a single quantity).

LIQUID MEASURES

Oral Work

To the Teacher. — A gallon, a quart, and a pint measure should be brought into class. Pupils should measure, and thus learn the relative capacities.

1. Memorize this table:

2 pints	**= 1 quart (qt.)**
4 quarts	**= 1 gallon (gal.)**
31$\frac{1}{2}$ gallons	**= 1 barrel (bbl.)**
63 gallons	**= 1 hogshead (hhd.)**
1 gallon	**= 231 cubic inches**
1 gal. = 4 qt. = 8 pt. = 32 gi.	

2. 1 gal. = —— qt. = —— pt.

3. 4 gal. = —— qt.

4. 5 gal. = —— pt.

5. 6$\frac{1}{2}$ gal. = —— qt.

6. 16 qt. = —— gal.

7. 8 pt. = —— qt.

8. 8 pt. = —— gal.

Written Work

1. Change 3 gal. 3 qt. 1 pt. to pints.

$$3 \text{ gal.} = 3 \times 8 \text{ pt.} = 24 \text{ pt.}$$
$$3 \text{ qt.} = 3 \times 2 \text{ pt.} = 6 \text{ pt.}$$
$$\underline{1 \text{ pt.} \qquad\qquad = 1 \text{ pt.}}$$
$$3 \text{ gal.} \ 3 \text{ qt.} \ 1 \text{ pt.} = 31 \text{ pt.}$$

Since there are 8 pints in 1 gallon, in 3 gallons there are 3 times 8 pints, or 24 pints. Since there are 2 pints in 1 quart, in 3 quarts there are 3 times 2 pints, or 6 pints. 24 pints + 6 pints + 1 pint = 31 pints. Hence 3 gal. 3 qt. 1 pt. = 31 pt.

Change :

2. 6 gal. 1 pt. to pints.   **4.** 4 gal. 3 qt. to pints.

3. 8 gal. 1 qt. 1 pt. to pints.   **5.** 5 gal. 2 qt. 1 pt. to pints.

6. Change $\frac{7}{8}$ gal. to pints.

     SOLUTION. — $\frac{7}{8}$ gal. = $\frac{7}{8}$ of 4 qt., or $3\frac{1}{2}$ qt.

             $\frac{1}{2}$ qt. = $\frac{1}{2}$ of 2 pt., or 1 pt.

             $\frac{7}{8}$ gal. = 3 qt. 1 pt.

Change:

7. $\frac{3}{4}$ gal. to quarts.   **9.** $\frac{3}{4}$ qt. to pints.   **11.** $3\frac{3}{4}$ qt. to pints.

8. $\frac{1}{2}$ gal. to pints.   **10.** $\frac{7}{8}$ qt. to pints.   **12.** $1\frac{7}{8}$ pt. to quarts.

13. Change 127 pt. to gallons, etc.

8 pt.)127 pt.(15, no. gal.
    120

2 pt.)7 pt.(3, no. qt.
      6
      1 pt.

Since 8 pt. = 1 gal., 127 pt. = 15 gal. with 7 pt. remaining. Since 2 pt. = 1 qt., 7 pt. = 3 qt. with 1 pt. remaining.

Change :

14. 376 pt. to gallons.   **21.** 8 qt. 4 pt. to pints.

15. 846 pt. to gallons, etc.   **22.** 15 gal. 5 qt. to quarts.

16. 278 qt. to gallons, etc.   **23.** 469 qt. to gallons, etc.

17. 675 pt. to gallons, etc.   **24.** 84 pt. to quarts, etc.

18. 12 gal. 1 qt. to quarts.   **25.** 64 qt. to gallons.

19. 10 qt. 2 pt. to pints.   **26.** 37 pt. to quarts, etc.

20. 7 gal. 3 qt. to quarts.   **27.** 100 pt. to gallons, etc.

28. Henry sold 100 qt. of fresh cider. How many pints did he sell ?

29. John sold 50 pt. of milk at 8 ¢ a quart. How many gallons and quarts remaining did he sell, and how much did he receive for the milk ?

30. A milkman received 5 gal. of milk in the morning, and made the following sales: 1 pt., 1 qt., 1 gal., 1½ pt., 3 qt., 1½ gal., 1 pt. How much did he sell, and how much milk had he left?

31. Mrs. Smith buys 2 gal. lamp oil. How many days will it last if she uses 1 pt. a day?

32. Find the cost of 2 gal. and 1 pt. of milk at 8 ¢ per quart.

33. How many gallons of milk will a family consume in 75 days, if they use 2 qt. 1 pt. daily?

34. James sold milk as follows: Monday, 1 gal. 1 qt.; Tuesday, 1 gal. 1 pt.; Wednesday, 3 qt. 1 pt.; Thursday, 16 pt.; Friday, 10 qt.; Saturday, 1 gal. 1 pt. How many pints did he sell during the week?

35. A dairyman owns a cow that averages 3 gal. 2 qt. 1 pt. of milk daily. If he sells the milk at $.06 per quart, how much will he realize from the cow during the month of May?

36. A tank in an oil field holds 7500 gal. of crude oil. How many barrels of 42 gal. each does it hold? If the oil is selling for $2.50 per barrel, what is its value?

37. Henry works in his father's sugar camp in the spring, and for his share of the work he is given 24 gal. of maple sirup, which he retails in quart bottles at 30 ¢ per quart. For how much does he sell the maple sirup?

38. What is my April milk bill, if I use 3 pt. every day, at 8 ¢ a quart?

39. A merchant sells linseed oil at 12 ¢ a pint that cost him 56 ¢ a gallon. Find his profits on 45 gal. 3 qt.

40. A milkman has a can of milk containing 10 gal. If he sells 5 qt. to his first customer, 4 qt. to the second customer, 4 gal. to the third customer, and 3 gal. 1 pt. to the fourth, how much has he left in the can?

DRY MEASURES

Oral Work

Dry measures are used in measuring grain, fruit, roots and other dry articles. Name five articles sold by the bushel.

1. Memorize this table:

2 pints (pt.)	= 1 quart (qt.)
8 quarts	= 1 peck (pk.)
4 pecks	= 1 bushel (bu.)
1 bushel	= 2150.42 cubic inches
1 bu. = 4 pk.	= 32 qt. = 64 pt.

2. 1 bu. = —— pt. = —— qt.

3. 5 bu. = — pt.   **6.** 16 pt. = — bu.   **9.** 32 qt. = — pt.

4. 4 bu. = — qt.   **7.** 64 qt. = — bu.   **10.** 14 pt. = — qt.

5. 6 pt. = — qt.   **8.** 64 pt. = — bu.   **11.** 48 pt. = — qt.

Written Work

Change:

1. 7 bu. 3 pecks to pints.   **9.** $1\frac{1}{2}$ bu. to pecks.

2. 3 pk. 3 qt. to quarts.   **10.** $3\frac{1}{2}$ pk. to quarts.

3. 5 bu. 1 qt. to quarts.   **11.** 18 pt. to quarts.

4. 4 bu. 3 qt. to quarts.   **12.** 128 pt. to bushels.

5. 10 pk. 3 qt. to quarts.   **13.** 64 pt. to pecks.

6. 14 bu. 4 pk. to pecks.   **14.** 32 qt. to bushels.

7. $\frac{7}{8}$ bu. to pecks.   **15.** 16 qt. to pecks.

8. $8\frac{3}{4}$ pk. to quarts.   **16.** 16 pk. to bushels.

17. James picked 20 qt. of cranberries. How many pecks did he pick? How many pints did he pick? He sold the cranberries at 8¢ per quart. How much did he receive for them?

18. John picked strawberries as follows: Monday, 16 qt.; Tuesday, 3 pk.; Wednesday, 2 pk. 3 qt.; Thursday, 1 bu.; Friday, 1 bu. 3 qt.; Saturday, 27 qt. Find the number of bushels and quarts he picked during the week.

19. James had a chestnut tree in the back yard. He gathered in all 1 bu. 3 qt. of chestnuts from the tree. How many pints did he gather?

20. The grocer sold to three customers as follows: to the first, 3 pk. of apples; to the second, ¼ pk. of apples; and to the third, 1 bu. of apples. Find the number of bushels and quarter pecks he sold in all.

21. Mary's mother gave her the apples on a certain tree to sell. Mary sold at six different times: 1 pk.; ½ pk.; ¼ pk.; ¼ pk.; 1¼ pk.; 1 bu. How many bushels and pecks of apples did she sell?

22. How many bushels of potatoes are necessary to plant 8⅝ acres, allowing 6 bu. 1 pk. to the acre?

23. Walter picked 4¼ bushels of blackberries and sold them to a grocer for 6¢ a quart. He took in exchange eggs at 24¢ a dozen. How many dozen did he receive?

MEASURES OF WEIGHT

To the Teacher. — Secure a scale and weights, and have pupils weigh articles of different kinds.

Avoirdupois weight is used in weighing heavy articles; as groceries, coal, grain, and metals, except gold and silver.

The long ton of 2240 lb. is used at the United States custom houses, and in wholesale transactions in coal and iron.

Oral Work

1. Memorize this table:

16 ounces (oz.)	**= 1 pound (lb.)**
100 pounds	**= 1 hundredweight (cwt.)**
20 hundredweight ⎫	**= 1 ton (T.)**
2000 pounds ⎭	
2240 lb.	**= 1 long ton**
1 T. = 20 cwt.	**= 2000 lb. = 32,000 oz.**

The avoirdupois pound contains 7000 grains, and the avoirdupois ounce, 437½ grains.

The unit of avoirdupois weight is the **pound.**

2. 1 T. = —— cwt. = —— lb. = —— oz.

3. 3 T. = — lb. **6.** 5 lb. = — oz. **9.** 64 oz. = — lb.

4. 32 oz. = — lb. **7.** 2 T. = — lb. **10.** 80 cwt. = — T.

5. 4 cwt. = — oz. **8.** 10 lb. = — oz. **11.** 400 lb. = — cwt.

NOTE. — A short method of estimating prices of hay is so much per pound. Hay at $10 per ton is ½ ⊄ a pound. Hay at $20 per ton is 1 ⊄ a pound, and hay at $30 per ton is 1½ ⊄ a pound.

Estimate mentally the cost of the following quantities of hay at $20 a ton:

12. 3 cwt. **16.** 1 T. 6 cwt. **20.** 950 lb.

13. 650 lb. **17.** 1275 cwt. 75 lb. **21.** 5 cwt.

14. 8 bales, 90 lb. each **18.** 2 bales, 70 lb. each **22.** 7 cwt.

15. 6 bales, 80 lb. each **19.** 540 lb. **23.** 360 lb.

Estimate the cost of the following at $30 a ton:

24. 6 bales, 90 lb. each. **27.** 1787 lb. **30.** 72 cwt.

25. 1½ T. **28.** 450 lb. **31.** 1750 lb.

26. 34 cwt. **29.** 54 cwt. **32.** 1570 lb.

Written Work

Change:

1. 6 lb. 5 oz. to ounces.

2. 3 T. 8 cwt. to pounds.

3. 4 cwt. 3 lb. to ounces.

4. $5\frac{1}{2}$ cwt. to ounces.

5. 3600 lb. to hundredweight.

6. 15 lb. 8 oz. to ounces.

7. 18 cwt. 25 lb. to pounds.

8. 2 T. 5 lb. to ounces.

9. 544 oz. to pounds.

10. 6000 lb. to tons.

11. 128 oz. to pounds.

12. 810 oz. to pounds, etc

13. $4\frac{4}{5}$ T. to pounds.

14. 16 T. 15 cwt. to pounds.

15. Change $\frac{5}{16}$ of a ton to hundredweight and pounds.

SOLUTION.—$\frac{5}{16}$ T. = $\frac{5}{16}$ of 20 cwt. = $6\frac{1}{4}$ cwt.

$\frac{1}{4}$ cwt. = $\frac{1}{4}$ of 100 lb. = 25 lb.

$\frac{5}{16}$ of a ton = 6 cwt. 25 lb.

Change:

16. $\frac{1}{2}$ cwt. to pounds.

17. $\frac{7}{8}$ T. to pounds.

18. $\frac{3}{4}$ lb. to ounces.

19. $\frac{7}{16}$ T. to cwt. and pounds.

20. At 65¢ per pound, how much will 10 oz. of tea cost?

21. 5 lb. of candy are how many ounces of candy?

Find the cost of the following amounts of hard coal at $6.00 a ton:

22. 16 cwt.

23. 10 cwt.

24. 30 cwt.

25. 40 cwt.

26. $\frac{3}{4}$ T.

27. $\frac{1}{4}$ T.

Potatoes, onions, and similar articles are frequently sold at so much per hundredweight.

28. Find the cost of 250 lb. of potatoes at $1.25 a hundredweight; of one bag of onions, containing 75 lb., at 90¢ a hundredweight.

29. Mr. Frank raised 1275 bu. of potatoes, and sold $33\frac{1}{3}\%$ of them at $40\cancel{c}$ a bushel, 10% of the remainder at $50\cancel{c}$ a bushel, and the remainder at $60\cancel{c}$ a bushel. For how much did he sell his potatoes?

30. Roy's father sold five loads of hay at the following prices: 750 lb. at $10 a ton; 2780 lb. at $15 a ton; 3890 lb. at $18 a ton. Estimate in the shortest possible way the total amount of the sales.

31. 27 cwt. equal how many tons and hundredweight over?

32. A wagon loaded with potatoes weighs 4200 lb. If the wagon weighs 1200 lb., how many tons and hundredweight of potatoes are there?

33. A wagon loaded with anthracite coal weighs 4940 lb. How many long tons and pounds are there, if a long ton of anthracite coal weighs 2240 lb.?

34. A man sells 30 bags of bran, each weighing 100 lb. How many tons does he sell?

35. A grain dealer buys 50 sacks of grain, averaging 90 lb. to a sack. How many hundredweight of grain does he buy? How many tons and hundredweight over does he buy?

36. Mary goes to the store and buys $3\frac{3}{4}$ lb. of butter. How many ounces of butter should she get?

37. A butcher sold $30\frac{1}{4}$ lb. of lard at $.12 a pound, and purchased with the money flour at $.03 a pound. How much flour did he buy?

38. 50 cwt. of starch was packed into boxes, each containing 10 lb. How much was received, if each box was sold for $12\cancel{c}$?

39. A Kentucky farmer clipped $241\frac{1}{2}$ lb. of mohair from 70 Angora goats. Find the average clip from each goat and its value at $.37\frac{1}{2}$ per pound.

WEIGHTS AT WHICH COMMON FOODS AND SEEDS ARE
USUALLY SOLD

	PER BUSHEL		PER BUSHEL
Wheat	60 lb.	Onions	50 lb.
Buckwheat	48 lb.	Potatoes	60 lb.
Oats	32 lb.	Sweet Potatoes	55 lb.
White Beans	60 lb.	Corn, in ear	70 lb.
Turnips	55 lb.	Corn, shelled	56 lb.
Blue Grass Seed	44 lb.	Rye	56 lb.
Clover Seed	60 lb.	Timothy Seed	44 lb.
Barley	48 lb.	Winter Apples	48 lb.

Oral and Written Work

Find the weight of:

1. ½ pk. of white potatoes.
2. ½ pk. of sweet potatoes.
3. ½ pk. of shelled corn.
4. ¼ pk. of onions.
5. 1 pk. of clover seed.
6. ¼ pk. of oats.
7. ½ pk. of barley.
8. ¼ pk. of Timothy seed.
9. ¾ pk. of white beans.
10. ¼ pk. of rye.
11. ½ pk. of winter apples.
12. ½ pk. of corn in ear.
13. 1 pk. of winter apples.
14. ¼ pk. of blue grass seed.
15. 8 pk. of clover seed.
16. 16 pk. of white beans.
17. 8 pk. of barley.
18. 1 pk. of shelled corn.

19. John took a load of potatoes to the market, and he found that the wagon and potatoes weighed 3000 lb. If the wagon weighed 1200 lb., how many bushels of potatoes did he have?

20. Mr. Long threshed 37 bu. per acre of buckwheat on 3¾ acres. What was the weight of the buckwheat? About how many loads of 1600 lb. each did he have?

21. A Dakota farmer had 60 acres in barley. He threshed 29 bushels to the acre. About how many loads of 3000 lb. each did he have to take to market?

22. Mr. Jackson of Moorhead, Minn., raised 260 bu. of potatoes to the acre on 40 acres. How many pounds of potatoes had he? If 80,000 lb. fill a car, how many car loads had he? (Call a fraction a full load.)

23. Estimate the number of bushels of shelled corn, corn in the ear, oats, wheat, or rye, that a car having 60,000 lb. capacity will hold.

24. Make problems from the above table relating to experiences in the home or on the farm.

MEASURES OF LENGTH

Oral Work

A tapeline 50 to 100 feet in length, marked in feet and tenths and hundredths of a foot, is commonly used for measuring short distances.

1. With a tapeline 50 ft. in length, measure the distance in rods and feet around a square or a field.

2. 20 city blocks, each 16 rods in length, are 320 rods long. This is called **one mile.** **1 mile = 320 rods.**

3. $320 \times 16\frac{1}{2}$ ft. = —— feet. (Why do we multiply $16\frac{1}{2}$ ft. by 320?)

4. 5280 ft. ÷ 3 = —— yards. (Why do we divide 5280 ft. by 3?)

5. Memorize this table :

12 inches (in.)	= 1 foot (ft.)
3 feet	= 1 yard (yd.)
5½ yards, or 16½ feet	= 1 rod (rd.)
320 rods	= 1 mile (mi.)
1760 yards = 1 mile	5280 feet = 1 mile

6. 1 mi. = —— rd. = —— yd. = —— ft. = —— in.

7. 3 ft. = — in.   **10.** 36 in. = — ft.   **13.** 640 rd. = — mi.

8. 4 rd. = — ft.   **11.** 12 ft. = — yd.   **14.** 4 ft.   = — in.

9. 2 mi. = — rd.   **12.** 33 ft. = — rd.   **15.** 5 yd.   = — ft.

Written Work

Change :

1. 1 rd. 2 ft. to feet.
2. 5 yd. 1½ ft. to feet.
3. 10 ft. 6 in. to inches.
4. 12 rd. 3 yd. to yards.
5. 4 rd. 3 ft. to feet.
6. 5⅛ yd. to feet.
7. 5 yd. 2 ft. to feet.
8. 1 rd. 6 ft. to feet.
9. 1 yd. 7 in. to inches.
10. 6 ft. 5 in. to inches.
11. 5 mi. 4 rd. to rods.
12. ¾ mi. to rods.
13. 10 rd. 5 yd. to yards.
14. 15 ft. 9 in. to inches.
15. 4 yd. 2½ ft. to feet.
16. 5 rd. 5 in. to inches.
17. 10 yd. 4 ft. to feet.
18. 25 ft. 17 in. to inches.
19. ½ yd. to inches.
20. 59 in. to feet, etc.
21. 145 in. to yards, etc.
22. 17 ft. to yards, etc.
23. 127 ft. to rods and feet.
24. 59 yd. to rods and yards.
25. 63 yd. to rods and yards.
26. 520 rd. to miles and rods.

27. 2100 yd. to miles and yards.

28. 1820 yd. to miles and yards.

29. In an automobile race the fastest machine ran ⅞ mi. in 36 sec. Find the number of feet it ran per second.

30. James lives 1.85 mi. from school. Find the number of feet he walks to school.

31. A bicycle wheel is 7 ft. 4 in. in circumference. How many turns will it make in going 6 mi. ?

32. How much fence will be needed to inclose a square field, each side of which is 22 rd. ?

33. How much are 5½ mi. of telegraph wire worth at $.005 a foot?

34. At 2 cents a foot find the length, in miles and rods, of a telephone wire that costs $4672.80.

MEASURES OF TIME

Oral Work

1. Memorize this table :

60 seconds (sec.) = 1 minute (min.)
60 minutes = 1 hour (hr.)
24 hours = 1 day (da.)
7 days = 1 week (wk.)
365 days = 1 year (yr.)

April, June, September, and November have each 30 days. All the others except February have 31 days each. February usually has 28 days. A year that has 366 days is called a **leap year.** In leap year February has 29 days.

NOTE. — Centennial years divisible by 400, and other years divisible by 4, are leap years. Thus, 1600 and 1912 were leap years, but not 1900.

2. Memorize this rime:

Change:

Thirty days have September,
April, June, and November.
All the rest have thirty-one,
Save February, which alone
Has twenty-eight; and one day more
We add to it one year in four.

3. 3 min. to seconds.

4. 6 da. to hours.

5. 7 hr. to minutes.

6. 3 da. 6 hr. to hours.

7. 10 wk. 6 da. to days.

8. How many days are there in January, February, and March? in May, June, and July? in August, September, and October?

9. October 1, 1913, falls on a Wednesday. On what day does November 1, 1913, fall?

10. At \$720 a year, find the rent of a cottage for 3 months.

11. 1 yr. = —— mo. = —— da. = —— hr. = —— mi.

12. 5 min. = —— sec.

13. 6 hr. = —— min.

14. 4 da = —— wk.

15. 120 sec. = —— min.

16. 48 hr. = —— da.

17. 2 yr. = —— da.

18. 5 wks. = —— da.

19. 2 da. = —— hr.

Written Work

Change:

1. 3 hr. 6 min. to minutes.

2. 5 da. 3 hr. to hours.

3. 12 wk. 6 da. to days.

4. 30 min. 20 sec. to seconds.

5. ¾ wk. to days.

6. 3 hr. 10 min. to minutes.

7. 3 da. 7 hr. to hours.

8. 5 da. 4 hr. to hours.

9. ¼ hr. to minutes.

10. 150 sec. to minutes, etc.

11. 30 da. to weeks, etc.

12. 72 mo. to years.

13. 50 hr. to days, etc.

14. ⅝ min. to seconds.

15. Make a calendar for November, 1913, abbreviating the days of the week.

16. A train goes 104 miles in 3 hr. 15 min. What is its rate per hour?

17. A machinist works 10 hr. per day in summer and 8⅘ hr. per day in winter. ' If his wages in summer are $3.35 per day, at the same rate find his wages per day in winter.

18. If a watch gains 18 seconds in a day, how much too fast will it be in three weeks?

19. An automobile runs 2⅞ mi. in 5 min. At that rate, find the distance in miles and rods it runs in 1 hr. 35 min.

20. Mary walks ¾ mi. to school each day. How many miles does she walk in going to and from school in 180 da.?

21. Henry walks 80 % of the distance Mary walks each day. How far does Henry walk in a term of 160 da.?

22. At 20 cents an hour, how much will a man earn in 26 da., working each day from 8 A.M. to 5 P.M., allowing 1 hr. for lunch?

23. If a flour mill grinds wheat at the rate of 1 pt. in 5 sec., in how many hours will it grind 21,600 bu.?

24. An 11-hour train from Newark to Pittsburgh, a distance of 480 miles, starts 1 hr. late. How much must it increase its speed to arrive on time?

On schedule time it averaged (480 mi. ÷ 11) per hour; when late, (480 mi. ÷ 10) per hour.

25. A train running 1320 mi. in 24 hr. starts ½ hr. late. How much must it increase its speed to arrive on time?

COUNTING TABLE

Oral and Written Work

1. Memorize this table.

12 things	= 1 dozen
12 dozen	= 1 gross
24 sheets paper	= 1 quire
20 quires	= 1 ream
20 things	= 1 score

2. How much will 1 gross of buttons cost at 15¢ a dozen?

3. A stationer bought 12 quires of paper for $1.20 and sold it at 1¢ a sheet. How much did he gain?

4. Find the cost of 40 lemons at 15¢ a dozen.

5. If a dozen oranges cost 40¢, how much does 1 cost?

6. The teacher gets a box of pens containing one gross. How many dozen does she get? How much are they worth at 5¢ a dozen?

7. A dealer buys pens at 80¢ a gross, and sells them at 1¢ apiece. Find his profit per dozen.

8. Mr. Jones is 4-score and 6 years old. How old is he?

9. A merchant buys eggs at 35¢ a dozen, and retails them at 48¢ a dozen. What is the gain on each egg?

10. How much will 1½ doz. tablets cost at 12¢ apiece? If the dealer sells them at 15¢ apiece, what is his profit?

11. How much must be paid for 15 gross of lead pencils at 35¢ a dozen?

ADDITION AND SUBTRACTION OF DENOMINATE MEASURES

To THE TEACHER. — Short lengths measured with a foot ruler and a yardstick are always expressed in feet and inches, or yards and feet. For this reason, pupils need much drill in measures that can be expressed in one denomination. Thus, 1 ft. 5 in. $= 1\frac{5}{12}$ ft.; 1 yd. 2 ft. $= 1\frac{2}{3}$ yd.

Written Work

1. Find the sum of 1 ft. 3 in. $+$ 2 ft. 1 in. $+$ 7 ft. 8 in.

$$1\frac{1}{4} \text{ ft.} + 2\frac{1}{12} \text{ ft.} + 7\frac{2}{3} \text{ ft.} = 11 \text{ ft.}$$

2. John is 5 ft. 6 in. and Mary 4 ft. 11 in. in height. Find the difference in their heights.

$$5\frac{1}{2} \text{ ft.} - 4\frac{11}{12} \text{ ft.} = \frac{7}{12} \text{ ft., or 7 in.}$$

Perform the operation indicated, and test by actual measurements with a foot ruler or a yardstick.

3. 5 ft. 9 in. $+$ 4 ft. 10 in.

4. 3 ft. 1 in. $-$ 1 ft. 3 in.

5. 6 ft. 1 in. $+$ 4 ft. 9 in.

6. $1\frac{2}{3}$ yd. $-$ 4 ft. 6 in.

7. $\frac{2}{3}$ yd. $+$ 1 ft. 9 in.

8. $5\frac{1}{2}$ yd. $-$ 10 ft. 4 in.

9. $3\frac{7}{12}$ ft. $+$ 10 in.

10. 1 yd. 1 ft. $-$ $\frac{2}{3}$ yd.

11. 6 ft. 9 in. $+$ $\frac{11}{12}$ ft.

12. 5 ft. 1 in. $-$ 9 in.

13. 6 yd. 2 ft. $+$ $\frac{1}{3}$ yd.

14. 1 ft. 8 in. $-$ 10 in.

15. 3 yd. 3 in. $+$ 1 yd. 8 in.

16. 2 yd. 4 in. $-$ 1 yd. 3 in.

17. 2 yd. 1 in. $+$ 1 yd. 4 in.

18. 2 yd. 6 in. $-$ 24 in.

19. $5\frac{1}{2}$ yd. $-$ 6 in.

20. $1\frac{7}{12}$ ft. $-$ 11 in.

21. $2\frac{3}{4}$ ft. $+$ 10 in.

22. 18 in. $-$ $1\frac{1}{4}$ ft.

23. 40 in. $+$ $\frac{2}{3}$ yd.

24. 60 in. $-$ $4\frac{2}{3}$ ft.

25. $\frac{2}{3}$ yd. $+$ 19 in.

26. $3\frac{3}{4}$ ft. $-$ 1 ft. 11 in.

27. 24 in. $+$ $\frac{1}{8}$ yd.

28. 29 in. $-$ $\frac{3}{4}$ yd.

29. 30 ft. $+$ $29\frac{1}{4}$ in.

30. $29\frac{3}{4}$ in. $-$ $\frac{2}{3}$ yd.

31. $2\frac{3}{8}$ ft. $+$ 7 in.

32. 3 ft. 7 in. $-$ 2.3 ft.

DIFFERENCE IN TIME

Finding the **difference in time** between two dates is the most practical application of **subtraction of compound denominate numbers.**

Written Work

1. Find the difference in time between November 15, 1911, and August 12, 1913.

yr.	mo.	da.
1913	8	12
1911	11	15
1	8	27

Aug. 12, 1913, is represented as the 12th day of the 8th month of 1913, and Nov. 15, 1911, as the 15th day of the 11th month of 1911.

1 mo., or 30 da., + 12 da. = 42 da.; 42 da. − 15 da. = 27 da.; 1 yr., or 12 mo., + 7 mo. = 19 mo.; 19 mo. − 11 mo. = 8 mo.; 1912 yr. − 1911 yr. = 1 yr.

Subtract:

	yr.	mo.	da.			yr.	mo.	da.
2.	1908	7	12		**3.**	1905	9	1
	1901	9	15			1890	8	15

4. General Robert E. Lee was born January 19, 1807, and General Ulysses S. Grant, April 27, 1822. How old was each at the close of the Civil War, April 9, 1865? How much older was General Lee than General Grant?

5. President Woodrow Wilson was born at Staunton, Va., Sept. 28, 1856. How old was he at the date of his inauguration, March 4, 1913?

6. Ex-Vice President Garret A. Hobart was born in Long Branch, N.J., June 3, 1844. How old was he at his death, Nov. 21, 1899?

7. Ex-President Theodore Roosevelt was born in New York City, Oct. 27, 1858. How old was he when he took the oath of office as President of the United States, Sept. 14, 1901?

PERCENTAGE

Per cent means by the hundred or **hundredths.** The sign for per cent is %.

We may express the per cent of a number either as a *common fraction* or as a *decimal*.

Thus, $6\% = \frac{6}{100} = .06$; 6 % of 500 means $\frac{6}{100}$ of 500, which equals 30. Or, .06 of 500 = 30.

2 % of a number means $\frac{2}{100}$, or .02, of the number.

25 % of a number means $\frac{25}{100}$, or .25, of the number.

Oral Work

1. What term in common fractions corresponds to the number before the sign % ? to the sign % ?

2. In the following expressions, what represents the *numerator?* the *denominator?*

1 %	20 %	40 %	90 %
6 %	30 %	75 %	100 %

3. Find 6 % of 100.

SOLUTION. — $\frac{6}{100}$ of 100 = 6.   Or, .06 × 100 = 6.

Find :

4. 5 % of 100	**10.** 10 % of 100	**16.** 8 % of 75
5. .05 of 100	**11.** 25 % of 100	**17.** .08 of 75
6. $\frac{5}{100}$ of 100	**12.** .25 of 400	**18.** $\frac{8}{100}$ of 75
7. 6 % of 150	**13.** 3 % of 60	**19.** $33\frac{1}{3}\%$ of 300
8. .06 of 150	**14.** .03 of 60	**20.** $\frac{33\frac{1}{3}}{100}$ of 300
9. .10 of 100	**15.** $\frac{3}{100}$ of 60	**21.** $33\frac{1}{3}\%$ of 300

247

Changing per cents to equivalents.

Oral Work

Since $5\% = .05 = \frac{5}{100} = \frac{1}{20}$, these expressions are called **equivalents.**

1. Give the fractional and decimal equivalents of 10%; 6%; 4%; 20%; 25%.

Read the following equivalents:

2. $\frac{20}{100}$, 20%, $.20$, $\frac{1}{5}$ **5.** $\frac{37\frac{1}{2}}{100}$, $37\frac{1}{2}\%$, $.37\frac{1}{2}$, $\frac{3}{8}$

3. $\frac{12\frac{1}{2}}{100}$, $12\frac{1}{2}\%$, $.12\frac{1}{2}$, $\frac{1}{8}$ **6.** $\frac{80}{100}$, 80%, $.80$, $\frac{4}{5}$

4. $\frac{40}{100}$, 40%, $.40$, $\frac{2}{5}$ **7.** $\frac{87\frac{1}{2}}{100}$, $87\frac{1}{2}\%$, $.87\frac{1}{2}$, $\frac{7}{8}$

8. Change the fractions $\frac{1}{5}$, $\frac{2}{5}$, $\frac{3}{5}$, $\frac{4}{5}$ to hundredths and per cents; also $\frac{1}{8}$, $\frac{3}{8}$, $\frac{5}{8}$, $\frac{7}{8}$.

$\frac{1}{5} = 5)\overline{1.00}$ $\frac{1}{8} = 8)\overline{1.00}$

   .20, or 20%   .12$\frac{1}{2}$, or 12$\frac{1}{2}$%

$\frac{2}{5} = 2 \times .20 = .40$, or 40% $\frac{3}{8} = 3 \times .12\frac{1}{2} = .37\frac{1}{2}$, or 37$\frac{1}{2}$%

$\frac{3}{5} = 8 \times .20 = .60$, or 60% $\frac{5}{8} = 5 \times .12\frac{1}{2} = .62\frac{1}{2}$, or 62$\frac{1}{2}$%

$\frac{4}{5} = 4 \times .20 = .80$, or 80% $\frac{7}{8} = 7 \times .12\frac{1}{2} = .87\frac{1}{2}$, or 87$\frac{1}{2}$%

Change to hundredths and per cents:

9. $\frac{1}{2}$	**13.** $\frac{3}{4}$	**17.** $\frac{8}{10}$	**21.** $\frac{2}{5}$	**25.** $\frac{3}{8}$
10. $\frac{1}{5}$	**14.** $\frac{1}{8}$	**18.** $\frac{7}{10}$	**22.** $\frac{3}{5}$	**26.** $\frac{5}{8}$
11. $\frac{3}{5}$	**15.** $\frac{5}{8}$	**19.** $\frac{9}{10}$	**23.** $\frac{4}{5}$	**27.** $\frac{7}{8}$
12. $\frac{1}{4}$	**16.** $\frac{1}{10}$	**20.** $\frac{4}{5}$	**24.** $\frac{1}{8}$	**28.** $\frac{1}{16}$

Give the products rapidly:

29. $2 \times .33\frac{1}{3}$	**32.** $5 \times .12\frac{1}{2}$	**35.** $4 \times .12\frac{1}{2}$	**38.** $4 \times .04\frac{1}{4}$
30. $5 \times .16\frac{2}{3}$	**33.** $7 \times .12\frac{1}{2}$	**36.** $6 \times .12\frac{1}{2}$	**39.** $3 \times .16\frac{2}{3}$
31. $3 \times .12\frac{1}{2}$	**34.** $3 \times .8\frac{1}{3}$	**37.** $2 \times .15$	**40.** $4 \times .16\frac{2}{3}$

41. Memorize the following table:

$$\frac{1}{2} = 50\% \qquad \frac{1}{5} = 20\% \qquad \frac{5}{6} = 83\frac{1}{3}\% \qquad \frac{1}{12} = 8\frac{1}{3}\%$$
$$\frac{1}{3} = 33\frac{1}{3}\% \qquad \frac{2}{5} = 40\% \qquad \frac{1}{8} = 12\frac{1}{2}\% \qquad \frac{5}{12} = 41\frac{2}{3}\%$$
$$\frac{2}{3} = 66\frac{2}{3}\% \qquad \frac{3}{5} = 60\% \qquad \frac{3}{8} = 37\frac{1}{2}\% \qquad \frac{1}{16} = 6\frac{1}{4}\%$$
$$\frac{1}{4} = 25\% \qquad \frac{4}{5} = 80\% \qquad \frac{5}{8} = 62\frac{1}{2}\% \qquad \frac{1}{20} = 5\%$$
$$\frac{3}{4} = 75\% \qquad \frac{1}{6} = 16\frac{2}{3}\% \qquad \frac{7}{8} = 87\frac{1}{2}\% \qquad \frac{1}{25} = 4\%$$

Name rapidly the fractional equivalents, in lowest terms, of the following per cents:

42. 50%	**47.** 20%	**52.** $37\frac{1}{2}\%$	**57.** 90%
43. $33\frac{1}{3}\%$	**48.** 40%	**53.** $62\frac{1}{2}\%$	**58.** $12\frac{1}{2}\%$
44. $66\frac{2}{3}\%$	**49.** 60%	**54.** $87\frac{1}{2}\%$	**59.** $16\frac{2}{3}\%$
45. 25%	**50.** 16%	**55.** 10%	**60.** 80%
46. 75%	**51.** $83\frac{1}{3}\%$	**56.** 30%	**61.** 70%

Write the equivalents of the following in hundredths. Thus, $1\% = .01$; $32\% = .32$; $\frac{1}{2}\% = .00\frac{1}{2}$; etc.

62. 1%	**68.** $\frac{5}{6}\%$	**74.** 50%	**80.** 13%
63. 32%	**69.** 3%	**75.** $\frac{2}{3}\%$	**81.** $13\frac{1}{3}\%$
64. $\frac{1}{2}\%$	**70.** 11%	**76.** 6%	**82.** $\frac{5}{8}\%$
65. 2%	**71.** $\frac{1}{4}\%$	**77.** $\frac{3}{4}\%$	**83.** 100%
66. $16\frac{1}{3}\%$	**72.** 4%	**78.** 7%	**84.** 123%
67. $\frac{1}{8}\%$	**73.** $43\frac{1}{2}\%$	**79.** $\frac{7}{8}\%$	**85.** 127%

86. May received 80 % in arithmetic, 100 % in spelling, and 90 % in grammar. What was her average per cent?

87. The attendance was 80 % of the enrollment. What fractional part of the enrollment was it?

Finding a given per cent of a number.

Oral Work

Find results decimally:

NOTE.— Think of $66\frac{2}{3}\%$ as $\frac{2}{3}$; of $83\frac{1}{3}\%$ as $\frac{5}{6}$, etc.

1. $66\frac{2}{3}\%$ of 18.
2. $33\frac{1}{3}\%$ of 90.
3. 50% of \$500.
4. 25% of \$2000.
5. 75% of 16 inches.
6. 20% of 100 yards.
7. 40% of 60 feet.
8. 60% of 40 miles.
9. 80% of 75 gallons.
10. $16\frac{2}{3}\%$ of \$6000.
11. $83\frac{1}{3}\%$ of \$1200:

12. $37\frac{1}{2}\%$ of \$7200.
13. $12\frac{1}{2}\%$ of \$6400.
14. 75% of \$4800.
15. $66\frac{2}{3}\%$ of \$999.
16. 80% of 60 sheep.
17. 60% of 75 horses.
18. 40% of 90 miles.
19. $87\frac{1}{2}\%$ of \$160.
20. $62\frac{1}{2}\%$ of \$240.
21. $37\frac{1}{2}\%$ of \$880.
22. $12\frac{1}{2}\%$ of 24.

23. John earns \$21.60 per month, and spends 75% of his earnings for clothes. How much do his clothes cost him?

24. Mary earns \$50 per month as a stenographer and pays 40% for board and room. How much do her board and room cost her?

25. A boy spends \$8 for an overcoat and $37\frac{1}{2}\%$ of that sum for shoes. How much does he spend for shoes?

26. John earns \$50 during his vacation, and Margaret 25% as much as John. How much does Margaret earn?

27. What is $33\frac{1}{3}\%$ of 24? of \$4.80? of \$66.90?

28. A man's salary is \$150 per month. He spends 40% of it for clothing and other expenses. How much does he save?

29. John sells \$20 worth of berries for Mr. Smith and gets 10% for selling. How much does he receive?

30. There are 1000 pupils in school, and 40% are males. How many are males?

31. If a man buys a horse for $150 and sells it at a profit of 20%, how much does he gain?

The **base** is that number of which some per cent is to be taken; as, 5% of $200 (*base*).

The **rate** is the number of hundredths taken; as, 5% (*rate*) of 80 horses; that is, $\frac{5}{100}$ of 80 horses.

The **percentage** of a number is the result obtained by taking any per cent of it; as, 10% of 200 acres is $\frac{10}{100}$ of 200 acres, or 20 acres (*percentage*).

In problems involving percentage, discard less than $\frac{1}{2}$ cent in the answer and count $\frac{1}{2}$ cent or a greater fraction of a cent as an extra cent.

Written Work

1. What is 75% of $5.12?

MULTIPLIER	MULTIPLICAND	PRODUCT
Rate	Base	Percentage
75% of	$5.12 =	( )

Decimal Method

75% = .75.

$5.12, base

.75, rate

———

2560

3584

———

$3.8400, percentage

Fractional Method

75% = $\frac{3}{4}$

$\frac{3}{4}$ of $5.12 = $3.84

Study of Problem

What is the base? $5.12. What is the rate? 75%.

To what do the *base* and *rate* correspond in simple multiplication? *Multiplicand* and *multiplier*.

To what does *percentage* correspond in simple multiplication? *Product*.

How is the *product* found in simple multiplication? *Multiplier* × *multiplicand*.

How is the *percentage* found? *Rate* × *base*.

The **percentage of a number** *equals the product of the base by the rate.*

Find, using a short method when possible:

2. 16 % of $200.

3. 33⅓ % of 11 months.

4. 60 % of 30 days.

5. 12 % of 150 acres.

6. 15 % of 600 acres.

7. 18 % of 400.

8. 27 % of 400 horses.

9. 3⅓ % of 99.

10. 18 % of 150 lb.

11. 1½ % of $75.

12. A house costs $2500, and the damage by fire is 8 %. Find the amount of the damage.

13. John owed his tailor $80, and paid 37½ % of the debt. How much does he still owe?

14. Mary spells 90 % of 80 words correctly. How many does she miss?

15. A boy buys apples at $1 per bushel, and sells them at a profit of 20 %. How much profit is that per bushel?

16. 6¼ % of 3680 equals what number?

17. A man buys a farm for $2500, and sells it for 25 % more than it cost him. For how much does he sell the farm?

18. A man earns $180 per month, and puts 33⅓ % of it in the savings bank. What is his deposit each month?

19. If 37½ % of a man's farm is in timber, and the total area is 240 acres, how much timber land has he?

20. A teacher who earned $1200 a year spent 66⅔ % of her salary. How much did she save?

21. Mr. Scott's horse is valued at $250, and Mr. Hill's at 60 % of this. What is the value of Mr. Hill's horse?

22. The population of a town of 9672 inhabitants increased 12½ % in a year. What was the increase in population?

Finding what per cent one number is of another number.

Oral Work

1. What part of 8 is 4 ? What % of 8 is 4 ?

SOLUTION.—4 is ½ of 8, or 50 % of 8.

What per cent of :

2. 10 is 5?	**6.** 30 is 10?	**10.** 40 is 20?
3. 12 is 6?	**7.** 15 is 5?	**11.** 30 is 20?
4. 20 is 5?	**8.** 12 is 10?	**12.** 12½ is 6¼?
5. 40 is 8?	**9.** 20 is 10?	**13.** 27½ is 5½?

Written Work

1. What per cent of 80 is 12 ?

$12 \div 80 = \frac{12}{80} = \frac{3}{20}$, or 15 %
 Test : .15 × 80 = 12

$12 \div 80 = \frac{12}{80} = .15$, or 15 %

To **find what per cent one number is of another,** *make the number of which the per cent is to be found the numerator, and the other number the denominator, of a fraction, and express the fraction as a per cent.*

What per cent of :

2. 50 is 16?	**5.** 200 is 65?	**8.** 150 is 75?
3. 80 is 24?	**6.** 125 is 45?	**9.** 160 is 50?
4. 60 is 18?	**7.** 240 is 90?	**10.** 320 is 25.6?

11. Frank's salary for a month is $75 and Samuel's is $25. What per cent of Frank's salary is Samuel's?

12. I earn $600 a year and spend $300. What per cent of my money do I spend?

13. A farmer who raised 350 bu. of potatoes marketed only 280 bu. What per cent of his potatoes did he market?

14. If I spell 20 words out of 25 words correctly, what per cent do I get on the test?

15. What per cent of a quart is a pint?

16. What per cent of a yard is a foot?

17. What per cent of a gallon is a quart? a pint?

18. What per cent of a gross is a dozen?

19. What per cent of a pound is an ounce?

20. What per cent of a foot is an inch?

21. Tom earned 20 ¢ on Saturday, and Joe earned 50 ¢. What fractional part of Joe's money is Tom's money? what per cent?

22. Mary's hat cost $2.00, and Martha's hat cost $3.00. What per cent of the cost of Mary's hat did Martha's hat cost?

23. Nellie attends school 19 days out of 20 days. What is her per cent of attendance for the month?

24. James, Robert, and Paul are respectively 20, 15, and 10 years of age. What per cent of James's age is each of the other two?

25. In a school of 400 pupils, 220 are girls. What per cent of the school are girls? what per cent are boys?

26. The wool from Mr. Jones's sheep averaged $1.20 per head. He afterwards sold the sheep for $4.00 per head. What per cent of the value of the sheep is the wool?

27. Walter sold 60 out of 120 chickens that he raised during the summer. What per cent of his chickens did he sell?

28. John buys a sled for $3.00 and sells it for $3.25. What per cent of profit does he make?

HINT. — His profit is the relation between 25 ¢ and $3.00.

29. A huckster buys eggs at 20 ¢ a dozen, and sells them at 24 ¢ a dozen. What is his per cent of profit on the cost of the eggs?

30. In 1912, 80 pupils were graduated from Franklin School, and in 1913, 40 more pupils were graduated than in 1912. What was the per cent of increase? How many pupils were graduated in both years?

31. A turnip that weighs 10 oz. contains 9 oz. of water. What per cent of the turnip is water?

32. I bought a suit of clothes marked $10 for $8. What per cent of discount did the clerk give?

33. In an orchard there are 40 apple trees, 10 pear trees, 5 plum trees, and 5 cherry trees. What per cent of the whole number of trees are the apple trees? the plum trees? the pear trees? the cherry trees?

34. Frank works 8 hours on Saturday, and 2 hours on each of the other days in the week. What per cent of the number of hours that he works on Saturday does he work on each of the other days?

35. A discount of $4 is offered on a piece of furniture marked $20. What per cent of discount is offered?

36. Out of 40 chickens that James had on May 1, he succeeded in raising only 25. What per cent of chickens did he lose?

37. Mary spelled 20 words correctly out of 25. What per cent did she misspell?

38. A man bought from Mr. Grant a farm valued at $5000 for $4000 cash. What per cent of the value of the farm did he save by paying cash?

39. A bag of potatoes that weighs 100 lb. in September weighs 95 lb. in April after drying out. What per cent of the weight of the potatoes is lost in drying out?

40. John's salary in September was $40 a month, and in October it was raised $10 per month. What per cent of increase in salary did he receive?

41. I buy a house for $2500 and sell it for $3000. What is the per cent of my gain?

42. I buy a horse for $170 and sell it for $270. What is the per cent of my gain?

43. I buy a house and lot for $3000 and sell it for $2800. What is the per cent of my loss?

HINT.—The per cent of loss is the relation that $200 bears to $3000.

44. I buy a bill of goods for $1000 and by paying cash I get $100 discount. Find the per cent of discount.

45. My tuition in a Business College is $100, but by paying cash I get it for $80. What per cent discount do I save by paying cash?

In the following quotations to-day as compared with those of two weeks ago, find the per cent of increase or decrease:

		TO-DAY	TWO WEEKS AGO	PER CENT OF INCREASE OR DECREASE
46.	Sirloin steak, lb.	22 ¢	20 ¢	?
47.	Round steak, lb.	18 ¢	20 ¢	?
48.	Tenderloin, lb.	25 ¢	24 ¢	?
49.	Veal, lb.	25 ¢	20 ¢	?
50.	Cutlets, lb.	30 ¢	25 ¢	?
51.	Pork chops, lb.	20 ¢	16 ¢	?
52.	Leg of lamb, lb.	16 ¢	18 ¢	?
53.	Ham, whole, lb.	17 ¢	16 ¢	?
54.	Bacon, lb.	18 ¢	17 ¢	?

55. A workman who received $3.00 a day had his wages reduced 25 ¢ per day. What was the per cent of the reduction?

56. A man received $20 for selling $2000 worth of goods. What per cent did the man receive for selling the goods?

AMOUNT RECEIVED FOR SELLING		AMOUNT SOLD	PER CENT FOR SELLING
57.	$ 10.00	$ 100.00	?
58.	50.00	1000.00	?
59.	20.00	200.00	?
60.	6.25	100.00	?
61.	80.00	4000.00	?
62.	20.00	1000.00	?
63.	75.00	2000.00	?
64.	2.50	50.00	?
65.	50.00	5000.00	?
66.	5.25	100.00	?
67.	60.00	3000.00	?

68. Ruth earned $25 in picking berries and fruits, and put $16.75 in the savings bank. What per cent of her earnings did she save?

69. A grocer bought strawberries for $3.20 a bushel, and was compelled to sell them at a loss of $.80 a bushel. What was the per cent of his loss?

70. Strawberries were sold in the market at 10¢ per basket on Monday and at 8¢ per basket on Tuesday. What was the per cent of reduction from Monday to Tuesday?

71. Peaches were quoted at $1.50 per bushel on Monday, and at $2.00 per bushel on Saturday. What was the per cent of increase in price from Monday to Saturday?

COMMERCIAL DISCOUNT

Wholesale merchants and manufacturers usually publish printed price lists of their goods. The prices in these lists are higher than the wholesale prices and are subject to deductions called **trade discounts** or **commercial discounts.**

A **discount** is any deduction from a fixed price.

Sometimes several discounts are allowed. The first is a discount from the list price; the second, a discount from the remainder, etc.

The **net price** is the price less all trade discounts.

Oral Work

Find the selling price of goods marked:

1. $15, less 20 %.
2. $20, less 40 %.
3. $6, less 50 %.
4. $25, less 20 %.
5. $7.50, less 20 %.
6. $12.50, less 40 %.

7. $40, less 60 %.
8. $48, less 25 %.
9. $6.80, less 25 %.
10. $4.50, less $33\frac{1}{3}$ %.
11. $9.60, less $16\frac{2}{3}$ %.
12. $4.80, less $37\frac{1}{2}$ %.

Written Work

Find the selling price of goods marked:

1. $168.75, less 25 %.
2. $1374, less $16\frac{2}{3}$ %.
3. $1872, less $33\frac{1}{3}$ %.
4. $278.40, less $37\frac{1}{2}$ %.
5. $3030, less 40 %.
6. $225.65, less 20 %.

7. $875.50, less 30 %.
8. $278.90, less 10 %.
9. $2378.50, less 4 %.
10. $6775.20, less 5 %.
11. $8888.88, less $12\frac{1}{2}$ %.
12. $4255.75, less 40 %.

13. Find the net price of a bill of goods for $75.40, trade discounts 20 %, 10 %.

List price, $75.40

Less 20 %,   15.08   Observe that the *second* discount is

First remainder, 60.32   reckoned on the *first* remainder. As there

Less 10 %,   6.03   are only two discounts, the second remain-

Net price, $54.29   der is the net price.

Find the net price of articles listed at:

14. $400, less 20 %, 10 %.   **17.** $10.75, less 40 %, 5 %.

15. $375.50, less 25 %, 5 %.   **18.** $6.80, less 25 %, 10 %.

16. $290.80, less 40 %, 10 %. **19.** $12.75, less 33⅓ %, 10%.

Find the net price of the following bills of goods :

20. 36 dozen boys' caps @ $6, discounts 25 %, 20 %.

21. 50 buggies @ $120, discounts 20 %, 15 %.

22. 75 sets harness @ $40, discounts 30 %, 10 %.

23. 25 grain drills @ $95, discounts 40 %, 5 %.

24. 12 rubber hose, each 50 feet long, at 15 ¢ per foot, discounts 30 %, 15 %.

25. A merchant buys 12 stoves listed at $45, less 40 %, 10 %. Find the net amount of the bill. Compare this with the net amount of the bill with only one discount of 50 %.

26. Compare the net price of an article listed at $500, discounts of 20 %, 10 %, with the net price of a similar article listed at $500, discounts of 10 %, 20 %.

Discounting bills.

Merchants and manufacturers frequently offer an extra discount from their bills if cash is paid at the time of purchase or within a certain time. This is called a **cash discount.** Cash discounts are always mentioned in connection with the *terms of payment.*

Oral Work

Find the net price:

	List Price	Rate of Discount	Net Price		List Price	Rate of Discount	Net Price
1.	$ 80.00	25 %	?	**10.**	$ 350.00	25 %	?
2.	100.00	50 %	?	**11.**	25.00	20 %	?
3.	450.00	11⅛ %	?	**12.**	64.00	12½ %	?
4.	600.00	33⅓ %	?	**13.**	500.00	20 %	?
5.	12.00	16⅔ %	?	**14.**	360.00	25 %	?
6.	32.00	12½ %	?	**15.**	56.00	22½ %	?
7.	50.00	20 %	?	**16.**	100.00	20 %	?
8.	150.00	20 %	?	**17.**	990.00	33⅓ %	?
9.	54.00	16⅔ %	?	**18.**	1080.00	11⅑ %	?

The W. V. Snyder Company in Newark advertised the following goods. Find the per cent of discount.

	List Price	Net Price	Per Cent of Discount
19.	25 ¢ cambric	20 ¢	?
20.	12½ ¢ ribbon	10 ¢	?
21.	$ 10.00 boys' suits	$ 7.50	?
22.	$ 3.00 girls' trimmed hats	$ 2.00	?
23.	$ 3.50 girls' shoes	$ 2.75	?
24.	50 ¢ dress goods	37½ ¢	?
25.	$ 8.00 remnant	$ 4.00	?
26.	40 ¢ ribbon	30 ¢	?
27.	$ 12.00 girls' suits	$ 9.00	?
28.	$ 150.00 piano player	$ 150.00	?
29.	$ 4.00 shoes	$ 3.00	?

Written Work

1. Find the net cash price of the sectional bookcases as given in the following bill at $33\frac{1}{3}\%$ and 10 % trade discounts. Terms: 60 days net; 5% cash in 10 days.

GRAND RAPIDS, MICH., *Jan. 4,* 1913.

Messrs. *Geo. H. Alexander & Co.,*
300 Wood Street.

Bought of THE KENSINGTON MFG. CO.
GRAND RAPIDS, MICH.

TERMS : 60 days net ; 5% cash in 10 days.

Jan. 4	13	12 Bookcase Sections $28	$14	40		
		12 " Tops $21	9			
		6 " Bases $25	27			
			50	40		
		Less ⅛, ⅒,	20	16		
		Net price,	30	24		
		Cash in 10 days, less 5%	1	51	$28	73
		Received payment, Jan. 10,	1913			
		Kensington Mfg. Co.				
		Per R.				

The sign #, when placed before a number, is read " number." Thus, # 12 is read "number 12."

Make out bills showing net cost price for the following:

2. Mr. King bought for cash from Banister and Polard, Newark, N.J., 4 doz. Acme lawn mowers @ $30, 50 lb. lawn seed @ 15¢, 2½ doz. brushes @ 40¢. Trade discounts: 20 %, 10 %. Terms: 30 days net; 2 % cash in 10 days.

3. Jamison and Redmond, Elizabeth, N.J., bought, for cash, from the Acme Buggy Co., Cincinnati, O., 72 buggies @ $105, 50 sets harness @ $45, 15 sleighs @ $60, 40 robes @ $20. Trade discounts: 30%, 15%. Terms: 30 days net; 3% cash in 10 days.

4. James Cubbison, Greenville, O., bought for cash from Arbuthnot, Stevenson & Co., Camden, N.J., 5 doz. handkerchiefs @ $3.60; 5 bolts muslin, 40 yd. each, @ 8¢; 5 bolts prints, 42 yd., @ 7¢. Trade discount: 33⅓%. Terms: 30 days net; 2% cash in 10 days.

5. S. H. Gardner Co., piano dealers, Newark, N.J., order from the Harmonic Piano Co., Chicago, Ill., 2 Harmonic pianos #266 @ $600, less 40%, 10% trade discount. Terms: 90 days net; 10% off 10 days. Find the net cash price. Find the net price if paid in 30 days.

6. M. L. Smith, tailor, Brockton, Mass., orders from Bender & Co., New York, importers, 3 pieces suiting, 22 yd. each, @ $3.15. Terms: 30 days net; 2% off 10 days. Make out and receipt bill if paid within 10 days.

7. A $2400 automobile was sold for $1800. For what per cent of the cost was it sold?

8. I sold a bill of goods for $1200 that cost me $1050. What was my gain per cent?

HINT. — My gain was $150. My per cent of gain was the relation between what numbers?

COMMISSION

An **agent** is a person who transacts business for another. **Commission** is the sum charged by an agent for his services

Oral Work

Find the commission on:

1. $2000 at 5%	**3.** $2500 at 4%	**5.** $1000 at 8%
2. $5000 at 10%	**4.** $3200 at 2%	**6.** $4000 at 5%

Written Work

An agent made the following sales. Find his commission for each day at 5%:

1. Monday, $1800
2. Tuesday, $1594
3. Wednesday, $1954
4. Thursday, $1400.80
5. Friday, $1528
6. Saturday, $2370.60

7. Find his total commission for the week.

8. A real estate agent sells a house and lot for $6750, charging 2% commission. Find his commission.

9. A traveling salesman sold $50,000 worth of goods in a year, at a commission of 8%. If his expenses for the year were $2200, how much had he left?

10. An agent rents 12 houses at $40 per month. If he receives 5% for collecting the rents, how much is remitted to the owners each month?

11. James and John are employed in a factory. John works 307 days in the year at $1.95 a day, and James works 296 days in the year at $2.15 a day. John deposits 25% of his earnings in the savings bank, and James deposits 30% of his earnings. Find their total deposit for the year in the savings bank.

12. Mr. Daniels promised his son George and his daughter Mary to add 20% to their total savings each year. George saved $159 for the year, and Mary $124. How much money did the father have to add to the earnings of each?

13. William earns $1.80 a day, and puts 10% of his earnings in the savings bank. If he works 302 days in the year, how much does he deposit in the savings bank?

14. Curtis works in vacation for $50 a month. If he works 2 mo., and saves 60% of his money, how much money does he save?

SIMPLE INTEREST

Oral Work

1. Mr. Johnston pays the liveryman $6 for the use of a horse and buggy for two days. What does he get in exchange for the $6?

2. Mr. Daniels pays $6 for the right to pasture his cow in a field for two months. What does he get in exchange for the $6?

3. Mr. Watson pays $6 for the use of $100 for one year. What does he get in exchange for the $6?

4. In the first two examples money is paid for the use of something that is not money. For what does Mr. Watson pay the money in the last example?

Interest is money paid for the use of money.

5. How much does Mr. Watson pay for the use of the money? What is the $6 called?

6. On what is the interest reckoned? The $100 is called the *principal*.

The **principal** is the sum on which the interest is paid.

The **rate of interest** is a certain number of hundredths of the principal paid for the use of the principal for *one year*.

Time is always a factor in *interest*. Interest, then, is the product of three factors: **principal**, **rate**, and **time**.

The **amount** is the sum of the principal and the interest.

Interest for years and months.

Oral Work

1. What part of a year are 6 months? 4 months? 3 months? 2 months? 1 month?

2. If the interest for a year is $100, what should it be for 6 months? for 4 months? for 3 months? for 2 months? for 1 month?

Find the interest on :

3. $300 for 1 year at 5 %

4. $800 for 2 years at 6 %

5. $150 for $3\frac{1}{2}$ years at 6 %

6. $700 for $4\frac{1}{4}$ years at 4 %

Written Work

1. What is the interest on $200 for $2\frac{1}{2}$ years at 6 % ?

$200, principal
.06, rate
$12.00, interest for one year
$2\frac{1}{2}$,
$30, interest for $2\frac{1}{2}$ years

The interest for 1 year is .06 of the principal, or $12. The interest for $2\frac{1}{2}$ years is $2\frac{1}{2} \times$ $12, or $30.

Multiply the principal by the rate and the product by the number of years.

The year is usually considered as 360 days, that is, 12 months of 30 days each.

Find the interest of :

2. $250 for $1\frac{1}{2}$ years at 4 %

3. $75 for 2 years at 5 %

4. $100 for $3\frac{2}{3}$ years at 7 %

5. $80 for $4\frac{1}{2}$ years at 5 %

6. $40 for $2\frac{1}{2}$ years at 6 %

7. $500 for $2\frac{1}{8}$ years at 4 %

8. $960 for 9 mo. at 6 %

9. $900 for $2\frac{3}{4}$ years at 7 %

10. $654 for $\frac{3}{4}$ year at 6 %

11. $220 for $\frac{7}{8}$ year at 8 %

Find the interest and amount of:

    12. $660 at 6% for 1 yr. 2 mo.

    13. $457.75 at 6% for 2 yr. 8 mo.

    14. $675 at 6% for 1 yr. 4 mo.

    15. $1200 at 6% for 2 yr. 8 mo.

    16. $84.50 at 6% for 3 yr. 3 mo.

    17. $90.75 at 6% for 2 yr. 6 mo.

    18. $530 at 5% for 3 yr. 6 mo.

    19. $150 at 4% for 4 yr. 1 mo.

    20. $275 at 6% for 1 yr. 10 mo.

    21. $120 at 5% for 2 yr. 8 mo.

    22. $625 at 4% for 8 mo.

    23. $400 at 6% for 3 yr. 7 mo.

    24. $275 at 5% for 2 yr. 11 mo.

Find the interest at 6% on:

25. $100 for 6 mo.   **28.** $624 for 120 da.

26. $500 for 4 mo.   **29.** $170 for 8 mo.

27. $150 for 2 yr. 2 mo.   **30.** $355 for 180 da.

31. Find the amount necessary to pay a loan of $250 at 6% from Jan. 4, 1911, to March 4, 1913.

32. Find the interest at 6% on $375 borrowed Sept. 1, 1911, and paid March 1, 1913.

PRACTICAL MEASUREMENTS

MEASURES OF SURFACE

Oral and Written Work

Review Areas of Rectangles, pp. 194–196.

1. Draw a diagram on a scale of 1 in. to 3 ft., to represent a rectangle 24 ft. long and 18 ft. wide. Find its area.

Draw diagrams, on suitable scales, and find the surfaces of the following rectangles:

2. A rectangle 20 ft. by 24 ft.

3. A flower bed 16 ft. by 8 ft.

4. A floor 16 ft. long and 14 ft. wide.

5. A wall 15 yd. long and 5 yd. high.

6. By actual measurement find the number of square feet in the floor, the door, the blackboard, and the walls of the schoolroom.

7. In what denominations did we find the lengths and widths of the problems just given?

Land is measured in *acres, square rods, square feet,* or in acres and parts of an acre.

8. Measure a square rod on your playground. How long is it? how wide?

9. Measure the length and width of your school grounds in rods and feet.

10. Since $16\frac{1}{2}$ feet equal 1 rod, how many yards equal 1 rod? How many square yards equal 1 square rod?

11. Since 16½ feet equal 1 rod, how many square feet equal 1 square rod?

12. A field is 70 rd. long and 40 rd. wide. How many square rods are there in it? how many acres?

13. Memorize this table:

144 square inches (sq. in.)	= 1 square foot (sq. ft.)
9 square feet	= 1 square yard (sq. yd.)
30¼ square yards	= 1 square rod (sq. rd.)
160 square rods	= 1 acre (A.)
640 acres	= 1 square mile (sq. mi.)
1 A. = 160 sq. rd. = 4840 sq. yd. = 43,560 sq. ft.	

Written Work

Change:

1. 2700 sq. yd. to sq. ft.

2. 50 sq. ft. to sq. in.

3. 1600 sq. rd. to A.

4. 1⅝ A. to sq. rd.

5. 800 sq. yd. to sq. rd.

6. 5¾ A. to sq. ft.

7. A farm is 90 rd. long and 60 rd. wide. Find the number of acres in it. Find its cost at $60 per acre.

8. A lot 100 ft. square has a house 36 ft. by 42 ft. located on it. The remaining space is lawn. Find the number of square feet of lawn. Draw a diagram.

9. A concrete sidewalk in front of the lot is 4 ft. wide. Find its cost at 19 ¢ per square foot.

10. Find the cost of a flagstone walk, 135 ft. long and 6 ft. wide, at 21 ¢ per square foot.

11. City lots are sometimes sold by the square foot. Find the cost of a lot in Newark, N.J., 21 ft. by 70 ft. at $27.50 per square foot.

12. A farm 160 rd. long and 120 rd. wide is sold in two pieces, ⅔ of it at $60 per acre, and the remainder at $50 per acre. Find the amount of the entire sale.

13. A New Jersey farmer owns a farm a mile square. How many acres has he? Find its value at $85 per acre.

14. A western wheat field 100 rd. long and 80 rd. wide yields 880 bu. of wheat. Find the average yield per acre.

15. City lots are usually sold by the front foot. Find the cost, at $20 per foot front, of a lot 25 ft. front by 120 ft. deep. Find the cost per square foot.

16. A four-room school building has a slate blackboard 24 ft. by 4 ft. in each room. Find the total cost of the blackboard at 23¢ per square foot.

17. The area of a field in the form of a rectangle is 8 A. If one side is 32 rd., what is the other side?

These diagrams represent pieces of land. The dimensions are given in rods, and the corners are all square.

18. Divide the first piece into 3 rectangles and find (1) how many square rods there are in each; (2) the perimeter of each; (3) the area of the entire piece in acres.

19. Divide the second piece into rectangular lots, and find (1) the perimeter of each; (2) the area of each; (3) the area of the entire piece.

20. Find the value of a lot on Broadway, New York, 28' × 56' at $40.50 per square foot.

21. Mr. Franklin's farm is in the shape of the diagram. The dimensions are given in rods. What is the size of the farm in acres? On what scale is the farm represented in the diagram?

22. Find by the scale the number of acres in oats; in corn; in pasture; in wheat.

23. The wheat yields 23 bushels to the acre and is sold for $1.12½ per bushel. Find the amount of the sale of wheat.

24. The corn yields 85 bushels of ears per acre, worth 40¢ per bushel. Find the amount of the corn crop, not counting the fodder.

THE RIGHT TRIANGLE

Oral and Written Work

1. Draw on the blackboard a rectangle 12 inches long and 8 inches wide. Connect the opposite corners by a straight line.

This line is called the **diagonal** of the rectangle. It divides the rectangle into two triangles.

A **triangle** is a surface bounded by three straight lines.

A **right triangle** is a triangle having one right angle.

The **base** of a triangle is the side on which it is assumed to stand.

The **altitude** of a triangle is the line that meets the base line at a right angle.

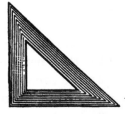

To THE TEACHER. — As an aid in drawing, have each pupil, if possible, get a right triangle, like the one here represented.

2. Point out the base and the altitude in the triangles at the right.

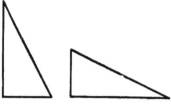

3. Fold a rectangular piece of paper, as *ABCD*, on its diagonal. Observe:

(1) That the rectangle *ABCD* and the triangle *ABD* have the same base and altitude.

(2) That the area of the triangle is just ½ the area of the rectangle.

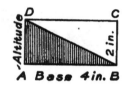

Hence the area is ½ of $4 \times 2 \times 1$ *sq. in.*, or 4 sq. in.

The **area of a right triangle** *equals the unit of measure multiplied by ½ the product of the base and altitude.*

Draw on a suitable scale and find the area of the following right triangles in square inches:

4. Base 10 in., altitude 8 in. **6.** Base 25 in., altitude 18 in.

5. Base 12 in., altitude 6 in. **7.** Base 36 in., altitude 24 in.

8. Observe the dimensions of the school building. What are the heights of the sides of the building?

9. Find the number of square feet of siding needed for the sides and the two ends of the same height as the sides, making no allowance for openings.

10. The triangular parts at the top of the house, in front and in back, are called *gables*. Each gable can be

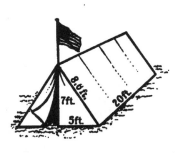

divided by a line through the center of its base into two right triangles. How many square feet of siding are necessary for the two gables?

11. Find the cost of painting the siding on the house, at 10 ¢ per square yard.

12. John has a tent 20 ft. in length and 10 ft. in width, as shown in the illustration. The top of the tent is 8.6 ft. from the ground. Find the area of each end of the tent, and of each side of the tent.

13. Find the entire surface of the gable and the end of the house in this picture.

MEASURES OF VOLUME

Oral Work

1. How many cubic inches equal 1 cubic foot?

2. How many cubic feet equal 1 cubic yard?

3. How many cubic yards equal a cart load of earth?

4. How many surfaces has a cube?

5. Show that all the surfaces of an inch cube are the same in area; of a 2-inch cube; of a 9-inch cube.

Review pp. 198–202.

6. Examine carefully the figure on p. 273. Observe:

(1) That the surface of the face upon which it rests contains 9 square inches.

(2) That the first layer of units of volume contains 9 cubic inches.

(3) That the whole solid, if 6 inches high, contains 6 × 9 cubic inches, or 54 cubic inches.

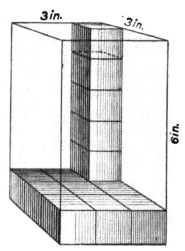

7. How many 1-inch cubes are there in the first layer? how many in the solid?

8. What is the shape of the surfaces of the solid? Is each surface a rectangle?

A **rectangular solid** is a solid whose surfaces are all rectangles.

9. Observe that the number of inch cubes in the solid is equal to the product of its three dimensions.

10. What is the unit of measure in this solid?

Observe that 3 × 3 × 6 × 1 *cu. in.* = 54 cu. in.

The **contents,** *or* **volume, of a rectangular solid** *is found by multiplying the unit of measure by the product of its three dimensions, when expressed in like units.*

To the Teacher. — Secure 144 1-inch cubes.

11. Build a cube 2 inches on an edge.

12. Build a cube 4 inches on an edge.

13. Compare the 4-inch cube with the 2-inch cube.

14. Give the different units of measure of surface; of length; of contents.

15. Find the contents of a box 3 ft. long, 2 ft. wide, and 1½ ft. high.

Written Work

Excavations are estimated by the cubic yard.

1. Find the cost, at 30 ⊄ per cubic yard, of excavating a cellar 36 ft. in length, 24 ft. in width, and 4 ft. in depth.

36 × 24 × 4 × 1 *cu. ft.* = 3456 cu. ft., the contents of the cellar.
3456 cu. ft. ÷ 27 cu. ft. = 128, number of cu. yd.
128 × $.30 = $38.40, cost of excavation.

2. This diagram shows the outline of a cellar 5 ft. deep. Its dimensions are given in feet.   What is the scale of the drawing?

3. Find its area in square feet.

4. Find the length of its walls.

5. What is the cost of excavating it at 32 ⊄ per cubic yard?

6. How much will it cost to cement the floor at $.90 per square yard?

7. If a boy inhales 24 cu. in. of air at a breath, how many times must he breathe to inhale 1 cu. ft.?

8. 29 pupils and their teacher occupy a schoolroom 30 ft. in length, 24 ft. in width, and 12 ft. in height.   What is the average number of cubic feet of air for each person?

9. A city lot 37½ ft. by 120 ft. is to have a layer of earth 1 ft. thick over its surface.   Find the number of loads (cubic yards) needed and the cost at 25 ⊄ per load.

10. If the rainfall on a certain day was $2\frac{1}{4}$ inches, find the number of cubic inches that fell on a lot 25 feet wide and 100 feet long. Find the number of gallons.

11. Find the cost of digging a ditch, 60 rd. long, $3\frac{1}{2}$ ft. wide, and 6 ft. deep, at 60 ¢ per cubic yard.

12. Memorize this table:

1 gallon = 231 cu. in.	1 T. coal = about 35 cu. ft.
1 bushel = 2150.42 cu. in.	1 T. hay = about 500 cu. ft.
1 bushel = about $1\frac{1}{4}$ cu. ft.	

13. Compare a 3-inch cube with a 4-inch cube. If a 2-inch cube weighs 6 ounces, how much will a 4-inch cube of the same material weigh?

14. A bin is 8 ft. long, 6 ft. wide, and 4 ft. deep. Estimate quickly about the number of bushels of wheat or oats it will hold.

15. A farmer has a tank 12 ft. long, 8 ft wide, and 6 ft. deep. How many gallons of water will it hold?

16. Mr. Hoyt's coal bin is 12 ft. long, 8 ft. wide, and 5 ft. deep. Estimate quickly the number of tons of coal it will hold.

17. A hay mow is 20 ft. wide, 50 ft. long, and 16 ft. deep. Estimate quickly the number of tons of hay in the mow.

18. A barrel of apples ordinarily holds 3 bu. How many cubic feet are there in such a barrel?

19. Estimate quickly the number of bushels of apples in a wagon bin 4 ft. wide, $8\frac{1}{2}$ ft. long, and 2 ft. deep.

20. A cistern is 4 ft. long, 3 ft. wide, and 3 ft. deep. How many barrels of water will it hold?

REVIEW OF PRACTICAL MEASUREMENTS

1. How many tiles 12 in. square will be required to lay a floor 36 ft. by 15 ft.?

2. What is the length of a board walk that is 4 ft. 8 in. wide and contains 1350 sq. ft.?

3. How many cubic yards of earth must be removed in digging a cellar 36 ft. long, 26 ft. wide, and 8 ft. deep?

4. Find the number of square feet in a hall 45 ft. long and 30 ft. wide.

5. How many bushels can be put into a bin $12\frac{1}{2}$ ft. long, 10 ft. wide, and 10 ft. deep?

6. Find the cost of 30 boards 16 ft. long, 12 in. wide, at 5 ¢ a square foot.

7. At $.80 a bushel what is the value of a bin of wheat 16 ft. long, 8 ft. wide, and 4 ft. deep?

8. What is the number of gallons in a tank 12 ft. long, 10 ft. wide, and 8 ft. deep?

9. How much will it cost to cement the floor of a cellar 50 ft. long and 28 ft. wide, at $1.08 a square yard?

10. At 7 ¢ a square yard, how much will it cost to paint the four sides of a building 50 ft. long, 20 ft. wide, and 15 ft. high?

11. My farm is in the form of a rectangle, and contains 40 A.   What is its width, if its length is 128 rd.?

12. What will be the cost of plastering the ceiling of a room 22 ft. by 18 ft., at 18 ¢ a square yard?

13. A rectangular field contains 5 A.   If its length is 80 rd., what is its width?

14. How many cakes of soap 4 in. by 3 in. by 2 in. can be packed in a box whose inside dimensions are 2 ft., 3 ft., and 4 ft.?

15. Find the cost of digging a cellar 42 ft. long, 30 ft. wide, and 6 ft. 3 in. deep, at 40 ¢ a cubic yard.

16. The length of a field is 80 rd., and its width is 30 rd. How many acres are there in the field?

17. What is the number of bushels in a bin 20 ft. long, 16 ft. wide, and 8 ft. deep?

18. A tank 9 ft. square and 8 ft. deep contains how many gallons?

19. A building lot 100 ft. front contains 15,000 sq. ft. What is its depth?

20. A baseball ground 160 yd. by 170 yd. has a tight board fence around it 8 ft. high. How much will the painting of the outside of the fence cost at $5\frac{1}{2}$ cents a square yard?

21. The area of a right triangle is 560 sq. ft., and its altitude is 28 ft. What is the base of the triangle?

22. How much will it cost to excavate a street 800 ft. long and 50 ft. wide, to a depth of 18 in., at 36 cents a load?

23. A plot of ground in the form of a square is 100 ft. on each side. A straight walk 8 ft. wide divides it into two equal parts — a lawn for flowers and a garden for vegetables. In the lawn there is a flower bed 5 ft. by 8 ft. Draw the plot.

24. Find the perimeter of the plot; of the lawn; of the garden; of the flower bed; of the walk.

Find the area in square yards:

25. Of the plot.   **27.** Of the flower bed.

26. Of the lawn.   **28.** Of the walk.

29. How much will it cost to fence the plot at $3¼ per rod?

30. How much will it cost to pave the walk at $1.55 per square yard?

31. How much will it cost to spade the flower bed at 5¢ per square yard?

32. How much will it cost to sod the lawn, excluding the flower bed, at $.26 per square yard?

33. A room is 20 ft. long, 16 ft. wide, and 10 ft. high. How much will it cost to plaster the walls and ceiling at 20¢ a square yard?

34. How many gallons of water are there in a tank 12 ft. long, 8 ft. wide, and 6 ft. deep, if it is half full?

35. A city 5 mi. long and 3 mi. wide is equal in area to how many farms of 160 A. each?

36. How many sods 16 in. square will be required to turf a lawn 106 ft. 8 in. long and 50 ft. wide?

37. What will be the cost of painting the outside of a house 48 ft. long, 30 ft. wide, and 20 ft. high, at 18 cents a square yard?

38. Estimating that 300 cu. ft. of air is required for each person, how many persons should occupy a hall 40 ft. long, 30 ft. wide, and 12 ft. high?

HOME AND FARM PROBLEMS

1. A house is insured against fire in the Home Fire Insurance Company, for $2000, for three years at 2%. Write a check for the insurance.

2. A barn is insured for $1000 in the London Insurance Company, at 1% per year. Write the check for the insurance.

3. Mr. Jones pays 1¾% insurance on his store valued at $5000. Write the receipt for the insurance.

4. Mr. Warren bought an automobile in 1912 for $1875. The insurance for one year was 2½% and the repairs cost $150. Find the cost of the insurance and the repairs.

5. Mr. Jones had 4 fancy cows, — Flora, Bess, Queen, and Daisy. The following table shows the number of pounds of milk each gave in a certain week and the per cent of butter fat. Find the number of pounds of butter each cow produced that week.

NOTE. — Butter fat is the richest part of milk from which butter is made. A 5% milk contains 5 lb. of butter fat to every 100 lb. of milk.

	FLORA	BESS	QUEEN	DAISY
Pounds of milk	200	190	196	204
Per cent of butter fat	4.2	4.8	4.32	4.

6. Mr. Ames reckoned his profits on his farm and found that in 1910 he had a net profit of 12% on a valuation of $3000; in 1911, 5½% on a valuation of $3000; in 1912, 10¾% on a valuation of $3000. Find his profits for the three years.

279

7. Mr. Frank Barnett buys 6 T. 5 cwt. of coal at $6 per ton in September, and by paying cash gets a discount of 5 %. Write his check to the Acme Coal Company, Trenton, N.J., on your local bank in payment of the bill. Write the coal company's receipt.

8. My natural gas bill for the month of February was 19,000 cu. ft. at 30 ¢ per thousand cu. ft., less a discount of $2\frac{1}{2}$ % per thousand cubic feet if paid on or before March 2. Write the check on your local bank in payment of the bill if paid in ten days.

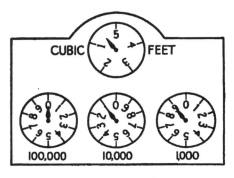

9. My artificial gas bill for the month of February was for 3000 cu. ft. at 80¢ per thousand cubic feet, less a discount of 10 % if paid before March 5. Write the check on your local bank in payment of the bill if paid March 4.

Electricity is measured by a unit called a watt. A **kilowatt** (Kw.) is 1000 watts.

A 16 candle power electric lamp consumes 56 watts of electrical energy per hour and 10 and 8 candle power electric lights consume, respectively, 40 and 32 watts per hour. "Watt" is here used in the sense of watt hour.

10. If my electric meter on Feb. 1 read 35,207 watts and I used during February 66 kilowatts, what was the reading on March 1? At 12 ¢ per kilowatt, less a discount of 2 %, find the cost of my electricity for February.

11. Read the meters in your own house or in a neighbor's house and make practical problems about gas and electric light.

12. My electric light bill for the month of February was $7.92. If I received a discount of 10 % on the bill, how much did I have to pay?

13. A gas meter reads 34,600 cu. ft. January 1, and 46,700 cu. ft. February 1. The price is 30 ¢ per thousand cubic feet, less 2 % discount, if paid before February 10. Find the amount of the bill. Write the check in payment of the bill February 9.

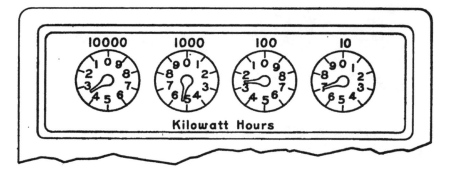

14. Mr. James's electric light meter reads as follows: December 1, current, 1636 kilowatt hours, January 1, current, 1742 kilowatt hours. At 10 ¢ per kilowatt, less 2 % discount in 10 days, make out Mr. James's bill and his receipt.

15. If a mile of railroad has a 2 % grade, how much higher is one end of the railroad than the other?

HINT. — 2% of 5280 ft. = ?

16. The top of a hill whose slope is 200 rd. is 100 ft. higher than the bottom. How many feet of rise are there to the rod?

SOLUTION. — 100 ft. to 200 rd. = ½ ft. to 1 rd.

17. On an old road Mr. Barnes's team could haul only 2500 lb. After the road was graded it could haul 40 % more. How many pounds could it haul on the graded road?

18. May lives 100 rd. uphill from Helen's house. If the grade is 8 %, how much higher on the hill is Mary's house than Helen's?

Find the differences in elevation between two points in a road running up a grade, indicated as follows:

	DISTANCE	PER CENT OF GRADE	DIFFERENCE IN ELEVATION
19.	40 rd.	1 %	?
20.	80 rd.	2 %	?
21.	100 rd.	3 %	?
22.	1 mi.	$3\frac{1}{2}$ %	?
23.	220 rd.	5 %	?
24.	50 rd.	4 %	?
25.	500 ft.	$1\frac{1}{2}$ %	?
26.	400 ft.	$2\frac{1}{2}$ %	?

GENERAL REVIEW

1. The remainder is 92,568 and the minuend is 202,660. Find the subtrahend.

2. The dividend is 364,450 and the quotient is 9850. What is the divisor?

3. Add 3.5, .035, 45.006, and 2.06.

4. Write decimally twenty-five and sixty-one thousandths; one hundred twenty-five and five tenths; and three hundred and two ten-thousandths.

5. What number multiplied by one hundred seventy-nine is equal to 848,818?

6. From 2.0011 take 1.9892.

7. Add $\frac{1}{2}$, $2\frac{3}{4}$, $\frac{1}{5}$, $\frac{3}{8}\frac{1}{8}$, and $4\frac{5}{8}$.

8. The multiplicand is 1325 and the multiplier is .0416. What is the product?

9. If 38 dozen eggs cost $11.40, what is the cost per dozen?

10. A building is 34 ft. 2 in. wide, and twice as long as wide. Find the distance around the walls.

11. A man bought 48 head of cattle, at $36 per head, and sold them at a gain of 25%. What was the total amount received for the cattle?

12. Find the interest on $370.50 for 4 yr. at 6%.

13. Divide $48\frac{1}{2}$ by $21\frac{1}{5}$.

14. Divide .65 by 6.5.

15. How much will it cost to ship a carload of wheat containing 42,000 lb. from Fargo, N.D., to Chicago, Ill., if the freight rate is $0.06 per bushel?

16. A train leaves Newark at 8.15 A.M. and arrives at Pittsburgh at 8.20 P.M. The distance is 430 miles. Find the number of miles per hour the train travels.

17. The steel rails on a railroad weigh 100 pounds to the yard. Find the number of tons necessary to lay 5 rd. of single track.

18. How much does an architect receive, at 4%, for the plans of a house that cost $8350?

19. A man's salary is $150 per month. He spends 40% of it a year. How much does he save in a year?

20. A man purchases 80 acres of land for $6400, and sells them at 25% gain. How much does he receive per acre?

21. A real estate man buys a house for $2800 and sells it for $3200. What is his gain per cent?

22. What is $33\frac{1}{3}$% of 24? of $4.80? of $62.50?

23. A piece of land 30 rd. wide and 480 rd. long was sold at $62.50 per acre. Find the amount of the sale.

24. If $\frac{4}{5}$ of a bushel of potatoes cost $.40, how much will $7\frac{1}{2}$ bu. cost?

25. A piece of land 40 rd. long in the form of a rectangle contains 5 A. Find its width in rods.

26. A dairyman owns a cow that averages 3 gal. 2 qt. 1 pt. of milk daily. If he sells the milk at $.06 per quart, how much will he realize from the cow during the month of May?

27. How much commission, at 4%, does a salesman receive for selling goods to the amount of $16,500?

28. In a certain city there are 12,600 persons of school age, but only 9450 are enrolled in the school. What per cent of the persons of school age are there in the school?

29. An auctioneer sold $16,200 worth of goods at 2% commission. How much did the owner realize from the sale?

APPENDIX

To THE TEACHER. — The following subjects, which were purposely omitted or condensed in the body of the book, are here presented for the convenience of teachers who think it desirable to teach them.

FACTORING

The **factors** of a number are the integers that, multiplied together, will produce that number. They are **exact divisors** of the number since they will divide it an integral number of times.

Thus, the factors of 4 are 2 and 2, since $2 \times 2 = 4$; the factors of 6 are 2 and 3, since $2 \times 3 = 6$; the factors of 8 are 4 and 2, since $4 \times 2 = 8$, or 2, 2, and 2, since $2 \times 2 \times 2 = 8$.

Oral Work

1. What two factors will produce 9? 15? 25? 32?

2. What three factors will produce 8? 12? 20? 42?

3. What number will divide 13, giving a whole number as quotient? Observe that 13 has no factors except itself and 1.

A **prime number** is an integer that has no other factors except itself and 1. All other integers are called **composite numbers**.

Thus, $5 = 5 \times 1$; $7 = 7 \times 1$. 5 and 7 are prime numbers. 4, 6, 8, 9, etc., are composite numbers.

4. What numbers between 0 and 30 are prime numbers? between 30 and 60?

Prime factors are factors that are prime numbers.

Thus, 3 and 7 are the prime factors of 21.

PRIME FACTORS

Oral Work

Find by inspection the prime factors of:

1. 12	**3.** 27	**5.** 35	**7.** 45	**9.** 54
2. 21	**4.** 32	**6.** 40	**8.** 52	**10.** 60

Written Work

1. Find the prime factors of 120.

$$\begin{array}{r|l} 2 & 120 \\ \hline 2 & 60 \\ \hline 2 & 30 \\ \hline 3 & 15 \\ \hline & 5 \end{array}$$

Divide 120 by the prime factor 2, which gives the quotient 60. Divide 60 and the succeeding quotients by the smallest prime factors that will divide each. The last quotient, 5, is a prime number. The prime factors of 120 are 2, 2, 2, 3, 5; that is, $120 = 2 \times 2 \times 2 \times 3 \times 5$

Every composite number is equal to the product of all its prime factors.

2. Find the prime factors of 700.

$$100 \overline{)700}$$
$$\overline{7}$$

$100 = 10 \times 10 = 2 \times 5 \times 2 \times 5$
$700 = 2 \times 5 \times 2 \times 5 \times 7$

We see at a glance that $700 = 100 \times 7$, or $10 \times 10 \times 7$. Since the factors of 10 are 2 and 5, the factors of 700 are 2, 5, 2, 5, and 7.

Find the prime factors of:

3. 96	**11.** 240	**19.** 600	**27.** 1701
4. 100	**12.** 360	**20.** 900	**28.** 2142
5. 108	**13.** 432	**21.** 1122	**29.** 2310
6. 120	**14.** 576	**22.** 1155	**30.** 2520
7. 125	**15.** 720	**23.** 1225	**31.** 3850
8. 160	**16.** 960	**24.** 1331	**32.** 4199
9. 180	**17.** 300	**25.** 1575	**33.** 4235
10. 210	**18.** 400	**26.** 1596	**34.** 5145

GREATEST COMMON DIVISOR

Oral Work

1. Name a number that will divide 8 and 12 exactly; 10 and 15; 12 and 16.

A **common divisor** (c. d.) of two or more numbers is a number that divides each of them exactly. Thus, 4 is a common divisor of 16 and 24.

The **greatest common divisor** (g. c. d.) of two or more numbers is the *greatest* number that will divide each of them exactly. Thus, 8 is the g. c. d. of 16 and 24.

2. Name the g. c. d. of 12 and 16; of 20 and 30; of 18 and 27; of 10 and 15; of 22 and 33; of 63 and 81.

The greatest common divisor of two or more numbers is the product of all their common prime factors.

Written Work

1. Find the greatest common divisor of 24, 36, and 48.

```
2)24  36  48
2)12  18  24
3) 6   9  12
   2   3   4
```

Divide all the numbers by a common prime factor. In the same way divide the quotients until they are prime to each other. The divisors 2, 2, and 3 are *all the common prime factors.* Hence the g. c. d. of 24, 36, and 48 is $2 \times 2 \times 3$, or 12.

2. Find the greatest common divisor of 165 and 210.

```
3)165  210
5) 55   70
   11   14
```

Dividing each number by 3 and the quotients by 5, you find that the only prime factors common to both are 3 and 5. Hence their product, 15, is the greatest common divisor of 165 and 210.

Find the greatest common divisor of .

3. 72 and 108	**5.** 391 and 697	**7.** 408 and 544
4. 36 and 90	**6.** 135 and 270	**8.** 190 and 570

NOTE. — The chief application of greatest common divisor is in reducing fractions to their lowest terms.

LEAST COMMON MULTIPLE

Oral Work

1. Name a number that contains 3 and 4 a whole number of times; 4 and 5.

A **common multiple** of two or more numbers is a number that can be divided exactly by each of them. Thus, 24 is a common multiple of 3 and 4.

2. Name the *least* number that can be divided exactly by 3 and 4; by 6 and 8; by 5 and 6.

The **least common multiple** (l. c. m.) of two or more numbers is the least number that can be divided exactly by each of them. Thus, 12 is the l. c. m. of 3 and 4.

3. Name the l. c. m. of 4 and 5; of 6 and 8; of 6 and 9.

Written Work

1. Find the l. c. m. of 16 and 24.

$16 = 2 \times 2 \times 2 \times 2$
$24 = 2 \times 2 \times 2 \times 3$

2 occurs 4 times as a factor in 16. It must, therefore, be used 4 times in the l. c. m. 3 occurs once as a factor in 24. It must, therefore, be used once in the l. c. m. The l. c. m., therefore, of 16 and 24 is $2 \times 2 \times 2 \times 2 \times 3$, or 48.

2. Find the l. c. m. of 6, 12, 16, and 24.

```
2)6  12  16  24
  2)  8  12
    2)4   6
        2   3
```

Since 6 and 12 are exact divisors of 24, a multiple of 24 is also a multiple of 6 and 12; hence 6 and 12 may be rejected. Divide the remaining numbers by any prime factor that will divide *two or more of them*. In the same way divide the quotients until *no two of them* have a common divisor. 2 occurs 4 times in 16; hence it must be used 4 times in the l. c. m. 3 occurs once in 24; hence it must be used once in the l. c. m. Therefore the l. c. m. of 6, 12, 16, and 24 is $2 \times 2 \times 2 \times 2 \times 3$, or 48.

The least common multiple of two or more numbers is the product of all their prime factors each used as often as it occurs in any one of the numbers. . . .

NOTE. — The chief application of least common multiple is in finding the lowest common denominator of several fractions, in order to make them similar.

Find the least common multiple of:

3. 15, 20, 30	**9.** 64, 72, 108	**15.** 27, 45, 63
4. 18, 24, 36	**10.** 72, 84, 120	**16.** 28, 40, 56
5. 14, 21, 42	**11.** 54, 81, 135	**17.** 36, 48, 64
6. 27, 54, 63	**12.** 8, 12, 16, 20	**18.** 32, 52, 65
7. 32, 48, 96	**13.** 16, 24, 32	**19.** 50, 60, 70
8. 36, 54, 63	**14.** 20, 35, 42	**20.** 55, 75, 88

PAINTING AND PLASTERING

Painting, plastering, and kalsomining are generally measured by the **square yard.** In some localities an allowance is made for doors and windows, but there is no uniform rule in practice.

Written Work

1. How much will it cost to paint a ceiling 18 ft. long and 15 ft. wide at 10 ¢ per square yard?

2. How much will it cost to kalsomine a hall 30 ft. long, 9 ft. wide, and 15 ft. high, at 5 ¢ per square yard? (Observe that the perimeter of the hall is 78 ft.)

3. How many square yards of plastering are there in a room 21 ft. long, 18 ft. wide, and 12 ft. high, making no allowance for openings?

4. How much will it cost, at 15 ¢ a square yard, to plaster a room 24 ft. by 19½ ft. by 15 ft.?

5. A hall is 120 ft. by 66 ft. by 22½ ft. How much will it cost to paint the walls and ceiling, at 10 ¢ per square yard?

MEASUREMENT OF LUMBER

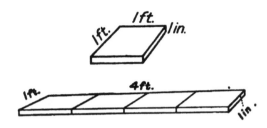

A board 1 foot square and 1 inch thick or less is a **board foot**.

A *board foot* is the *unit* in measuring lumber.

Oral and Written Work

1. Draw on the blackboard a figure to represent a board 4 feet long, 1 foot wide, and 1 inch thick.

2. Show that this board contains 4 board feet.

3. How many board feet are there in a sill 4 ft. long, 1 ft. wide, and 4 in. thick?

Observe:

(1) That the sill is equal to 4 boards 4 ft. long, 1 ft. wide, and 1 in. thick.

(2) That each board contains 4 board feet.

Hence the sill contains 4 × 4 × 1 *board foot* = 16 board feet.

The number of board feet in a piece of lumber *equals the number of board feet in one surface multiplied by the number of inches in thickness.*

Find the number of board feet in the following:

4. A plank 12 ft. long, 12 in. wide, and 2 in. thick.

5. A board 12 ft. long, 9 in. wide, and 1 in. thick.

6. A plank 15 ft. long, 12 in. wide, and 3 in. thick.

7. A plank 16 ft. long, 18 in. wide, and 2 in. thick.

8. A sill 20 ft. long, 10 in. wide, and 6 in. thick.

9. A sill 30 ft. long and 12 in. square.

Lumber is usually *sold* at so much per 1000 (M) *board feet.* Find the cost of:

10. 5000 ft. poplar at $40 per M.

11. 500 ft. hemlock at $24 per M.

12. 10,850 ft. Georgia pine at $24 per M.

13. 8000 ft. white pine at $50 per M.

Small bills of lumber are usually estimated at so much per *board foot.*

A pile of wood, of 4-foot sticks, 8 ft. in length and 4 ft. in height, is called a **cord of wood.**

$$4 \times 4 \times 8 \times 1 \; cu. \; ft. = 128 \; cu. \; ft. = 1 \; cord \; of \; wood.$$

14. How many cords are there in a pile of 4-foot wood, 160 feet long and 4 feet high?

15. Two men cut several piles of 4-foot wood that measure in all 640 feet in length and 4 feet in height. How many cords do they cut and how much do they receive for the work, at $5.50 per cord?

Wood is frequently cut for house purposes into short lengths from 16 in. to 2 ft. The price of such a cord varies according to the length of the sticks.

The **number of cords of short wood** *in a pile is found by dividing the number of square feet in one side by* 32.

16. At a school building there is a pile of 16-inch wood 80 ft. long and 4 ft. high. Find its cost at $1.50 per cord.

17. Two men cut 4 cd. of 2-foot wood each day for 16 da. Find the cost of the cutting at 70 ¢ per cord.

18. One side of a pile of 2-foot wood contains 400 sq. ft. Find the number of cords it contains.

PERCENTAGE

Finding a number when a per cent of it is given.

Oral Work

1. If $\frac{1}{2}$ of a number is 5, what is the number?

2. If 50 % of a number is 5, what is the number?

Solution. — 50 %, or $\frac{1}{2}$, of the number = 5; $\frac{1}{2}$ of the number = 2 × 5, or 10.

Find the number of which:

3. 10 is 25 %	8. 40 is 20 %	13. 75 is 75 %
4. 25 is 50 %	9. 10 is 5 %	14. 70 is 87$\frac{1}{2}$ %
5. 20 is 33$\frac{1}{3}$ %	10. 50 is 12$\frac{1}{2}$ %	15. 60 is 62$\frac{1}{2}$ %
6. 15 is 60 %	11. 30 is 66$\frac{2}{3}$ %	16. 90 is 37$\frac{1}{2}$ %
7. 60 is 75 %	12. 50 is 62$\frac{1}{2}$ %	17. 75 is 83$\frac{1}{3}$ %

Written Work

1. Find the number if 25 % of it is 128.

$$128 \div .25 = 512$$
Test: $.25 \times 512 = 128$ Since 128 is .25 times the number, the number = 128 ÷ .25, or 512.

Divide the given number by the rate per cent.

Find the number if:

2. 10 % of it is $35	8. 62$\frac{1}{2}$ % of it is $75
3. 20 % of it is $38	9. 75 % of it is $96
4. 30 % of it is $45	10. 60 % of it is $84
5. 12$\frac{1}{2}$ % of it is $56	11. 8 % of it is $24
6. 33$\frac{1}{3}$ % of it is $65	12. 35 % of it is $42
7. 37$\frac{1}{2}$ % of it is $72	13. 87$\frac{1}{2}$ % of it is $70

14. A boy pays $3, or 20% of his month's pay, for a pair of shoes. How much does he earn in a month?

15. A teacher pays $20 per month for board, and this is 40% of her salary. What is her monthly salary?

16. What is the cost of a house on which 25% is gained when it is sold at a profit of $2500?

17. If a profit of $150 is 5% of the cost, what is the cost?

18. A man drew out 24% of the money he had in bank to invest in an automobile costing $2880. How much money had he in the bank at first?

REFERENCE TABLES OF MEASURES

Liquid Measures

4 gills	= 1 pint
2 pints	= 1 quart
4 quarts	= 1 gallon
1 gal.	= 4 qt. = 8 pt.

The gill is now seldom used.
The standard unit of liquid measure is the gallon.

1 gallon	= 231 cubic inches
1 cubic foot	= nearly 7½ gallons
31½ gallons	= 1 barrel
63 gallons	= 1 hogshead

in measuring the capacity of cisterns and vats

1 gallon of water weighs nearly 8⅓ pounds
1 cubic foot of water weighs nearly 62½ pounds

Dry Measures

2 pints	= 1 quart
8 quarts	= 1 peck
4 pecks	= 1 bushel
1 bu.	= 4 pk. = 32 qt. = 64 pt.

Our standard unit, the Winchester bushel, used for measuring shelled grains, = 2150.42 cu. in., or nearly 1¼ cubic feet. In form it is a cylinder 18½ inches in diameter and 8 inches deep.

The dry gallon = 268.8 cu. in.

The heaped bushel, used for measuring corn in the ear, apples, potatoes, etc., = 2747.71 cu. in., or nearly 1⅗ cu. ft.

The standard English bushel = 2218.192 cu. in.

Measures of Length

12 inches	= 1 foot
3 feet	= 1 yard
5½ yards	= 1 rod
16½ feet	
320 rods	= 1 mile
5280 feet	

1 mi. = 320 rd. = 1760 yd. = 5280 ft. = 63360 in.

The standard unit of length is the yard.

A nautical mile (knot) = 6080.27 ft. or nearly 1.15 common miles. A league = 3 nautical miles; a fathom, used in measuring the depth of water, = 6 ft.; a hand, used in measuring the height of horses, = 4 in. A furlong = ⅛ mi.

Measures of Surface

$$144 \text{ square inches} = 1 \text{ square foot}$$
$$9 \text{ square feet} = 1 \text{ square yard}$$
$$30\tfrac{1}{4} \text{ square yards} = 1 \text{ square rod}$$
$$\left.\begin{array}{l} 160 \text{ square rods} \\ 43560 \text{ square feet} \end{array}\right\} = 1 \text{ acre}$$
$$640 \text{ acres} = 1 \text{ square mile}$$
$$1 \text{ mile square} = 1 \text{ section}$$
$$36 \text{ square miles} = 1 \text{ township}$$

The acre is not a square unit like the square foot, the square yard, etc. When in the form of a square, it is nearly 209 feet on a side.

Surveyors' Measures

Surveyors and engineers formerly used the *Gunter's Chain*. It is 66 feet long and divided into 100 links of 7.92 inches each. The tables are as follows:

Length	Surface
7.92 inches = 1 link	16 square rods = 1 square chain
100 links = 1 chain	10 square chains = 1 acre
80 chains = 1 mile	

They now generally use a steel tape 50 ft. to 100 ft. long divided into feet and tenths of a foot; or a chain 50 ft. to 100 ft. long having links each a foot in length, divided into tenths of a foot.

Land Measure is computed by dividing the number of square feet of surface by 43560, the number of square feet in an acre, and changing the decimal of an acre to square rods, etc.

Measures of Volume

$$1728 \text{ cubic inches} = 1 \text{ cubic foot}$$
$$27 \text{ cubic feet} = 1 \text{ cubic yard}$$

A cubic yard of earth is considered a load.

A cord of 4 foot wood is a pile of wood 8 feet long and 4 feet high, the sticks averaging 4 feet in length, making 128 cubic feet in the pile.

A cord of short wood is a pile of wood 8 feet long and 4 feet high, the number of cords in a pile being computed by multiplying the length of the pile in feet by the height in feet, and dividing the product by 32.

Avoirdupois Weight

16 ounces = 1 pound
100 pounds = 1 hundredweight
2000 pounds = 1 ton
2240 pounds = 1 *long* ton
   1 T. = 20 cwt. = 2000 lb. = 32000 oz.

The standard unit of weight is the **pound** = 7000 grains.
1 Av. oz. = $437\frac{1}{2}$ grains

*60 pounds = 1 bu. of wheat or of potatoes
*56 pounds = 1 bu. of shelled corn or of rye
*70 pounds = 1 bu. of corn in ear
*32 pounds = 1 bu. of oats
*48 pounds = 1 bu. of buckwheat or of barley
196 pounds = 1 bbl. of flour
200 pounds = 1 bbl. of beef or pork

*In most states.

The long ton is used in the United States custom houses and in the wholesale transactions in coal and iron. The long cwt. = 112 lb.

Measures of Time

60 seconds = 1 minute
60 minutes = 1 hour
24 hours = 1 day
7 days = 1 week
12 months } = 1 common year
365 days }
366 days = 1 leap year
10 years = 1 decade
100 years = 1 century

Thirty days have September,
April, June, and November.
All the rest have thirty-one
Save February, which alone
Has 28, and one day more
We add to it one year in four.

The true solar year is 365 days, 5 hr., 48 min., 46 sec. The standard unit of time is the **day**, which is divided into 24 hours, counting from midnight to midnight. In business transactions, 30 days are considered a month, and 12 months are regarded as a year.

The centennial years divisible by 400, and all other years divisible by 4 are **leap years**.

United States Money

10 mills = 1 cent
10 cents = 1 dime
10 dimes = 1 dollar
10 dollars = 1 eagle

INDEX

ANSWERS

FIFTH GRADE

Page 13. — **18.** 2489. **19.** 2243. **20.** 2940. **21.** 2554. **22.** 2726. **23.** 2735. **24.** 9450. **25.** 54,130. **26.** 26,083. **27.** 233,301.

Page 14. — **28.** 54,603. **29.** 20,427. **30.** 7844. **31.** 20,025. **32.** 26,399. **33.** $174.77. **34.** $1345.11. **35.** $1247.82. **36.** $18,934.63. **37.** $267.94. **38.** $3195.20. **39.** $2120.46. **40.** $3039.49. **41.** $741.57. **42.** $553.85. **43.** $505.25. **44.** $790.17.

Page 15. — **45.** $1599.98. **46.** 2,082,724. **47.** $1696.22. **48.** 2,689,411. **49.** $620.48. **50.** 3,211,787. **51.** $660.52. **52.** 2,965,515.

Page 17. — **22.** 158,219. **23.** 357,511. **24.** 349,197. **25.** 81,485. **26.** 51,008. **27.** 133,003. **28.** 701,898. **29.** 171,951. **30.** 107,803.

Page 18. — **32.** 16,909. **33.** 62,062. **34.** 5500. **35.** $953.86. **36.** $936.088. **37.** $963.98. **38.** $2959.96. **39.** $17,062.83. **40.** $309.16. **41.** $6530.43. **42.** 440,441. **43.** 373,068. **44.** 346,229. **45.** $1844.12. **46.** $4833.22. **47.** $5442.91. **48.** $2856.17. **49.** $2408.91. **50.** $72,669.65. **51.** $10,009.89.

Page 20. — **3.** 2835 bu. **4.** 2523 ft. **5.** 9890 doz. **6.** $25,428. **7.** $25,198. **8.** $13,195. **9.** 59,073. **10.** 83,895. **11.** 47,192. **12.** 2925. **13.** 4368. **14.** 7020. **15.** 2600. **16.** 3738. **17.** 3900. **18.** 1710. **19.** 3015. **20.** 3216. **21.** 2278. **22.** 9400. **23.** 30,150. **24.** 38,286. **25.** 68,100. **26.** 6104. **27.** 14,335. **28.** 14,652. **29.** $266.49. **30.** $273. **31.** $244.67 **32.** $672.75. **33.** $4087.70. **34.** $6644.40. **35.** $6496.10. **36.** $6678.12. **37.** $5151.90. **38.** $3131.22. **39.** $1520. **40.** $1821.76. **41.** $1844.10.

Page 21. — **45.** 3,083,750. **46.** 743,330. **47.** 1,239,200. **48.** 1,401,972. **49.** 4,919,502. **50.** 2,752,122. **51.** 5,251,350. **52.** 3,215,300. **53.** 4,866,432. **54.** 3,511,200. **55.** 6,730,470. **56.** 1,192,276. **57.** 2,432,500. **58.** 1,241,884. **59.** 2,955,960. **60.** 5,552,820. **61.** 4,266,600. **62.** 899,001. **63.** 5,040,000. **64.** 7,188,384. **65.** 4,524,660. **66.** 3,458,400. **67.** 3,154,050. **68.** 3,479,211. **69.** 2,837,420. **70.** 5,438,111. **71.** 579,579. **72.** 7,281,809. **73.** 4,648,581. **74.** 2,955,591. **75.** 3,390,000. **76.** 2,700,450. **77.** 5,296,980. **78.** 3,584,560. **79.** 101,000. **80.** 250,700. **81.** 5,664,708. **82.** 3,354,000. **83.** 5,492,700. **84.** 7,488,832. **85.** 3,451,890. **86.** 5,420,430. **87.** 343,060. **88.** 3,857,142. **89.** 677,039. **90.** a. 1,574,125; b. 2,210,200; c. 3,623,700; d. 4,793,050; e. 5,165,700; f. 6,174,425; g. 5,583,325; h. 5,114,800; i. 6,219,400; j. 5,769,650. **91.** a. 250,880; b. 352,256; c. 577,536; d. 763,904; e. 823,296; f. 984,064; g. 889,856; h. 815,104; i. 991,232; j. 919,552. **92.** a. 2,136,400; b. 2,999,680;

c. 4,918,080 ;   *d.* 6,505,120 ;   *e.* 7,010,880 ;   *f.* 8,379,920 ;   *g.* 7,577,680 ;
h. 6,941,120 ;   *i.* 8,440,960 ;   *j.* 7,830,560.   **93.** *a.* 2,364,740 ;   *b.* 3,320,288 ;
c. 5,443,728 ;   *d.* 7,200,392 ;   *e.* 7,760,208 ;   *f.* 9,275,572 ;   *g.* 8,387,588 ;
h. 7,682,992 ;   *i.* 9,343,136 ;   *j.* 8,667,496.   **94.** *a.* 2,122,925 ;   *b.* 2,980,760 ;
c. 4,887,060 ;   *d.* 6,464,090 ;   *e.* 6,966,660 ;   *f.* 8,327,065 ;   *g.* 7,529,885 ;
h. 6,897,340 ;   *i.* 8,387,720 ;   *j.* 7,781,170.   **95.** *a.* 1,934,030 ;   *b.* 2,715,536 ;
c. 4,452,216 ;   *d.* 5,888,924 ;   *e.* 6,346,776 ;   *f.* 7,586,134 ;   *g.* 6,859,886 ;
h. 6,283,624 ;   *i.* 7,641,392 ;   *j.* 7,088,812.   **96.** *a.* 2,073,925 ;   *b.* 2,911,960 ;
c. 4,774,260 ;   *d.* 6,314,890 ;   *e.* 6,805,860 ;   *f.* 8,134,865 ;   *g.* 7,356,085 ;
h. 6,738,140 ;   *i.* 8,194,120 ;   *j.* 7,601,570.   **97.** *a.* 1,885,275 ;   *b.* 2,647,080 ;
c. 4,339,980 ;   *d.* 5,740,470 ;   *e.* 6,186,780 ;   *f.* 7,894,895 ;   *g.* 6,686,955 ;
h. 6,125,220 ;   *i.* 7,448,760 ;   *j.* 6,910,110.   **98.** *a.* 2,064,125 ;   *b.* 2,898,200 ;
c. 4,751,700 ;   *d.* 6,285,050 ;   *e.* 6,773,700 ;   *f.* 8,096,425 ;   *g.* 7,321,325 ;
h. 6,706,300 ;   *i.* 8,155,400 ;   *j.* 7,565,650.   **99.** *a.* 2,321,620 ;   *b.* 3,259,744 ;
c. 5,344,464 ;   *d.* 7,069,096 ;   *e.* 7,618,704 ;   *f.* 9,106,436 ;   *g.* 8,234,644 ;
h. 7,542,896 ;   *i.* 9,172,768 ;   *j.* 8,509,448.

Page 23. — **2.** $4969\frac{2}{4}$.   **3.** $22994\frac{4}{8}$.   **4.** $36,357\frac{3}{4}$.   **5.** $5506\frac{44}{8}$.   **6.** $7240\frac{44}{8}$.
7. 2201.   **8.** 1357.   **9.** $13,938\frac{47}{8}$.   **10.** $12,583\frac{5}{8}$.   **11.** $13,397\frac{77}{8}$.
12. $9564\frac{44}{8}$.   **13.** $5787\frac{47}{8}$.   **14.** $4494\frac{44}{8}$.   **15.** $3185\frac{44}{8}$.   **16.** $2223\frac{44}{8}$.
17. 987.   **18.** 805.   **19.** 11,336.   **20.** $6247\frac{44}{8}$.   **21.** 105.   **22.** 109.

Page 24. — **23.** $6457\frac{16}{15}$.   **24.** $3181\frac{49}{60}$.   **25.** $2768\frac{444}{8}$.   **26.** $3097\frac{28}{85}$.
27. $1425\frac{44}{8}$.   **28.** $861\frac{47}{8}$.   **29.** $6827\frac{42}{8}$.   **30.** \$407.   **31.** \$409.   **32.** \$70$\frac{44}{8}$.
33. \$83$\frac{44}{8}$.   **34.** \$152$\frac{58}{8}$.   **35.** \$50$\frac{44}{8}$.   **36.** \$22$\frac{68}{50}$.   **37.** 1503$\frac{49}{8}$.
38. 1001$\frac{44}{8}$.   **39.** 542$\frac{41}{17}$.   **40.** 478$\frac{49}{7}$.   **41.** 885$\frac{35}{05}$.   **42.** 135$\frac{44}{17}$.
43. 308$\frac{47}{8}$.   **44.** *a.* 24,120$\frac{44}{8}$;   *b.* 18,280$\frac{44}{8}$;   *c.* 9562$\frac{29}{78}$;   *d.* 8528$\frac{47}{8}$;
e. 7473$\frac{44}{8}$; *f.* 6569$\frac{444}{8}$; *g.* 8193$\frac{44}{775}$; *h.* 7206$\frac{44}{8}$; *i.* 12,700$\frac{41}{8}$; *j.* 7263$\frac{44}{8}$.
45. *a.* 29,278$\frac{44}{8}$; *b.* 22,165$\frac{49}{8}$; *c.* 11,607$\frac{44}{8}$; *d.* 10,351$\frac{47}{8}$; *e.* 9071$\frac{44}{8}$;
f. 7974$\frac{44}{8}$;   *g.* 9945$\frac{44}{8}$;   *h.* 8747$\frac{47}{8}$;   *i.* 15,416$\frac{16}{505}$;   *j.* 8816$\frac{44}{8}$.
46. *a.* 21,882$\frac{174}{77}$; *b.* 16,566$\frac{44}{8}$; *c.* 8675$\frac{44}{8}$; *d.* 7736$\frac{49}{8}$; *e.* 6779$\frac{44}{8}$;
f. 5959$\frac{44}{8}$;   *g.* 7432$\frac{49}{8}$;   *h.* 6537$\frac{44}{8}$;   *i.* 11,521$\frac{44}{8}$;   *j.* 6589$\frac{44}{8}$.
47. *a.* 32,258$\frac{127}{8}$; *b.* 24,421$\frac{197}{8}$; *c.* 12,788$\frac{444}{8}$; *d.* 11,405$\frac{44}{8}$; *e.* 9994$\frac{44}{8}$;
f. 8785$\frac{49}{8}$;   *g.* 10,957$\frac{44}{8}$;   *h.* 9638$\frac{55}{8}$;   *i.* 16,984$\frac{44}{8}$;   *j.* 9713$\frac{47}{8}$.
48. *a.* 35,913$\frac{44}{8}$; *b.* 27,188$\frac{410}{8}$; *c.* 14,237$\frac{49}{8}$; *d.* 12,697$\frac{44}{8}$; *e.* 11,127$\frac{17}{185}$;
f. 9781$\frac{44}{8}$;   *g.* 12,198$\frac{49}{8}$; *h.* 10,730$\frac{42}{87}$;   *i.* 18,909$\frac{49}{8}$;   *j.* 10,$\frac{91444}{8}$;
49. *a.* 29,045$\frac{40}{105}$; *b.* 21,988$\frac{44}{8}$; *c.* 11,514$\frac{49}{878}$; *d.* 10,269$\frac{98}{8}$; *e.* 8998$\frac{44}{8}$;
f. 7910$\frac{49}{8}$;   *g.* 9865$\frac{44}{8}$;   *h.* 8677$\frac{44}{8}$;   *i.* 15,292$\frac{473}{8}$;   *j.* 8746$\frac{49}{8}$.
50. *a.* 25,411$\frac{215}{8}$; *b.* 19,238$\frac{44}{8}$; *c.* 10,074$\frac{440}{8}$; *d.* 8984$\frac{44}{8}$; *e.* 7873$\frac{44}{8}$;
f. 6921$\frac{449}{8}$;   *g.* 8631$\frac{44}{8}$;   *h.* 7592$\frac{49}{887}$;   *i.* 13,379$\frac{44}{8}$;   *j.* 7652$\frac{44}{8}$.
51. *a.* 29,492$\frac{49}{8}$; *b.* 22,327$\frac{447}{8}$; *c.* 11,692$\frac{44}{878}$; *d.* 10,427$\frac{49}{8}$; *e.* 9137$\frac{449}{8}$;
f. 8032$\frac{447}{8}$;   *g.* 10,017$\frac{44}{8}$;   *h.* 8811$\frac{44}{8}$;   *i.* 15,528$\frac{473}{8}$;   *j.* 8880$\frac{44}{8}$.
52. *a.* 18,160$\frac{49}{8}$;   *b.* 13,748$\frac{44}{8}$;   *c.* 7109$\frac{411}{8}$;   *d.* 6420$\frac{44}{8}$;   *e.* 5626$\frac{49}{8}$;
f. 4946$\frac{181}{81}$;   *g.* 6168$\frac{44}{8}$;   *h.* 5425$\frac{49}{8}$;   *i.* 9561$\frac{44}{8}$;   *j.* 5468$\frac{44}{8}$.
53. *a.* 12,069$\frac{75}{105}$;   *b.* 9137$\frac{9}{54}$;   *c.* 4784$\frac{44}{8}$;   *d.* 4267$\frac{44}{8}$;   *e.* 3789$\frac{44}{8}$;
f. 3287$\frac{44}{8}$; *g.* 4099$\frac{49}{8}$; *h.* 3805$\frac{44}{8}$; *i.* 6354$\frac{44}{8}$; *j.* 3634$\frac{197}{8}$.   **3.** \$1.12.
4. \$.85.   **5.** \$.09.   **6.** 40 pads.   **7.** 15 caps.   **8.** 45 plates.

Page 26. — **1.** 26.   **2.** 32.   **3.** 18.   **4.** 17.   **5.** 29.   **6.** 28.   **7.** 957.
8. 702.   **9.** 608.   **10.** 918.   **11.** 63.   **12.** 49.   **13.** 57.   **14.** 1656.
15. 31.   **16.** 44.   **17.** 1443.   **18.** 71.   **19.** 2496.   **20.** 64.

Page 30. — 5. $\frac{1}{2}$; $\frac{1}{3}$; $\frac{7}{8}$. 7. $1\frac{1}{2}$; 2; $2\frac{1}{4}$. 8. $\frac{4}{5}$; $\frac{1}{2}$; $\frac{1}{3}$; $\frac{1}{8}$. 9. $6\frac{1}{2}$.
10. $15\frac{2}{3}$. 11. $16\frac{1}{4}$. 12. $15\frac{2}{3}$. 13. $17\frac{1}{4}$. 14. $5\frac{1}{4}$; $15\frac{1}{4}$. 15. $31\frac{1}{4}$; 94.
16. $2\frac{1}{4}$; $8\frac{1}{4}$. 17. $6\frac{1}{4}$; $18\frac{2}{3}$. 18. $24\frac{1}{4}$; $103\frac{1}{2}$. 19. $1\frac{1}{4}$; $18\frac{1}{4}$. 20. $19\frac{1}{4}$;
$36\frac{1}{4}$. 21. $2\frac{1}{4}$; $51\frac{1}{4}$. 22. 6; 13. 23. $13\frac{1}{4}$; $25\frac{1}{4}$. 24. $9\frac{1}{4}$; $28\frac{1}{4}$. 25. $1\frac{1}{4}$;
$9\frac{1}{4}$. 26. $20\frac{1}{4}$; $52\frac{7}{8}$. 27. $20\frac{4}{5}$; $60\frac{7}{8}$. 28. $6\frac{1}{4}$; $11\frac{1}{4}$. 29. $7\frac{1}{4}$. 30. $73\frac{1}{4}$.
31. $72\frac{7}{8}$. 32. $33\frac{7}{8}$. 33. $62\frac{1}{4}$. 34. $\frac{1}{4}$ yd. 35. $1\frac{1}{4}$ yd. 36. $\frac{1}{4}$ yd.
37. $1\frac{1}{4}$ lb.

Page 32. — 5. $\frac{2}{3}$. 6. $\frac{7}{12}$. 7. $8\frac{2}{3}$. 8. $4\frac{1}{4}$. 9. $\frac{1}{3}$. 10. $\frac{5}{12}$. 11. $2\frac{7}{12}$.
12. $1\frac{1}{4}$. 13. $25\frac{7}{12}$; $18\frac{1}{4}$. 14. $19\frac{1}{4}$; $6\frac{7}{12}$. 15. $22\frac{1}{4}$; $11\frac{1}{4}$. 16. $26\frac{7}{12}$; $5\frac{7}{12}$.
17. $23\frac{1}{4}$; $2\frac{5}{12}$.

ag **33.** — 18. $15\frac{2}{3}$; $7\frac{1}{4}$. 19. $11\frac{7}{12}$; $1\frac{5}{12}$. 20. $58\frac{5}{12}$; $25\frac{1}{4}$. 21. $28\frac{1}{3}$; $4\frac{1}{4}$;
22. P $21\frac{1}{12}$; $10\frac{1}{12}$. 23. $\frac{11}{12}$ doz.; 11 buttons. 24. $5\frac{1}{4}$ hr.; $2\frac{1}{4}$ more hour.
25. $2\frac{1}{4}$.

Page 35. — 16. $5\frac{3}{4}$ in. 17. $13\frac{3}{4}$ yd. 18. $2\frac{1}{4}$ ft. 19. $10\frac{1}{2}$ ft. 20. 1 ft.
21. $1\frac{1}{4}$ yd. 22. $1\frac{1}{2}$ ft. 23. $5\frac{1}{4}$ ft. 1. $18\frac{3}{4}$. 2. $13\frac{3}{4}$. 3. $21\frac{1}{4}$. 4. $18\frac{1}{4}$. 5. $20\frac{1}{4}$. 6. $61\frac{7}{8}$. 7. $54\frac{1}{4}$. 8. $27\frac{2}{3}$.
9. 100. 10. 98. 11. $12\frac{1}{4}$. 12. $14\frac{7}{8}$. 13. $15\frac{1}{4}$. 14. $155\frac{7}{12}$. 15. $83\frac{1}{4}$.

Page 36. — 16. $8\frac{1}{2}$. 17. $12\frac{7}{12}$. 18. $3\frac{1}{4}$. 19. $3\frac{1}{4}$. 20. $3\frac{7}{8}$. 21. $73\frac{1}{4}$.
22. $9\frac{1}{4}$. 23. $21\frac{1}{4}$. 24. $2\frac{1}{4}$. 25. $4\frac{1}{2}$. 26. $64\frac{7}{12}$. 27. $29\frac{1}{4}$. 28. $1\frac{1}{4}$.
29. $29\frac{1}{4}$. 30. $4\frac{1}{4}$. 31. $41\frac{1}{4}$. 32. $4\frac{1}{4}$. 33. $9\frac{2}{3}$. 34. $2\frac{1}{4}$. 35. $26\frac{1}{4}$.
36. $5\frac{3}{4}$; $5\frac{1}{4}$; $5\frac{1}{4}$.

Page 37. — 1. $\$5\frac{9}{10}$. 2. $\$16\frac{1}{2}$. 3. $29\frac{7}{10}$ mi. 4. $9\frac{1}{4}$. 5. $38\frac{7}{15}$ mi.
6. $\$61\frac{17}{20}$. 7. $90\frac{1}{4}$. 8. $31\frac{7}{10}$ da. 9. $\$1\frac{7}{10}$. 10. $12\frac{2}{15}$ hr. 11. $\$4\frac{9}{10}$.
12. $36\frac{4}{15}$ min. 13. $5\frac{1}{4}$ hr.

Page 38. — 14. $\frac{1}{12}$ hr. longer; 5 min. 15. $\frac{1}{4}$ of a dozen; $\frac{1}{4}$ of a foot.
16. $\frac{9}{10}$ of a dollar; 90 cents. 17. $5\frac{3}{4}$. 18. $6\frac{1}{4}$. 19. $5\frac{9}{10}$. 20. $12\frac{7}{10}$.
21. $4\frac{1}{4}$. 22. $7\frac{4}{15}$. 23. $6\frac{4}{15}$. 24. $6\frac{1}{10}$. 25. $3\frac{7}{10}$. 26. $2\frac{1}{10}$.
27. $1\frac{9}{10}$. 28. $1\frac{4}{15}$.

Page 39. — 10. $10\frac{11}{16}$ ft. 11. $\$54\frac{5}{8}$. 12. $37\frac{1}{16}$. 13. $22\frac{7}{16}$.
14. $\$4\frac{7}{16}$. 15. $5\frac{9}{16}$ yd. 16. $4\frac{5}{16}$ mi. 17. $16\frac{7}{16}$. 18. $15\frac{1}{4}$ ft. 19. 54 rd.
20. $3\frac{1}{4}$ yd. 21. $2\frac{7}{16}$ in. 22. $\$8\frac{1}{4}$.

Page 41. — 1. $1\frac{1}{2}$; $1\frac{1}{2}$; $1\frac{1}{5}$; 2; $1\frac{1}{4}$; $1\frac{2}{3}$; $1\frac{1}{2}$; $1\frac{1}{15}$; $1\frac{1}{4}$. 2. $107\frac{1}{15}$.
3. $99\frac{1}{4}$ ft. 4. $44\frac{1}{4}$. 5. $75\frac{11}{16}$ mi. 6. $105\frac{1}{4}$. 7. 50 bu. 8. $24\frac{7}{15}$.
9. $35\frac{7}{12}$ da. 10. $2\frac{1}{4}$. 11. $2\frac{1}{4}$ in. 12. $2\frac{1}{4}$. 13. $2\frac{7}{15}$ bu. 14. $2\frac{7}{15}$.
15. $4\frac{1}{4}$ lb. 16. $2\frac{1}{12}$. 17. $10\frac{1}{4}$. 18. $52\frac{1}{4}$ ft.; $1\frac{1}{2}$ ft. longer. 19. $14\frac{1}{4}$ in.
20. $18\frac{5}{12}$ hr. 21. $\frac{1}{4}$ mi. farther; $5\frac{3}{4}$ mi.

Page 42. — 1. $77\frac{7}{16}$ in. 2. $83\frac{11}{16}$ bu. 3. $71\frac{9}{16}$. 4. $89\frac{9}{16}$.
5. $2\frac{7}{16}$ yd. 6. $7\frac{1}{4}$ da. 7. $15\frac{5}{16}$. 8. $31\frac{7}{16}$.

Page 45. — 8. a. $\frac{9}{10}$; b. $\frac{11}{12}$; c. $\frac{9}{11}$; d. $\frac{17}{12}$; e. $\frac{4}{5}$; f. $\frac{7}{8}$. 9. a. $\frac{5}{11}$;
b. $\frac{4}{5}$; c. $\frac{7}{10}$; d. $\frac{1}{2}$; e. $\frac{11}{12}$; f. $\frac{11}{14}$. 10. a. $\frac{4}{5}$; b. $\frac{11}{12}$; c. $\frac{4}{5}$; d. $\frac{4}{7}$;
e. $\frac{9}{11}$; f. $\frac{11}{14}$.

Page 47. — 2. $\frac{47}{7}$. 3. $\frac{71}{11}$. 4. $\frac{111}{8}$. 5. $\frac{140}{9}$. 6. $\frac{217}{8}$.
7. $\frac{216}{5}$. 8. $\frac{563}{8}$. 9. $\frac{725}{12}$. 10. $\frac{593}{7}$. 11. $\frac{701}{8}$. 12. $\frac{1141}{8}$.
13. $\frac{409}{10}$. 14. $\frac{1385}{13}$. 15. $\frac{712}{9}$. 16. $\frac{1927}{17}$. 17. $\frac{372}{20}$. 18. $\frac{1151}{11}$.
19. $\frac{4011}{50}$. 20. $\frac{1901}{25}$. 21. $\frac{1074}{13}$. 22. $\frac{471}{8}$. 23. $\frac{1317}{20}$. 24. $\frac{712}{8}$.
25. $\frac{211}{8}$.

Page 48. — 26. $1\frac{11}{13}$. 27. $2\frac{14}{15}$. 28. $\frac{452}{15}$. 29. $4\frac{23}{25}$. 30. $2\frac{11}{14}$.
31. $1\frac{91}{10}$. 32. $\frac{307}{12}$. 33. $2\frac{17}{11}$. 34. $\frac{412}{15}$. 35. $4\frac{21}{12}$. 36. $1\frac{121}{13}$.
37. $4\frac{17}{13}$. 38. $2\frac{07}{14}$. 39. $4\frac{211}{30}$. 40. $4\frac{21}{10}$. 41. $\frac{445}{15}$.

Page 50. — 2. $\frac{4}{5}$, $\frac{7}{8}$. 3. $\frac{8}{12}$, $\frac{7}{12}$. 4. $\frac{4}{10}$, $\frac{8}{10}$. 5. $\frac{7}{12}$, $\frac{5}{12}$. 6. $\frac{12}{16}$.
7. $\frac{2}{16}$, $\frac{6}{16}$. 8. $1\frac{4}{15}$, $1\frac{7}{15}$. 9. $\frac{2}{10}$, $\frac{4}{10}$. 10. $\frac{5}{12}$, $\frac{8}{12}$. 11. $\frac{10}{16}$.
$\frac{8}{16}$, $\frac{2}{16}$. 12. $1\frac{6}{10}$, $\frac{4}{10}$. 13. $1\frac{5}{16}$, $\frac{10}{16}$, $1\frac{9}{16}$. 14. $\frac{8}{15}$, $\frac{6}{15}$, $\frac{7}{15}$. 15. $\frac{7}{12}$.
$\frac{9}{16}$, $\frac{2}{16}$. 16. $\frac{5}{10}$, $\frac{8}{16}$. 17. $\frac{2}{16}$, $1\frac{5}{16}$. 18. $\frac{5}{12}$, $1\frac{4}{12}$, $\frac{3}{12}$. 19. $\frac{7}{10}$, $\frac{6}{10}$, $\frac{4}{10}$.
$\frac{9}{16}$, $\frac{2}{16}$, $1\frac{7}{16}$. 21. $\frac{7}{10}$, $\frac{4}{10}$, $\frac{6}{10}$. 22. $\frac{70}{10}$, $\frac{104}{12}$, $\frac{41}{12}$. 23. $\frac{4}{12}$, $\frac{7}{12}$, $\frac{6}{12}$.
24. $1\frac{9}{16}$, $1\frac{4}{16}$. 25. $\frac{5}{10}$, $\frac{10}{16}$, $\frac{4}{12}$. 26. $\frac{4}{12}$, $\frac{7}{12}$, $\frac{6}{12}$. 27. $1\frac{11}{16}$, $1\frac{4}{16}$, $1\frac{9}{16}$.
28. $1\frac{11}{16}$, $1\frac{10}{16}$, $\frac{41}{10}$. 29. $\frac{7}{10}$, $\frac{4}{10}$, $\frac{6}{10}$.

Page 51. — 2. $1\frac{12}{16}$. 3. $1\frac{11}{12}$. 4. $1\frac{14}{16}$. 5. $1\frac{5}{12}$. 6. $1\frac{7}{8}$. 7. $1\frac{11}{14}$.
8. $\frac{9}{11}$. 9. $1\frac{11}{15}$. 10. $1\frac{11}{14}$. 11. $2\frac{5}{13}$. 12. $\frac{9}{11}$. 13. $2\frac{3}{10}$. 14. $2\frac{7}{14}$.
15. $1\frac{14}{15}$. 16. $1\frac{11}{14}$. 17. $1\frac{14}{16}$. 18. $2\frac{11}{13}$. 19. $2\frac{17}{16}$.

Page 52. — 2. $16\frac{1}{4}$. 3. $219\frac{1}{8}$. 4. $99\frac{1}{4}$. 5. $92\frac{11}{16}$. 6. 192.
7. $618\frac{1}{4}$. 8. $382\frac{7}{16}$. 9. $257\frac{7}{16}$. 10. $92\frac{11}{14}$. 11. $5\frac{11}{16}$. 12. $29\frac{5}{16}$.
13. $17\frac{11}{16}$. 14. $18\frac{17}{10}$. 15. $32\frac{7}{16}$. 16. $6\frac{17}{11}$. 17. $9\frac{7}{8}$. 18. $19\frac{11}{16}$.
19. $18\frac{7}{11}$. 20. $6\frac{14}{14}$. 21. $8\frac{1}{2}$ mi. 22. $\$67\frac{7}{10}$. 23. $\$33$.

Page 53. — 24. $4\frac{1}{4}$ A. 25. $28\frac{1}{4}$ in. 26. $63\frac{7}{16}$ mi. 27. $\$35\frac{5}{16}$.
28. $\$23\frac{7}{16}$. 29. $386\frac{7}{16}$ lb. 30. $213\frac{1}{4}$ rd. 31. $160\frac{5}{16}$ ft. 32. $106\frac{1}{4}$ ft.
33. $1409\frac{7}{16}$ mi.

Page 54. — 34. $21\frac{11}{16}$ bu. 35. $1\frac{1}{16}$. 36. $\$1\frac{1}{4}$. 37. $5\frac{1}{4}$ ft. 38. $\$4\frac{5}{16}$.
39. $\$2\frac{11}{16}$. 40. $8\frac{1}{4}$ T. 41. $12\frac{7}{16}$ in. 42. $29\frac{1}{4}$ mi. 43. $\$19\frac{9}{16}$.
44. $6\frac{11}{16}$ yd.

Page 56. — 2. $\frac{1}{6}$. 3. $\frac{1}{10}$. 4. $\frac{1}{13}$. 5. $\frac{1}{4}$. 6. $\frac{4}{15}$. 7. $\frac{3}{8}$. 8. $\frac{7}{11}$.
9. $\frac{1}{11}$. 10. $\frac{1}{15}$. 11. $\frac{1}{11}$. 12. $\frac{7}{12}$. 13. $\frac{1}{10}$. 14. $\frac{4}{15}$. 15. $\frac{5}{14}$. 16. $\frac{1}{7}$.
17. $\frac{3}{14}$. 18. $\frac{1}{10}$. 19. $\frac{1}{13}$. 20. $\frac{1}{10}$. 21. $\frac{5}{10}$. 22. $\frac{7}{13}$. 23. $\frac{1}{5}$ yd.
24. $\frac{1}{10}$ A. 25. $\frac{1}{10}$ of the distance farther.

Page 57. — 52. $1\frac{1}{4}$. 53. $3\frac{1}{4}$. 54. $7\frac{1}{4}$. 55. $8\frac{1}{4}$. 56. $7\frac{1}{4}$. 57. $13\frac{7}{16}$.
58. $13\frac{7}{15}$. 59. $53\frac{7}{16}$. 60. $62\frac{7}{11}$. 61. $55\frac{5}{16}$. 62. $66\frac{1}{4}$. 63. $66\frac{5}{8}$.
64. $15\frac{7}{10}$. 65. $13\frac{7}{16}$. 66. $48\frac{7}{11}$. 67. $48\frac{7}{15}$. 68. $10\frac{7}{15}$. 69. $12\frac{7}{13}$.
70. $40\frac{5}{16}$. 71. $7\frac{3}{4}$. 72. $102\frac{1}{4}$.

Page 58. — 74. $1\frac{1}{4}$. 75. $3\frac{1}{4}$. 76. $2\frac{17}{17}$. 77. $4\frac{4}{5}$. 78. $6\frac{1}{4}$.
79. $19\frac{11}{16}$. 80. $4\frac{7}{16}$. 81. $5\frac{11}{14}$. 82. $32\frac{1}{4}$. 83. $25\frac{7}{16}$. 84. $66\frac{11}{14}$.
85. $9\frac{9}{16}$. 86. $1\frac{11}{14}$ hr. 87. $\$6\frac{1}{4}$. 88. $\frac{7}{10}$ mi. 89. $\$2\frac{11}{16}$. 90. $\frac{3}{4}$ ft.
91. $7\frac{1}{4}$ ft. 92. $8\frac{1}{4}$ gal. 93. $\$24\frac{11}{16}$.

Page 59. — 1. $\frac{7}{14}$. 2. $4\frac{1}{4}$. 3. $5\frac{1}{4}$. 4. $1\frac{7}{8}$. 5. $8\frac{9}{16}$. 6. $9\frac{17}{16}$.
7. $2\frac{5}{16}$. 8. $17\frac{11}{14}$. 9. $\frac{1}{4}$. 10. $16\frac{1}{4}$. 11. $15\frac{5}{16}$. 12. $130\frac{11}{14}$.
13. $9\frac{17}{17}$. 14. $23\frac{7}{16}$. 15. $5\frac{4}{16}$. 16. $9\frac{7}{11}$. 17. $74\frac{11}{14}$. 18. $438\frac{17}{17}$.
19. $4\frac{11}{14}$. 20. $44\frac{109}{16}$. 21. $\$2\frac{1}{4}$. 22. $\$258$. 23. $\$55\frac{3}{4}$. 24. $\$209.25$.
25. $\frac{1}{4}$ da. 26. 201 rd.

Page 60. — 27. $\$7\frac{7}{10}$. 28. $\frac{7}{16}$. 29. $\$3\frac{11}{16}$. 30. $1\frac{1}{4}$ yd. 31. $1\frac{1}{4}$ hr.
32. $\frac{4}{15}$. 33. $2\frac{1}{4}$ lb. 34. $8\frac{1}{4}$ A. 35. $\frac{11}{16}$. 36. $424\frac{1}{4}$ gal. 37. $610\frac{9}{16}$ mi.

Page 61. — 38. $243\frac{3}{4}$ lb. 39. $43\frac{9}{10}$ hr.; $28\frac{11}{14}$ hr. 40. $21\frac{1}{4}$ mi.; $1\frac{5}{16}$
mi. farther. 41. 84 yd.; $9\frac{1}{4}$ yd. longer. 45. $\frac{1}{4}$ lb.; $\frac{5}{16}$ lb.; $\frac{1}{4}$ lb.; $\frac{1}{4}$ lb.;
$1\frac{1}{16}$ lb.

Page 62.—**47.** $7\frac{1}{4}$ T. **48.** $20\frac{11}{16}$ in. **49.** $2\frac{1}{16}$ in.; $3\frac{5}{16}$ in. **50.** $4\frac{3}{4}$ ft.; $3\frac{3}{4}$ ft.; $1\frac{1}{2}$ ft. **51.** $\frac{5}{16}$ da.; $\frac{1}{3}$ da.; $\frac{7}{33}$ da.; $2\frac{1}{11}\frac{1}{4}$ hr. **52.** $2\frac{9}{20}$ sec. **53.** $70\frac{1}{4}$ lb.; $98\frac{1}{4}$ lb.; $168\frac{3}{4}$ lb.; $27\frac{3}{4}$ lb. **54.** Paul, $\frac{1}{8}$ in.; Susan, $\frac{7}{16}$ in.

Page 63.—**55.** $6\frac{23}{24}$ yd. Addition. **56.** $6\frac{17}{40}$; $7\frac{11}{20}$; $11\frac{7}{15}$; $8\frac{9}{10}$; $10\frac{11}{16}$; $12\frac{7}{15}$. **57.** $10\frac{43}{48}$; $9\frac{17}{24}$; $9\frac{17}{20}$; $12\frac{49}{52}$; $14\frac{1}{24}$; $13\frac{5}{8}$. **58.** $20\frac{7}{10}$; 21; $23\frac{41}{60}$; $30\frac{13}{18}$; $40\frac{13}{48}$; $34\frac{13}{60}$. **59.** $49\frac{7}{8}$; $54\frac{7}{16}$; $58\frac{7}{10}$; $23\frac{11}{24}$; $69\frac{7}{8}$; $74\frac{11}{40}$. **60.** $83\frac{5}{16}$; $88\frac{17}{18}$; $67\frac{23}{40}$; $85\frac{17}{40}$; $76\frac{53}{60}$; $106\frac{51}{70}$. Subtraction. **56.** $\frac{43}{60}$; $1\frac{13}{40}$; $5\frac{11}{18}$; $2\frac{7}{10}$; $4\frac{19}{40}$; $6\frac{9}{16}$. **57.** $1\frac{21}{24}$; $\frac{21}{36}$; $\frac{11}{44}$; $4\frac{5}{72}$; $5\frac{7}{24}$; $4\frac{7}{8}$. **58.** $3\frac{1}{4}$; $4\frac{5}{8}$; $7\frac{1}{18}$; $14\frac{7}{80}$; $24\frac{7}{60}$; $17\frac{37}{70}$. **59.** $31\frac{1}{8}$; $35\frac{11}{18}$; $39\frac{11}{40}$; $5\frac{7}{10}$; $51\frac{3}{8}$; $56\frac{7}{60}$. **60.** $67\frac{1}{10}$; $72\frac{7}{45}$; $51\frac{7}{80}$; $69\frac{11}{40}$; $60\frac{11}{40}$; $90\frac{19}{110}$.

Page 64.—**3.** $9\frac{3}{20}$. **4.** $6\frac{1}{10}$. **5.** $46\frac{9}{10}$. **6.** $1\frac{31}{44}$. **7.** $11\frac{11}{14}$. **8.** $11\frac{23}{30}$. **9.** $8\frac{7}{9}$. **10.** $8\frac{11}{14}$. **11.** $88\frac{43}{80}$. **12.** $12\frac{7}{24}$. **13.** $28\frac{7}{15}$. **14.** $113\frac{11}{14}$. **15.** $9\frac{21}{40}$. **16.** $10\frac{1}{4}$. **17.** $10\frac{7}{16}$. **18.** $10\frac{1}{4}$. **19.** $44\frac{21}{44}$. **20.** $28\frac{11}{16}$. **21.** $18\frac{5}{8}$. **22.** $15\frac{1}{4}$. **23.** $6\frac{11}{14}$. **24.** $5\frac{11}{16}$. **25.** $20\frac{11}{14}$. **26.** $8\frac{7}{15}$. **27.** $7\frac{1}{4}$. **28.** $14\frac{1}{10}$ hr. **29.** $4\frac{1}{8}$ yd. **30.** $22\frac{1}{4}$ yr. **31.** $2\frac{1}{4}$ ft.

Page 67.—**2.** 36. **3.** 45. **4.** 24. **5.** 45. **6.** 36. **7.** 32. **8.** 102. **9.** 51. **10.** 18. **11.** $26\frac{1}{4}$. **12.** 45. **13.** 108. **14.** $579\frac{1}{4}$. **15.** 2685. **16.** 1826. **17.** $614\frac{2}{3}$. **18.** 1320. **19.** 360. **20.** $2843\frac{1}{4}$. **21.** $1779\frac{6}{7}$. **22.** 1470. **23.** $733\frac{1}{3}$. **24.** $1756\frac{4}{7}$. **25.** $368\frac{2}{3}$. **26.** 1425. **27.** $812\frac{1}{2}$. **28.** 3030. **29.** $150. **30.** $525. **31.** $3.75 cost; $2.25 gain.

Page 68.—**32.** $4. **33.** $.50.

Page 69.—**1.** $900. **2.** 1000 A.; 1000 A.; 500 A. **3.** 292 da. **4.** $160. **5.** $5460.

Page 70.—**6.** 9200 girls; 5520 boys. **7.** $35. **8.** $648. **9.** $7600.

Page 72.—**2.** 45. **3.** 138. **4.** 326. **5.** 4805. **6.** 10,805. **7.** 1100. **8.** 3927. **9.** 6006. **10.** 4284. **11.** 1704. **12.** 9729. **13.** 15,606. **14.** 9004. **15.** 3241. **16.** 2005. **17.** 2315. **18.** 2499. **19.** 3027. **20.** 8165. **21.** 15,005. **22.** 114. **23.** 144. **24.** 249. **25.** 645. **26.** 3996. **27.** 2012. **28.** 14,895. **29.** 2092. **30.** 6276. **31.** 5615. **32.** 2476. **33.** 4900. **34.** 470. **35.** 208. **36.** 7990. **37.** 4104. **38.** 348. **39.** 1020. **40.** 5971. **41.** 196. **42.** 357. **43.** 1687. **44.** 438 mi. **45.** $73.50. **46.** $24.75. **47.** $220; $110. **48.** $3.06.

Page 73.—**2.** $\frac{21}{32}$. **3.** $\frac{31}{64}$. **4.** $1\frac{3}{8}$. **5.** $\frac{11}{32}$. **6.** $\frac{29}{32}$. **7.** $\frac{27}{32}$. **8.** $\frac{15}{16}$. **9.** $\frac{21}{32}$. **10.** $\frac{29}{40}$. **11.** $\frac{7}{25}$. **12.** $1\frac{7}{8}$. **13.** $\frac{27}{50}$. **15.** $4\frac{1}{4}$. **16.** $8\frac{3}{4}$. **17.** $4\frac{3}{16}$. **18.** $10\frac{1}{2}$. **19.** $2\frac{17}{24}$. **20.** $5\frac{1}{4}$. **21.** $33\frac{7}{10}$. **22.** $120\frac{5}{8}$. **23.** $60\frac{11}{16}$. **24.** $52\frac{1}{4}$. **25.** $33\frac{1}{4}$. **26.** $156\frac{1}{4}$. **27.** $29\frac{1}{2}$ ¢. **28.** $74\frac{3}{8}$ ¢.

Page 74.—**2.** $2\frac{2}{3}$ ¢. **3.** $9\frac{1}{4}$. **4.** 12. **5.** $10\frac{1}{2}$. **6.** $1\frac{1}{4}$. **7.** $71\frac{1}{4}$. **8.** 105. **9.** 49. **10.** $3\frac{1}{4}$. **11.** 85. **12.** $33\frac{3}{4}$. **13.** $37\frac{1}{2}$. **14.** 54. **15.** $199\frac{1}{4}$. **16.** $5.60. **17.** $144. **18.** $354\frac{2}{3}$ mi. **19.** $34\frac{2}{3}$. **20.** $81\frac{2}{3}$.

Page 75.—**1.** $409\frac{11}{11}$. **2.** $682\frac{1}{2}$ lb. **3.** $8602.50. **4.** 115 hr.; $517\frac{1}{2}$ hr. **5.** 500 mi. **6.** $1525. **7.** 92 ¢. **8.** $268\frac{1}{4}$ rd. **9.** $7401\frac{1}{4}$ mi. **10.** $1409\frac{3}{16}$ T.

Page 78. — **3.** 78. **4.** $5\frac{5}{8}$. **5.** 50. **6.** $3\frac{1}{3}$. **7.** $4\frac{1}{4}$. **8.** $16\frac{2}{3}$. **9.** 2. **10.** 72. **11.** 72. **12.** $3\frac{3}{4}$. **13.** 20. **14.** $2\frac{1}{2}$. **15.** $2\frac{1}{3}$. **16.** $2\frac{1}{15}$. **17.** 24. **18.** 4. **19.** $7\frac{1}{3}$. **20.** $3\frac{1}{3}$.

Page 79. — **21.** 27. **22.** 40. **23.** 36. **24.** 49. **25.** 40. **26.** 77. **27.** 63. **28.** 72. **29.** 81. **30.** 100. **31.** 96. **32.** 117. **33.** 144. **34.** 108. **35.** 60. **36.** 225. **37.** 171. **38.** 910. **39.** 204. **40.** 405. **41.** 504. **43.** 12. **44.** 9. **45.** 10. **46.** 16. **47.** 16. **48.** 9. **49.** 20. **50.** 27. **51.** 24. **52.** 40. **53.** 45. **54.** 42. **55.** 100. **56.** 40. **57.** 45. **58.** 204. **59.** 63. **60.** 80. **61.** 5. **62.** 4. **63.** 3. **64.** 4. **65.** $6\frac{2}{7}$. **66.** $1\frac{2}{5}$. **67.** 2. **68.** 5. **69.** $1\frac{3}{10}$. **70.** $1\frac{22}{63}$. **71.** $\frac{42}{55}$. **72.** $\frac{10}{11}$. **73.** $\frac{1}{2}$. **74.** $1\frac{1}{11}$. **75.** $1\frac{2}{5}$. **76.** $2\frac{13}{21}$. **77.** $1\frac{1}{4}$. **78.** $1\frac{14}{25}$. **79.** $3\frac{3}{5}$. **80.** $3\frac{2}{5}$. **81.** $4\frac{5}{9}$. **82.** $3\frac{1}{5}$. **83.** $4\frac{1}{3}$. **84.** $3\frac{1}{5}$. **85.** $4\frac{1}{5}$. **86.** $5\frac{5}{8}$. **87.** $2\frac{1}{4}$.

Page 80. — **88.** 32¢. **89.** $5. **90.** 30 badges. **91.** 12¢. **92.** 7 pictures. **93.** 12 straw hats. **94.** $3. **95.** 7 pairs.

Page 81. — **2.** 105. **3.** 153. **4.** 880. **5.** 260 ft. **6.** 840 bu. **7.** 600 lb. **8.** $175.

Page 84. — **39.** $17.50. **40.** $3. **41.** $37.50. **42.** $9. **43.** $\$3\frac{1}{3}$. **44.** $31.25. **45.** $40. **46.** $14. **47.** $7. **48.** $\$5\frac{5}{8}$. **49.** 80¢. **50.** $\$3\frac{1}{3}$. **51.** $66\frac{2}{3}$¢. **52.** $4. **53.** $12. **54.** $5.

Page 87. — **12.** 28 qt. **13.** 4 pk. **14.** 7 pk. **15.** 32 pt. **16.** 7 pt. **17.** 99 in. **18.** 18 ft. **19.** 12 qt. **20.** 22 qt. **21.** 104 oz. **22.** 108 oz. **23.** 56 qt. **24.** 30 hr. **25.** 6 qt. **26.** 3500 lb. **27.** 90 sec. **28.** 7 qt. **29.** 26 qt. **30.** 162 sq. ft. **31.** $24\frac{3}{4}$ ft. **32.** 324 sq. in. **33.** 7 qt. **34.** 12 pt. **35.** 80 oz. **36.** 126 in. **37.** 112 pt. **38.** $2\frac{1}{4}$ ft. **39.** $30\frac{1}{18}$ ft. **40.** 400 rd. **41.** 3888 cu. in.

Page 88. — **43.** $30\frac{3}{4}$ yd. **44.** $4\frac{1}{2}$ gal. **45.** 6 bu. **46.** 8 lb. **47.** 8 gal. **48.** 2 da. **49.** 10 rd. **50.** 43 yd. **51.** 41 pk. **52.** 2 T. **53.** 20 lb. **54.** $4\frac{1}{2}$ ft.

Page 89. — **2.** $3.50. **3.** $3.20. **4.** $2.

Page 90. — **5.** 63¢. **6.** $9.35. **7.** $1. **8.** $1.20. **9.** $4.50. **10.** 75¢. **12.** 60¢. **13.** $1.70. **14.** $5.20. **15.** $1.96. **16.** $11.52. **17.** $2.04. **18.** $6.96.

Page 93. — **1.** 182,501. **2.** 166,326. **3.** 234,136. **4.** 291,438.

Page 94. — **5.** 364,888. **6.** 322,279. **7.** 385,851. **8.** 464,647. **9.** $639.51. **10.** $298.18. **11.** $230.10. **12.** $213.14. **13.** $297.04. **14.** $177.58. **15.** $264.37. **16.** $220.38. **17.** $342.95. **18.** $316.01. **19.** $355.82. **20.** $341.63.

Page 95. — **21.** $317.15. **22.** $361.84. **23.** $429.82. **24.** $1019.80. **25.** $395.72. **26.** $2167.46. **27.** $324.30. **28.** $544.39.

Page 96. — **1.** *a.* 15; *b.* 353; *c.* 478; *d.* 118; *e.* 105; *f.* 55. **2.** *a.* 6507; *b.* 211; *c.* 3299; *d.* 98; *e.* 1672; *f.* 4858. **3.** *a.* 1163; *b.* 1373; *c.* 454; *d.* 1535; *e.* 573; *f.* 1224. **4.** *a.* 159; *b.* 1425; *c.* 2216; *d.* 1126; *e.* 202; *f.* 485. **5.** *a.* 1245; *b.* 3065; *c.* 4069; *d.* 1911; *e.* 1543; *f.* 1514.

Page 97. — **6.** *a.* 10,273; *b.* 16,008; *c.* 20,444; *d.* 11,179; *e.* 2166.
7. *a.* 79,857; *b.* 168,876; *c.* 175,892; *d.* 28,353; *e.* 11,845. **8.** *a.* $242.68;
b. $828.14; *c.* $2.733; *d.* $21. **9.** *a.* $213.66; *b.* $229.09; *c.* $8.01;
d. $2066.32. **10.** *a.* $324.89; *b.* $1809.59; *c.* $93.89; *d.* $125.102.
11. *a.* $1141.09; *b.* $1519.21; *c.* $7158.89; *d.* $293.744. **12.** *a.* $1619.71;
b. $201.92; *c.* $679.18; *d.* $212.62; *e.* $365.91. **13.** *a.* $3520.62;
b. $2341 02; *c.* $7135.85; *d.* $1182.78; *e.* $6021.52. **14.** *a.* $2151.02;
b. $2143.71; *c.* $739.42; *d.* $2046.43; *e.* $1071.46. **15.** *a.* $149.25;
b. $6216.68; *c.* $1134.76; *d.* $4809.01; *e.* $2219.14. **16.** *a.* $90.11;
b. $2231.27; *c.* $7069.82; *d.* $8130.89; *e.* $3320.13.

Page 98. — **17.** *a.* $1354.39; *b.* $1811.75; *c.* $1849.11; *d.* $2363.02;
e. $1008.13. **18.** *a.* $1200.84; *b.* $3428.35; *c.* $3646.56; *d.* $2150.80;
e. $6159.21.
1. 945. **2.** 1833. **3.** 3312. **4.** 1554. **5.** 3283. **6.** 3286. **7.** 2886.
8. 1863.

Page 99. — **9.** 5248. **10.** 7872. **11.** 3366. **12.** 3420. **13.** 4872.
14. 6566. **15.** 784. **16.** 2816. **17.** 1332. **18.** 9702. **19.** 4875.
20. 4050. **21.** 5628. **22.** 4116. **23.** 782. **24.** 4256. **25.** 3060.
26. 22,050. **27.** 20,672. **28.** 10,300. **29.** 32,400. **30.** 36,216.
31. 29,000. **32.** 34,476. **33.** 34,650. **34.** 50,292. **35.** 59,202.
36. 11,088. **37.** 18,122. **38.** 29,430. **39.** 66,196. **40.** 54,936.
41. 35,298. **42.** 12,500. **43.** 18,791. **44.** 44,958. **45.** 3717.
46. 6370. **47.** 3600. **48.** 98,901. **49.** 16,800. **50.** 37,500. **51.** 24,000.
52. 26,780. **53.** 105,000. **54.** 550. **55.** 600. **56.** 800. **57.** 855.
58. 630. **59.** 425. **60.** 20,451. **61.** 102,714. **62.** 64,134.
63. 1,848,754. **64.** 1,220,305. **65.** 4,800,156. **66.** 3,153,392.
67. 2,830,143. **68.** 9,989,001. **69.** 555,500. **70.** 1,873,472.
71. 2,263,527. **72.** 3,295,072. **73.** 1,440,285. **74.** 9,259,888.
75. 2,796,480. **76.** 7,351,680. **77.** 3,662,650. **78.** 2,178,583.
79. 4,802,880. **80.** 808,202. **81.** 495,110. **82.** 7,223,760.
83. 5,646,207.

Page 100. — **1.** 475. **2.** $112\frac{2}{9}$. **3.** $106\frac{6}{7}$. **4.** $280\frac{1}{4}$. **5.** $51\frac{7}{9}$.
6. $124\frac{4}{6}$. **7.** $108\frac{3}{4}$. **8.** $23,391\frac{1}{6}$. **9.** $6383\frac{4}{9}$. **10.** $491\frac{1}{2}$. **11.** $11,865\frac{1}{4}$.
12. $1557\frac{7}{8}$. **13.** $16,626\frac{1}{4}$. **14.** $1044\frac{3}{4}$. **15.** $348\frac{1}{6}$. **16.** $23,425\frac{7}{10}$.
17. 51,305. **18.** $1902\frac{1}{4}$. **19.** $6535\frac{1}{6}$. **20.** 7075. **21.** $8534\frac{5}{6}$.
22. $10,315\frac{9}{10}$. **23.** $2339\frac{7}{10}$. **24.** $5061\frac{7}{10}$. **25.** $29,674\frac{9}{10}$. **26.** $23,606\frac{4}{9}$.
27. 248. **28.** $23,658\frac{1}{10}$. **29.** 2035. **30.** $1642\frac{2}{6}$. **31.** 3020. **32.** 3655.
33. $1052\frac{9}{10}$. **34.** 10,100. **35.** 1420. **36.** 204. **37.** $90\frac{45}{360}$. **38.** $2080\frac{31}{40}$.
39. $1932\frac{5}{360}$. **40.** $21\frac{1}{12}$. **41.** $101\frac{148}{360}$. **42.** $90\frac{1}{360}$. **43.** $139\frac{4}{44}$.
44. $87\frac{10}{12}$. **45.** $84\frac{1}{44}$. **46.** $73\frac{1}{44}$. **47.** $125\frac{75}{360}$. **48.** $57\frac{1}{17}$. **49.** $388\frac{1}{48}$.
50. $28\frac{88}{66}$. **51.** $96\frac{24}{44}$. **52.** $976\frac{101}{144}$. **53.** $80\frac{150}{168}$. **54.** $500\frac{47}{77}$.

Page 103. — **1.** 4000 packages. **2.** 32,000 oz. **3.** $1\frac{1}{2}$ T.; $21.
4. $51\frac{1}{2}$ bu. **5.** 4600 packages. **6.** $1093\frac{1}{4}$ yd. **7.** 755,040 ft. **8.** 60 ft.;
$3.60. **9.** 42 lb. **10.** $60.12. **11.** $277\frac{1}{2}$ ft. **12.** 42 bu. 13 qt.

Page 106. — **1.** $54\frac{11}{12}$. **2.** $214\frac{1}{4}$. **3.** $63\frac{1}{4}$. **4.** $204\frac{13}{16}$. **5.** $150\frac{1}{12}$.
6. $281\frac{11}{16}$. **7.** $218\frac{11}{16}$. **8.** $267\frac{1}{4}$. **9.** $8\frac{1}{4}$. **10.** $61\frac{1}{4}$. **11.** $23\frac{5}{12}$. **12.** $13\frac{5}{12}$.
13. $24\frac{5}{12}$. **14.** $14\frac{1}{12}$. **15.** $59\frac{1}{12}$. **16.** $22\frac{1}{11}$. **17.** $10\frac{1}{11}$. **18.** $35\frac{5}{24}$. **19.** $7\frac{11}{14}$.

20. $16\frac{9}{11}$. 21. $55\frac{13}{15}$. 22. 4 . 23. $11\frac{1}{15}$. 24. $14\frac{13}{30}$. 25. $42\frac{1}{4}$. 26. $29\frac{5}{6}$. 27. $36\frac{17}{60}$. 28. $59\frac{14}{15}$. 29. $\frac{37}{30}$. 30. $2\frac{47}{90}$; $\frac{13}{90}$. 31. $2\frac{1}{12}$; $\frac{11}{12}$. 32. $1\frac{5}{16}$; $\frac{7}{16}$. 33. $2\frac{11}{12}$; $2\frac{1}{12}$. 34. $1\frac{1}{16}$; $\frac{13}{16}$.

Page 107. — 35. $1\frac{5}{12}$; $\frac{1}{12}$. 36. $2\frac{1}{72}$; $\frac{17}{72}$. 37. $2\frac{1}{4}$; $1\frac{5}{12}$. 38. $5\frac{7}{8}$; $\frac{5}{8}$. 39. $9\frac{11}{24}$; $\frac{19}{24}$. 40. $2\frac{1}{56}$; $\frac{15}{56}$. 41. $3\frac{17}{24}$. 42. $8\frac{25}{36}$. 43. $3\frac{1}{45}$. 44. $2\frac{3}{16}$. 45. $9\frac{4}{15}$. 46. $8\frac{31}{72}$. 47. $9\frac{31}{45}$. 48. $4\frac{1}{48}$. 49. $2\frac{61}{120}$. 50. $2\frac{1}{12}$. 51. $\frac{13}{24}$. 52. $8\frac{53}{144}$. 53. $4\frac{7}{8}$. 54. $2\frac{47}{60}$. 55. $6\frac{4}{7}$. 56. $12\frac{71}{80}$. 57. $10\frac{5}{36}$. 58. $6\frac{109}{180}$. 59. $16\frac{1}{4}$. 60. $7\frac{13}{18}$. 61. $1\frac{23}{120}$. 62. $10\frac{53}{84}$.

Page 108. — 2. 6. 3. 21. 4. 18. 5. 20. 6. 60. 7. 44. 8. $9\frac{1}{2}$. 9. $12\frac{5}{8}$. 10. $11\frac{1}{3}$. 11. $8\frac{1}{2}$. 12. $13\frac{1}{3}$. 13. 42. 14. $4\frac{4}{5}$. 15. 26. 16. 63. 17. $5\frac{5}{8}$. 18. $20\frac{5}{6}$. 19. 35. 20. 48. 21. [illegible]. 22. $\frac{1}{4}$. 23. $\frac{1}{2}$. 24. $\frac{2}{11}$. 25. $\frac{4}{11}$. 26. $\frac{5}{11}$. 27. $\frac{2}{15}$. 28. $\frac{2}{5}$. 29. $\frac{2}{3}$. 30. $\frac{2}{15}$. 31. $\frac{10}{119}$. 32. $\frac{9}{13}$. 33. $\frac{7}{13}$. 34. $\frac{11}{13}$. 35. $\frac{2}{5}$. 36. $\frac{14}{25}$. 37. $20\frac{1}{4}$. 38. 44. 39. 96. 40. $94\frac{3}{5}$. 41. $49\frac{1}{3}$. 42. $59\frac{1}{2}$. 43. $24\frac{2}{5}$. 44. $150\frac{2}{3}$. 45. $198\frac{6}{7}$. 46. $234\frac{2}{5}$. 47. $327\frac{3}{4}$. 48. $158\frac{1}{3}$. 49. $134\frac{2}{5}$. 50. $256\frac{1}{2}$. 51. $339\frac{1}{2}$.

Page 109. — 53. $2\frac{1}{4}$. 54. 2. 55. $2\frac{2}{7}$. 56. $\frac{5}{8}$. 57. [illegible]. 58. $1\frac{3}{16}$. 59. $2\frac{1}{7}$. 60. $\frac{45}{1024}$. 61. $\frac{3}{10}$. 62. $\frac{2}{3}$. 63. $\frac{1}{90}$. 64. $\frac{1}{5}$. 65. $1\frac{3}{200}$. 66. $\frac{27}{70}$. 67. $1\frac{7}{20}$. 68. $\frac{3}{25}$. 69. $\frac{15}{512}$. 70. $1\frac{4}{15}$. 71. $103\frac{59}{128}$. 72. $\frac{3}{80}$. 73. $\frac{9}{160}$. 74. $\frac{25}{64}$. 75. $\frac{4}{25}$. 76. $\frac{1}{32}$. 77. $2\frac{7}{15}$. 78. $1\frac{2}{3}$. 79. $\frac{1}{4}$. 80. $\frac{3}{16}$. 81. $\frac{1}{10}$. 82. $\frac{1}{160}$. 83. $\frac{61}{400}$. 84. $\frac{1}{15}$. 85. $1\frac{11}{24}$. 86. $\frac{3}{5}$. 87. $\frac{3}{32}$. 88. $\frac{1}{16}$. 89. $\frac{1}{64}$. 90. $1\frac{9}{25}$. 91. $18\frac{9}{16}$. 92. $\frac{7}{20}$. 93. $2\frac{5}{24}$. 94. $337\frac{7}{8}$. 95. $3\frac{3}{8}$. 96. $2\frac{1}{4}$.

Page 110. — 1. 12. 2. $1\frac{4}{5}$. 3. $6\frac{2}{3}$. 4. $1\frac{5}{16}$. 5. $1\frac{4}{7}$. 6. $\frac{20}{21}$. 7. $1\frac{2}{5}$. 8. $\frac{15}{128}$. 9. $\frac{6}{35}$. 10. $\frac{11}{72}$. 11. $\frac{4}{33}$. 12. $\frac{4}{32}$. 13. $1\frac{3}{4}$. 14. $2\frac{19}{98}$. 15. $1\frac{17}{36}$. 16. $1\frac{1}{14}$. 17. $\frac{3}{4}$. 18. $1\frac{4}{57}$. 19. $1\frac{19}{55}$. 20. $1\frac{7}{52}$. 21. $2\frac{6}{35}$. 22. $1\frac{3}{5}$. 23. $9\frac{1}{7}$. 24. $8\frac{3}{4}$. 25. $\frac{6}{7}$. 26. $1\frac{23}{27}$. 27. $1\frac{1}{5}$. 28. $1\frac{7}{45}$. 29. $\frac{1}{20}$. 30. $\frac{11}{196}$. 31. $\frac{9}{121}$. 32. $\frac{15}{512}$. 33. $\frac{11}{392}$. 34. $1\frac{159}{175}$. 35. $3\frac{11}{15}$. 36. $\frac{73}{128}$. 37. $2\frac{4}{27}$. 38. $12\frac{7}{64}$. 39. 256. 40. $2\frac{43}{105}$. 41. 33. 42. 10. 43. 25. 44. 5. 45. 28. 46. $2\frac{2}{9}$. 47. $2\frac{4}{55}$. 48. $\frac{15}{16}$. 49. $\frac{47}{80}$. 50. $\frac{21}{44}$. 51. $\frac{4}{13}$. 52. $\frac{1}{90}$. 53. $\frac{1}{57}$. 54. $\frac{1}{50}$. 55. $2\frac{4}{7}$. 56. $3\frac{8}{35}$. 57. $3\frac{7}{75}$. 58. $8\frac{83}{124}$. 59. $5\frac{4}{7}$. 60. $5\frac{1}{4}$.

[A fraction of a cent is counted as an additional cent.]

Page 112. — 1. \$7350. 2. \$424.05. 3. \$165. 4. \$26$\frac{2}{5}$. 5. \$45. 6. \$720. 7. \$100. 8. \$87$\frac{1}{2}$.

Page 113. — 9. $22\frac{1}{2}$ mi. 10. \$11,340. 11. 70 ribbons. 12. \$1.35. 14. $\frac{3}{4}$. 15. $\frac{11}{72}$. 16. 50 ft. 17. \$60. 18. $20\frac{1}{2}$ rd. 19. $25\frac{23}{75}$ mi. 20. $434\frac{1}{72}$ lb. 21. 2112 steps. 22. $929\frac{13}{16}$ lb.

Page 118. — 2. $\frac{3}{20}$. 3. $\frac{33}{40}$. 4. $\frac{9}{10}$. 5. $\frac{13}{40}$. 6. $\frac{3}{4}$. 7. $\frac{1}{40}$. 8. $\frac{1}{8}$. 9. $\frac{17}{40}$. 13. $\frac{9}{20}$. 14. $\frac{5}{8}$. 15. $\frac{3}{4}$. 16. $\frac{13}{20}$. 17. $\frac{3}{8}$. 18. $\frac{4}{5}$. 19. $\frac{1}{5}$. 20. $\frac{3}{8}$. 21. $\frac{29}{40}$. 22. $\frac{9}{10}$. 23. $\frac{24}{25}$. 24. $\frac{11}{25}$. 25. $\frac{2}{5}$. 26. $\frac{13}{20}$. 27. $\frac{3}{20}$. 28. $\frac{1}{8}$. 29. $\frac{7}{8}$. 30. $\frac{2}{25}$. 31. $\frac{3}{40}$. 32. $\frac{1}{400}$. 33. $\frac{3}{25}$. 34. $\frac{9}{40}$. 35. $\frac{7}{10}$. 36. $\frac{4}{5}$.

Page 119. — 2. 1.015. 3. .221. 4. .256. 5. 9.1. 6. 13.189. 7. .0311. 8. 18.027. 9. .0922. 10. 3.603. 11. 6.052. 12. 2.777. 13. 10.856. 14. 15.728. 15. 13.356. 16. 99.05. 17. 3.505. 18. 44.85. 19. 2.9.

Page 120. — **20.** 17.165. **21.** 143.192. **22.** 33.798. **23.** 137.768.
24. 26.676. **25.** 2171.812. **26.** 14.806. **27.** 225.303. **28.** 47.97.
29. 2.75 bu. **30.** $4.94. **31.** 82.1 mi. **32.** 5.25 lb. **33.** 23.98 mi.
34. 2277.225.

Page 121. — **2.** 5.79. **3.** 12.954. **4.** .305. **5.** 180.892.
6. 16.02. **7.** 72.927. **8.** 695.725. **9.** 18.835. **10.** 131.745.
11. 38.615. **12.** 32.996. **13.** 108.967. **14.** $3.75. **15.** 5.875 mi.
16. 55.75 A.

Page 122. — **17.** $2.20. **18.** 10.86 ft. **19.** 28.46 ft. **20.** $2.05.
21. 6.75 lb. **22.** 5.125 lb. **23.** .375 lb. **24.** 27.65 T. **25.** 6.75
yd. less. **26.** $6.32. **27.** $38.55.

Page 124. — **4.** .12. **5.** .045. **6.** 3.06. **7.** 5.175. **8.** 246.4.
9. 487.5. **10.** 613.2. **11.** 1088.1 **12.** 1414.4. **13.** 1774.8.
14. 3060. **15.** 105.3. **16.** 2.24. **17.** 151.04. **18.** .596.
19. 1261.6. **20.** 3184.72. **21.** 3.59. **22.** 2.928. **23.** 3.225.
24. 1.981. **25.** 432.904. **26.** 228.03. **27.** 1282.413. **28.** 679.06.
29. 262.116. **30.** 257,352.83. **31.** 5638.844. **32.** 123.6.
33. 1756.125. **34.** 12.432. **35.** 607.92. **36.** 25,647.221. **37.** $5.74.
38. 1980 ft. **39.** 108 sq. in.

Page 125. — **40.** $8.40. **41.** 478.5 ft. **42.** $94.90. **43.** $14.49.
44. 1750 lb. **45.** 248.5 mi. **46.** 6 ; 8 ; 8.4 ; 8.6 ; 7.6 ; 6.5 ; 5.4 ; .05.
47. 30 ; 20 ; 2.5 ; 210 ; 235 ; 21,150 ; 28,350 ; 239 ; 4300 ; 4900. **48.** 60 ;
80 ; 84 ; 95 ; 86 ; 76 ; 6 ; 4 ; .5 ; 423 ; 5670 ; 47.8 ; 860 ; 980. **49.** 2970 ;
29,700 ; 297,000 ; 35 ; 350 ; 3500 ; 15,700 ; 12,500 ; 12.5. **50.** $3.
51. $4.46. **52.** $21. **53.** $3.50. **54.** $.75. **55.** $.50.
56. $480. **57.** $300. **58.** $2.34. **59.** $49.50. **60.** $.75.
61. $156.25. **62.** $1.60.

Page 126. — **63.** $20. **64.** 55.44 in. **65.** 354 ft. **66.** $36.
67. $2073.60.
4. .11. **5.** .32. **6.** .101. **7.** .102. **8.** .107. **9.** .212.
10. 1.121. **11.** 1.156. **12.** 1.112.

Page 127. — **15.** 3.04. **16.** 2.11. **17.** 6.1. **18.** 22.3. **19.** .101.
20. 3.4. **21.** .039. **22.** .124. **23.** .027. **24.** 1.16. **25.** .022.
26. .036. **27.** .14. **28.** .0017. **29.** .029. **30.** .032. **31.** .143.
32. 6.14. **33.** 6.04. **34.** .089. **35.** .475. **36.** .605. **37.** .065.
38. .0904. **39.** .25. **40.** .04. **41.** .0005. **42.** .125. **43.** .035.
44. .0001. **45.** .0001. **46.** .0003. **47.** .0015. **48.** .0888.
49. 2.5. **50.** .21. **51.** .0015. **52.** .026. **53.** .0003. **54.** .023.
55. .065. **56.** 2.03. **57.** 7.2. **58.** .36. **59.** .017. **60.** .045.
61. .8¼. **62.** .042. **63.** .018. **64.** .75. **65.** .0781¼. **66.** .651½.
67. 4.45. **68.** 4.43.

Page 128. — **73.** .075. **74.** .075. **75.** 1.045. **76.** 1.056. **77.** .8.
78. .3. **79.** .02. **80.** .001. **81.** .002. **82.** .3. **83.** .3. **84.** .03.
85. .170¼. **86.** .034. **87.** .003. **88.** .12. **89.** 1.005. **90.** .009.
91. 1.13. **92.** .072. **93.** 1.036. **94.** .04. **95.** .225. **96.** 1.04. **97.** .05.
98. .007. **99.** .011. **100.** .006. **101.** .39125 A.

ANSWERS

Page 129. — **102.** 188.58\frac{1}{3}$. **103.** 1.225 mi. **104.** 782.66\frac{2}{3}$. **2.** .26$\frac{2}{3}$. **3.** .4. **4.** .10$\frac{10}{99}$. **5.** .08. **6.** .44$\frac{4}{9}$. **7.** .5. **8.** .2. **9.** .5. **10.** .5. **11.** .25. **12.** .125. **13.** .012$\frac{543}{1111}$.

Page 130. — **3.** .2. **4.** .8. **5.** .375. **6.** .875. **7.** .625. **8.** .3. **9.** .35. **10.** .24. **11.** .5625. **12.** .6875. **13.** .52. **14.** .55. **15.** .45$\frac{5}{11}$. **16.** .58$\frac{1}{3}$. **17.** .73$\frac{1}{3}$. **18.** .85$\frac{5}{7}$. **19.** .55$\frac{5}{9}$. **20.** .83$\frac{1}{3}$. **21.** 5.25. **22.** 4.5. **23.** 3.125. **24.** .2. **25.** 3.66$\frac{2}{3}$. **26.** 7.25. **27.** 8.1. **28.** 6.25. **29.** 3.4. **30.** 5.875. **31.** 3.75. **32.** 2.875. **33.** 2.5. **34.** 3.2. **35.** 8.11$\frac{1}{9}$. **36.** 2.4. **37.** 2.75. **38.** 1.375. **39.** 1.625. **40.** 7.83$\frac{1}{3}$. **41.** 2.15. **42.** 2.125. **43.** 2.25. **44.** 1.35. **45.** 8.1875.

Page 133. — **16.** 540 ft. **17.** 1291$\frac{1}{3}$ ft. **18.** 228 ft. **19.** 5940 ft. **20.** 810 yd.; 950 yd. less. **21.** 16 trips.

Page 139. — **1.** $5.25. **2.** $4.95. **3.** $3. **4.** $10. **5.** $3. **6.** $9.38. **7.** $1.82. **8.** $.25. **9.** $3.40. **10.** $1.90. **11.** $5.25. **12.** $.38. **13.** $.29. **15.** $1.75. **16.** $.75. **17.** $.38. **18.** $1.10. **19.** $1.25. **20.** $3.90. **21.** $.90. **22.** $75.

Page 140. — **23.** $11.20. **24.** $2.50. **25.** $4.05. **26.** $1.32. **27.** $12. **28.** $.25. **29.** $.75. **30.** $.25. **31.** $1.24. **32.** $.50. **33.** $3.07. **34.** $1.80. **35.** $4.90. **36.** $3.60. **37.** $3.49. **38.** $15.12. **39.** $3.50. **40.** $6.65. **41.** $4.50. **42.** $1.80. **43.** $7.50. **44.** $1.55. **45.** $2.55. **46.** $2.75. **47.** $1.32. **48.** $1.05. **49.** $1.25. **50.** $.75. **51.** $.50. **52.** $34.20. **53.** $33.60. **54.** $1.70. **55.** $2.85. **56.** $1.38. **57.** $.44. **58.** $1.65.

Page 141. — **60.** $21.88. **61.** $1.26. **62.** $3.58. **63.** $2.24. **64.** $15.75. **65.** $59.40. **66.** $2. **67.** $7. **68.** $1.35. **69.** $16.88. **70.** $1.96. **71.** $14.63. **72.** $5.70. **73.** $1.05.

Page 142. — **74.** $.77. **75.** $2.70. **76.** $1.32. **77.** $3. **78.** $4.38. **79.** $2.10. **80.** $2.25. **81.** $1.68. **82.** $3.94. **83.** $2.10. **84.** $1.62. **85.** $2.67. **86.** $3.43. **87.** $96.23. **88.** $3.74.

Page 145. — **3.** $1672.50.

Page 146. — **4.** 6 pencils. **5.** 8 yd. **6.** 9 qt. **7.** $42. **8.** $1.35. **9.** *a.* $80; *b.* $90. **11.** $1.60. **12.** 60¢. **13.** $4.80.

Page 147. — **14.** $.63$\frac{9}{32}$ per bu. **15.** $17.50. **16.** $28.40. **17.** $7.50. **18.** 72. **19.** $15,000. **20.** $5680. **21.** $1.32. **22.** 268$\frac{1}{3}$ rd. **23.** 7401$\frac{1}{4}$ mi. **24.** 1409$\frac{3}{16}$ T. **25.** $38.21. **26.** $6.87.

Page 148. — **27.** $64.26. **28.** $45. **29.** $720. **30.** 14,910 pads. **31.** $417.10. **32.** $14.40. **33.** $2237.20. **34.** $19.20. **35.** Children, $786; adults, $3160.25; one-horse vehicles, $485.80; two-horse vehicles, $398. Total receipts, $4830.05. **36.** $2.28. **37.** $15.84. **38.** $6.52. **39.** $5.76.

Page 149. — **40.** $1. **41.** 75¢. **42.** $6.85. **43.** $22.80. **44.** $130.50. **45.** $1.90, gain. **46.** $5.40. **47.** $40, gain. **48.** 66 ft. **49.** $4.62. **50.** $76.96.

Page 150. — 51. $180.35. **52.** $180.75. **53.** Farmer owes $4.77.
54. $12,317.50.

SIXTH GRADE

Page 153. — 1. a. 290; b. 280; c. 269; d. 298; e. 331; f. 321; g. 292.
2. a. 310; b. 320; c. 298; d. 332; e. 318; f. 303; g. 337.

Page 154. — 3. a. 350; b. 394; c. 348; d. 332; e. 257; f. 289; g. 208.
4. a. 389; b. 368; c. 348; d. 337; e. 358; f. 340; g. 345. **5.** 27,479.
6. 20,360. **7.** 33,040. **8.** 37,457. **9.** 31,464. **10.** 24,777. **11.** 34,070.
12. 24,892. **13.** 24,677. **14.** 33,733. **15.** 26,546. **16.** 18,600.

Page 155. — 17. 21,363. **18.** 422,840. **19.** 281,524. **20.** 342,721.
21. 25,397. **22.** 411,368. **23.** 357,490. **24.** 422,320. **25.** 30,092.
26. 388,455. **27.** 332,442. **28.** 422,359. **29.** 24,990. **30.** 414,802.
31. 332,193. **32.** 453,283. **33.** $194\frac{1}{2}$. **84.** $199\frac{7}{12}$. **35.** 104.
36. $216\frac{7}{24}$. **37.** $170\frac{11}{12}$. **38.** $187\frac{37}{44}$.

Page 158. — 1. 29,678. **2.** 11,073. **3.** 4315. **4.** 199. **5.** 29,823.
6. 34,292. **7.** 29,331. **8.** 13,809. **9.** 5585. **10.** 2799. **11.** 300.
12. 9486. **13.** 16,088. **14.** 16,448. **15.** 28,099. **16.** 28,889.

Page 159. — 17. 9796. **18.** 991. **19.** 15,598. **20.** 19,988.
21. 12,818. **22.** 31,272. **23.** 1304. **24.** 32,397. **25.** 40,176.
26. 8999. **27.** 42,849. **28.** 11,199. **29.** 50,484. **30.** 31,104.
31. 41,108. **32.** 11,197. **33.** 27,724. **34.** 3026. **35.** 19,927.
36. 38,965. **37.** 11,993. **38.** 32,082. **39.** 21,129. **40.** 8112.
41. 4902. **42.** 91. **43.** 19,992. **44.** 19,099. **45.** 32,030. **46.** 23,106.
47. 11,198. **48.** 43,116. **49.** $9\frac{5}{12}$. **50.** $18\frac{11}{12}$. **51.** $42\frac{11}{12}$. **52.** $3\frac{7}{12}$.
53. $17\frac{7}{10}$. **54.** $6\frac{43}{48}$. **55.** $1\frac{5}{24}$. **56.** $26\frac{37}{44}$. **57.** $22\frac{4}{15}$. **58.** $10\frac{11}{18}$.
59. $25\frac{7}{10}$. **60.** $35\frac{11}{16}$. **61.** $11\frac{11}{12}$. **62.** $25\frac{5}{24}$. **63.** $7\frac{10}{13}$.

Page 161. — 1. a. 5700; 31,900; 24,875; $45,023\frac{1}{4}$. b. 6840; 38,280;
29,850; $54,028\frac{1}{4}$. c. 7980; 44,660; 34,825; $63,033\frac{1}{4}$. d. 9120; 51,040;
39,800; 72,038. e. 13,680; 76,560; 59,700; 108,057. **2.** a. 9920; 70,112;
47,280; 20,911. b. 11,160; 78,876; 53,190; $23,524\frac{2}{5}$. c. 8680; 61,348;
41,370; $18,297\frac{1}{2}$. d. 16,120; 113,932; 76,830; $33,980\frac{2}{5}$. e. 17,360; 122,696;
82,740; $36,594\frac{1}{4}$. **3.** a. 167,649; 257,752; 283,272; $199,856\frac{2}{5}$. b. 219,678;
337,744; 371,184; $261,880\frac{1}{4}$. c. 329,517; 506,616; 556,776; $392,821\frac{1}{2}$.
d. 554,976; 853,248; 937,728; $661,593\frac{3}{5}$. e. 450,918; 693,264; 761,904;
$537,544\frac{1}{2}$. **4.** a. 2,555,860; 3,578,580; 4,652,000; $4,282,013\frac{1}{4}$.
b. 4,894,200; 6,852,600; 8,908,200; $8,199,600$. c. 2,710,000; 3,807,000;
4,949,000; $4,555,933\frac{1}{3}$. d. 4,839,820; 6,776,460; 8,809,220; $8,108,493\frac{1}{4}$.
e. 4,388,466; 6,144,498; 7,987,686; 7,352,308. **5.** a. 5,403,904; 3,994,560;
4,135,008; $4,913,146\frac{1}{4}$. b. 6,301,592; 4,658,130; 4,821,909; $5,729,810\frac{1}{2}$.
c. 5,155,040; 3,810,600; 3,944,580; $4,686,883\frac{1}{4}$. d. 6,834,872; 5,052,330;
5,229,969; $6,214,160\frac{4}{5}$. e. 7,617,016; 5,630,490; 5,828,457; $6,925,274\frac{1}{2}$.
6. a. 5970; $4439\frac{1}{2}$; $3569\frac{1}{2}$; $3644\frac{12}{13}$. b. 21,890; $16,277\frac{1}{2}$; $13,087\frac{1}{4}$; $13,363\frac{33}{43}$.
c. 43,780; $32,554\frac{1}{2}$; $26,174\frac{1}{4}$; $26,727\frac{9}{17}$. d. $57,046\frac{2}{3}$; $42,419\frac{1}{4}$; $34,106\frac{1}{4}$;
$34,826\frac{11}{13}$. e. 33,432; $24,859\frac{1}{4}$; $19,987\frac{2}{5}$; $20,410\frac{1}{4}$. **7.** a. $2591\frac{1}{4}$; $2593\frac{2}{3}$;
$2182\frac{1}{4}$; $1358\frac{5}{10}\frac{1}{4}$. b. $219,866\frac{1}{4}$; $19,886\frac{1}{4}$; $16,729\frac{5}{8}$; $10,491\frac{11}{17}$. c. $42,227\frac{7}{8}$;
$42,270\frac{5}{8}$; 5,[6] ; $300\frac{45}{...}$. d. 24,876; $24,901\frac{1}{2}$; $20,948\frac{2}{3}$; $13,137\frac{1}{5}$.
e. $25,336\frac{2}{3}$; $25,362\frac{1}{3}$; $21,336\frac{1}{4}$; $13,380\frac{11}{17}$.

Page 162. — **8.** *a.* $7221\frac{1}{5}$; $6357\frac{3}{5}$; $8018\frac{3}{5}$. *b.* $8123\frac{4}{5}$; 7152; $9021\frac{3}{5}$; *c.* $9284\frac{3}{7}$; $8173\frac{5}{7}$; 10,310. *d.* 12,998; $11,443\frac{1}{5}$; 14,434. **9.** *a.* $525\frac{49}{70}$; $943\frac{9}{70}$; $1378\frac{59}{70}$. *b.* $427\frac{23}{56}$; $767\frac{17}{56}$; $1122\frac{15}{56}$. *c.* $375\frac{19}{38}$; $673\frac{35}{38}$; $984\frac{7}{38}$. *d.* $694\frac{8}{53}$; $1245\frac{24}{53}$; $1820\frac{50}{53}$. **10.** *a.* $354\frac{9}{44}$; $2958\frac{2}{44}$; $3495\frac{19}{44}$. *b.* $1893\frac{5}{45}$; $1577\frac{38}{45}$; $1864\frac{19}{45}$. *c.* $2837\frac{17}{35}$; $2448\frac{2}{35}$; $2892\frac{23}{35}$. *d.* $1521\frac{14}{56}$; $1267\frac{49}{56}$; $1498\frac{2}{56}$. **11.** *a.* $1521\frac{29}{50}$; 1660; $1491\frac{23}{150}$. *b.* 895; $976\frac{49}{85}$; $877\frac{33}{85}$. *c.* $800\frac{59}{95}$; $873\frac{65}{95}$; $785\frac{8}{95}$. *d.* $1135\frac{55}{67}$; $1238\frac{34}{67}$; $1113\frac{7}{67}$. **12.** *a.* $1009\frac{19}{98}$; 152; $555\frac{54}{98}$. *b.* $1545\frac{31}{54}$; $232\frac{45}{54}$; $850\frac{44}{54}$. *c.* $3532\frac{5}{25}$; 532; $1944\frac{13}{25}$. *d.* $2197\frac{38}{45}$; $331\frac{1}{45}$, $1209\frac{38}{45}$. **13.** *a.* $199\frac{319}{450}$; 100; $143\frac{301}{500}$. *b.* $111\frac{515}{805}$; $55\frac{725}{805}$; $80\frac{403}{3220}$. *c.* $115\frac{179}{780}$; $57\frac{449}{780}$; $82\frac{721}{1040}$. *d.* $599\frac{29}{150}$; 300; $430\frac{1}{300}$. **14.** *a.* $108\frac{131}{720}$; $106\frac{399}{720}$; $56\frac{5471}{5760}$. *b.* $119\frac{541}{650}$; $118\frac{19}{650}$; $63\frac{431}{5200}$. *c.* $110\frac{341}{705}$; $108\frac{579}{705}$; $58\frac{911}{5640}$. *d.* $162\frac{131}{480}$; $159\frac{329}{480}$; $85\frac{1531}{3840}$. **15.** *a.* $9108\frac{4}{17}$; $21,992\frac{8}{17}$; $3136\frac{16}{85}$. *b.* $6912\frac{1}{2}$; $16,690\frac{5}{7}$; $2380\frac{23}{56}$. *c.* $10,557\frac{3}{11}$; $25,491\frac{8}{11}$, $3635\frac{52}{110}$. *d.* $7201\frac{27}{43}$; $17,389\frac{17}{43}$; $2480\frac{12}{215}$. **16.** *a.* 46,024; $40,804\frac{13}{17}$; $17,788\frac{33}{85}$. *b.* $17,963\frac{24}{49}$; $15,926\frac{16}{49}$; $6942\frac{228}{245}$. *c.* $16,765\frac{11}{35}$; $14,864\frac{4}{7}$; $6480\frac{13}{175}$. *d.* $14,489\frac{1}{27}$; $12,845\frac{25}{27}$; $5600\frac{1}{135}$.

Page 168. — **1.** .34. **2.** .0675. **3.** 16.000075. **4.** 400.045. **5.** 6006.0066. **6.** 89.005. **7.** 700.0046. **8.** 900.000084. **9.** .005095. **10.** 8.0017. **11.** .000125. **12.** 896.00301. **13.** 1000.001. **14.** 18,051.957. **15.** 97.0008. **16.** .009864. **17.** 2135.000032. **18.** 1.000001. **19.** 1,000,000.1. **20.** 90,000.071. **21.** 1830.11684. **22.** 429,000.0046. **23.** 7035.97. **24.** 67,375.00035. **25.** .05815. **26.** 375.069.

Page 169. — **3.** 25,016.567. **4.** 410.002. **5.** 22.1925. **6.** 22.4965. **7.** 58.8468.

Page 170. — **8.** 11.205. **9.** 37.544. **10.** 26.725. **11.** 27.1933. **12.** 41.6767. **13.** 2715.05451. **14.** 65.28829. **15.** 2171.2623. **16.** 2.075. **17.** 3.799. **18.** 39.032. **19.** 8.999. **20.** 170.075. **21.** 46.0756. **22.** 256.2175. **23.** 81.0081. **24.** 6.994. **25.** 106.9194. **26.** 5.006. **27.** 105.3. **28.** 5.38875. **29.** 99.0001. **30.** 3.449. **31.** .141. **32.** 1.4002. **33.** 87.008. **34.** 192.935 yd.

Page 171. — **35.** $112.25. **36.** 104.353. **37.** 6.4 T. **38.** 7.875 yd. **39.** $ 38.05. **40.** 15,135.625 ft. **41.** 21.25 lb. **42.** Monday, .5 of a degree; Wednesday, 4.1 degrees. **43.** 27.1 in. **44.** 3.71 in. **45.** 2.005 A.

Page 172. — **46.** $ .057; $ .0115; $ $.006\frac{1}{4}$. **47.** 152.5 ft. **48.** 21.17 lb.

Page 173. — **3.** .216. **4.** .225. **5.** .0384. **6.** 9.75. **7.** 53.58. **8.** 14.445. **9.** .5792. **10.** .00462. **11.** .0855. **12.** .812. **13.** .696. **14.** 1.71. **15.** .0031. **16.** 90.9. **17.** .468.

Page 174. — **18.** 2.24. **19.** 3.75. **20.** 2.52. **21.** 6.51. **22.** 5.44. **23.** 3.48. **24.** 3.6. **25.** 4.55. **26.** 4.56. **27.** .03. **28.** .0308. **29.** .0625. **30.** .1376. **31.** .2255. **32.** .090625. **33.** .034592. **34.** .03075. **35.** .248814. **36.** .003404. **37.** .70632. **38.** 12.4476. **39.** .456. **40.** 8.272. **41.** .00072. **42.** .002928. **43.** .003225. **44.** .00581. **45.** .432904. **46.** .033696. **47.** .001212. **48.** 12.5664. **49.** .010304. **50.** .001089. **51.** 5.202. **52.** 2.3328. **53.** .040625. **54.** .002451. **55.** .00376. **56.** .119082. **57.** .045675. **58.** .000484. **59.** .05372. **60.** 231.8283. **61.** 1.616844. **62.** .0237888. **63.** .068625. **64.** .00046224. **65.** .979968. **66.** 1.629221. **67.** .00312. **68.** .01024. **69.** .000288. **70.** .000625. **71.** .083712. **72.** .0000016. **73.** .027775.

74. .0005832.   75. .0003416.   76. .000891.   77. .0111201.
78. .0465744.   79. .0014592.   80. .0003686.   81. 4.375.   82. 19.08.
83. 32.27875.   84. 184.6785.   85. 75.11502.   86. .840483.
87. 168.0336.   88. 1.82272.   89. 19.5168.   90. 70.70707.   91. 371.175.
92. 489.0756.   93. 19.70299.   94. 20445.6.   95. 171.342852.
93. 1113.111.   97. 265.972264.   98. 9.61996.   99. 266.30656.
100. 489.589848.   101. 369.738369.   102. 839.93161.   103. 4.29087.
104. 19.382137.

Page 175.—1. 45.45 lb.   2. 2481.93 sq. ft.   3. $6315.35.
4. 26,621.98 sq. ft.   5. 57.5 lb.   6. 15 lb.   7. 172.5 lb.   8. 4 cu. ft. ;
725 lb.   9. 8.755 lb.   10. 53.4375 mi.   11. $1195.06.   12. 464.58125 bu.
13. $2.35.

Page 176.—14. 6042.96875 lb.   15. 36 cu. ft. ; 270 gal.
16. 250.07136 ft.

Page 179.—4. 20.   5. 150.   6. 30.   7. 1000.   8. 200.   9. 640.
10. 48.   11. 4000.   12. 12,500.   13. 550.   14. 1500.   15. 2500.
16. 18.4.   17. 48.   18. 250.   19. 256.   20. 1000.   21. 10,000.   22. 25.
23. 3500.   24. 1200.   25. 16,000.   26. 150.   27. 3500.   28. 20.   29. 3.2.
30. .25.   31. 1.   32. .0122.   33. .0078.   34. 5.6.   35. 1.4.   36. 2.3625.

Page 180.—37. .016.   38. .009.   39. .01.   40. 1.7.   41. 1.5.
42. 5.   43. 6.   44. 2.9.   45. 40.   46. 60.   47. .08.   48. 4$\frac{2}{3}$.
49. 10.   50. 1.3.   51. 1.257.   52. 12.57.   53. .53.   54. 3.025.
55. 300.   56. 90.   57. .12.   58. 1270.   59. .365.   60. .1.   61. .1.
62. 100.   63. 3.5.   64. .53.   65. 2.1.   66. .575.   67. 3.375.
68. .1.   69. 10.   70. .1.   71. 290.   72. .35.   73. 43.6.   74. .384.
75. 28.3.

Page 181.—4. $\frac{7}{16}$.   5. $\frac{4}{15}$.   6. $\frac{1}{2}$.   7 $\frac{1}{3}$.   8. $\frac{4}{7}$.   9. $\frac{1}{6}$.   10. $\frac{5}{12}$.
11. $\frac{5}{6}$.   12. $\frac{2}{3}$.   13. $\frac{4}{7}$.   14. $\frac{4}{5}$.   15. $\frac{5}{6}$.   16. $\frac{4}{15}$.   17. $\frac{7}{24}$.   18. $\frac{7}{12}$.
19. $\frac{9}{3625}$.   20. $\frac{1}{4}$.   21. $\frac{1}{4}$.   22. $\frac{7}{12}$.   23. $\frac{21}{40}$.
3. .625.   4. .6.   5. .15.   6. .3125.   7. .875.   8. .44.   9. .5625.   10. .95.
11. .4375.   12. .68.   13. .66$\frac{2}{3}$.   14. .290$\frac{10}{17}$.   15. .629$\frac{11}{17}$.   16. .473$\frac{11}{17}$.
17. 1.29$\frac{7}{17}$.   18. .755$\frac{5}{15}$.   19. .01$\frac{1}{2}$.   20. .461$\frac{7}{13}$.   21. 1.588$\frac{4}{17}$.   22. .291$\frac{2}{3}$.

Page 182.—1. 75 books.   2. .00072.   3. 18.   4. 73.6.   5. $59.50.
6. .5 ; .75 ; .875 ; .5625.   7. $2250 ; $3750.   8. 40.   9. $184.45.
10. 100.   11. 75.5 bbl.   12. $10,381.25.   13. 12.75.   14. 671.875 lb.

Page 183.— 15. $\frac{4}{5}$.   16. Seventeen and forty-nine hundredths.
17. 121$\frac{1}{14}$ mi.   18. .008.   19. $3.   20. $32.   21. $62.   22. $8.
23. $90.   24. $108.   25. $17.   26. $105.   27. 48 yd.   28. 39 da.
29. $97.58.   30. 72 lb.   31. .015.   32. $24.50.

Page 184.—34. 30 da.   35. 5000 layers.   36. 920 yd.   37. 43,200
pencils.   38. 4000 pencils.   39. 30 yd.   40. 18$\frac{3}{4}$ doz.

Page 187.— 2. $106.65.   3. $66.92.   4. $126.48.   5. $8.444.
6. $4.54.   7. $31.71.   8. $8.20.   9. $200.40.   10. $10.03.   11. $35.035.
12. $453.75.   13. $741.85.   14. $3035.25.   15. $724.   16. $6867.84.
17. $157.50.   18. $33 more.   19. 594 mi.   20. $1027.   21. $50 a
month.   22. 403$\frac{1}{2}$ rd.

Page 188. — **23.** $886.90. **24.** $140. **25.** 1; 2; $2\frac{1}{4}$; 3; 4. **26.** $7.20.
27. $6. **28.** $10. **29.** $128. **30.** 225 sheep. **31.** 24 words.
33. $616.

Page 189. — **1.** $3.20. **2.** $12\frac{1}{4}$ pt. **3.** $2.56. **4.** 210 gal. **5.** $3.84.
6. $15\frac{1}{4}$ bu. **7.** 84¢. **8.** 6 pies. **9.** $2.

Page 190. — **10.** $2.10. **11.** 51¢. **12.** $8.96. **13.** 30¢. **14.** $2.
15. $9\frac{5}{14}$ lb. **16.** $25. **17.** 121 lb. **18.** $180. **19.** $9.60. **20.** $58.50.

Page 191. — **21.** 180 min.; 3 hr.; $5. **22.** $1.35. **24.** $3.02.
25. $6.20. **26.** $21. **27.** 208 pint cans. **28.** $10.08. **29.** $5.
30. $8.16. **31.** 12 T. 0 cwt. 41 lb. **32.** 1000 baskets.

Page 192. — **33.** $25. **34.** 1 gal. 1 qt. 1 pt. **38.** $4\frac{2}{3}$ ft. **39.** $1.10.

Page 196. — **27.** $11,760.
1. A, $2100; B, C, D, and E, $2800 each; F, $3500. **2.** $9750.

Page 197. — **3.** $440. **4.** $152. **5.** $400. **6.** 1 in. = 80 ft.; 30 ft.
7. 120 sq. in. **8.** $26\frac{2}{3}$ sq. ft. **9.** 3650 sq. ft. **10.** $36\frac{1}{4}$ sq. in. **14.** $24.
15. $42.63. **16.** $6\frac{1}{4}$ sq. ft.

Page 198. — **17.** $2.99. **18.** 20 in. × 12 in.; 110 sq. in. larger.
19. $37.50. **20.** $76.50. **21.** $192.50. **22.** $1

Page 202. — **1.** 19,200 cu. ft. **2.** $711\frac{1}{2}$ cu. yd. **3.** 12 cu. ft. **4.** 384
inch cubes. **5.** 40 sq. ft. **6.** 27,648 cu. in. **7.** 480 cu. ft. **8.** 385.7+
bu. **9.** 6000 lb. **10.** 320 cu. ft. **11.** $26.67. **12.** $13\frac{1}{4}$ loads; $4.17.

Page 209. — **1.** $\frac{4}{3}$. **2.** $\frac{7}{4}$. **3.** $\frac{11}{4}$. **4.** $\frac{11}{6}$. **5.** $\frac{21}{6}$. **6.** $\frac{21}{6}$. **7.** $\frac{54}{6}$.
8. $\frac{7}{6}$. **9.** $\frac{90}{7}$. **10.** $\frac{79}{6}$. **11.** $1\frac{11}{15}$. **12.** $\frac{84}{9}$. **13.** $1\frac{11}{15}$. **14.** $1\frac{11}{16}$.
15. $1\frac{9}{20}$. **16.** $\frac{27}{10}$. **17.** $\frac{203}{50}$. **18.** $4\frac{3}{10}$. **19.** $\frac{106}{25}$. **20.** $\frac{15}{50}$.
21. $2\frac{11}{14}$. **22.** $1\frac{2}{25}$. **23.** $1\frac{3}{50}$. **24.** $1\frac{22}{25}$. **25.** $1\frac{21}{50}$. **26.** $2\frac{11}{14}$.
27. $2\frac{7}{12}$. **28.** $1\frac{11}{24}$. **29.** $\frac{2}{5}$. **30.** $6\frac{1}{4}$. **31.** $6\frac{1}{2}$. **32.** $4\frac{1}{2}$. **33.** $9\frac{1}{4}$.
34. $9\frac{2}{5}$. **35.** $8\frac{1}{4}$. **36.** $8\frac{1}{4}$. **37.** $12\frac{2}{3}$. **38.** $9\frac{3}{4}$ oz. **39.** $12.50.
40. $8\frac{1}{4}$ lb. **41.** $11\frac{11}{12}$ hr. **42.** $11\frac{1}{12}$ min. **43.** $4\frac{1}{4}$. **44.** $10\frac{5}{9}$ mi. **45.** $7\frac{1}{2}$
rd. **46.** $8\frac{1}{2}$ bu. **47.** $12\frac{3}{4}$ in. **48.** $11\frac{1}{4}$ A. **49.** $\frac{1}{2}$. **50.** $\frac{2}{3}$. **51.** $1\frac{1}{4}$.
52. $\frac{7}{15}$. **53.** $\frac{7}{20}$. **54.** $\frac{4}{5}$. **55.** $\frac{7}{8}$. **56.** $\frac{4}{5}$. **57.** $\frac{11}{16}$. **58.** $\frac{13}{16}$. **59.** $\frac{1}{3}$.
60. $\frac{7}{15}$. **61.** $\frac{4}{5}$. **62.** $\frac{11}{15}$. **63.** $\frac{7}{15}$.

Page 210. — **64.** $\frac{4}{5}$, $\frac{3}{5}$. **65.** $\frac{3}{12}$, $\frac{2}{12}$. **66.** $\frac{4}{20}$, $\frac{5}{20}$. **67.** $\frac{3}{12}$, $\frac{2}{12}$. **68.** $\frac{9}{10}$,
$\frac{9}{10}$. **69.** $\frac{9}{10}$, $\frac{8}{10}$. **70.** $\frac{13}{14}$, $\frac{11}{14}$. **71.** $\frac{7}{12}$, $\frac{8}{12}$. **72.** $\frac{8}{15}$, $\frac{7}{15}$. **73.** $\frac{9}{10}$,
$\frac{7}{10}$, $\frac{11}{14}$. **74.** $\frac{2}{7}$, $\frac{4}{7}$, $1\frac{1}{7}$. **75.** $\frac{15}{16}$, $\frac{10}{16}$, $\frac{12}{16}$. **76.** $\frac{9}{40}$, $\frac{4}{40}$, $\frac{10}{40}$. **77.** $\frac{8}{5}$, $\frac{11}{5}$,
$\frac{11}{6}$, $\frac{11}{6}$. **78.** $\frac{15}{20}$, $\frac{56}{20}$. **79.** $\frac{20}{15}$, $\frac{10}{15}$. **80.** $\frac{30}{35}$, $\frac{15}{35}$, $\frac{21}{35}$. **81.** $\frac{7}{24}$, $\frac{21}{24}$, $\frac{22}{24}$. **82.** $\frac{8}{15}$,
$\frac{11}{15}$, $\frac{27}{30}$. **83.** $\frac{15}{25}$, $\frac{21}{25}$, $\frac{10}{25}$. **84.** $\frac{19}{25}$, $\frac{9}{25}$, $\frac{13}{25}$. **85.** $\frac{9}{40}$, $\frac{28}{40}$, $\frac{35}{40}$. **86.** $\frac{7}{15}$,
$\frac{1}{15}$. **87.** $\frac{9}{20}$, $\frac{10}{20}$, $\frac{9}{20}$. **88.** $\frac{54}{63}$, $\frac{42}{63}$, $\frac{11}{63}$. **89.** $\frac{44}{63}$, $\frac{11}{63}$. **90.** $\frac{11}{24}$, $\frac{14}{24}$, $\frac{1}{24}$.
91. $1\frac{13}{20}$. **92.** $1\frac{1}{12}$. **93.** $1\frac{23}{35}$. **94.** $1\frac{5}{12}$. **95.** $\frac{23}{30}$. **96.** $1\frac{11}{16}$. **97.** $\frac{15}{16}$.
98. $1\frac{21}{24}$. **99.** $1\frac{11}{12}$. **100.** $2\frac{5}{12}$. **101.** $1\frac{11}{16}$. **102.** $2\frac{1}{20}$. **103.** $2\frac{7}{24}$.
104. $2\frac{79}{200}$. **105.** $1\frac{11}{14}$. **106.** $1\frac{13}{16}$. **107.** $2\frac{11}{12}$. **108.** $2\frac{17}{24}$.

Page 210. — **109.** $5\frac{11}{16}$.   **110.** $28\frac{4}{11}$.   **111.** $17\frac{11}{13}$.   **112.** $18\frac{11}{17}$.
113. $32\frac{7}{16}$.   **114.** $6\frac{11}{13}$.   **115.** $9\frac{1}{4}$.   **116.** $19\frac{11}{13}$.   **117.** $18\frac{1}{13}$.   **118.** $6\frac{11}{13}$.
119. $\frac{7}{13}$.   **120.** $1\frac{5}{8}$.   **121.** $3\frac{1}{4}$.   **122.** $2\frac{7}{13}$.   **123.** $4\frac{4}{9}$.   **124.** $6\frac{1}{4}$.
125. $5\frac{1}{4}$.   **126.** $19\frac{1}{4}$.   **127.** $4\frac{7}{13}$.   **128.** $10\frac{7}{13}$.   **129.** $5\frac{11}{13}$.   **130.** $5\frac{4}{13}$.
131. $1\frac{11}{13}$.   **132.** $32\frac{1}{4}$.   **133.** $25\frac{7}{13}$.   **134.** $8\frac{11}{13}$.   **135.** $66\frac{11}{13}$.   **136.** $\frac{5}{13}$.

Page 211. — **137.** $\frac{7}{13}$.   **138.** $4\frac{1}{4}$.   **139.** $5\frac{1}{4}$.   **140.** $1\frac{1}{7}$.   **141.** $8\frac{1}{16}$.
142. $9\frac{1}{17}$.   **143.** $2\frac{7}{13}$.   **144.** $17\frac{11}{13}$.   **145.** $\frac{3}{4}$.   **146.** $16\frac{1}{4}$.   **147.** $15\frac{7}{13}$.
148. $180\frac{11}{13}$.   **149.** $9\frac{11}{17}$.   **150.** $23\frac{7}{13}$.   **151.** $5\frac{11}{13}$.   **152.** $9\frac{7}{13}$.   **153.** $74\frac{11}{13}$.
154. $438\frac{77}{136}$.   **155.** $4\frac{11}{17}$.   **156.** $44\frac{11}{13}$.   **157.** 6.   **158.** 21.   **159.** 18.
160. 20.   **161.** 4.   **162.** 44.   **163.** $9\frac{1}{4}$.   **164.** $12\frac{1}{8}$.   **165.** $11\frac{1}{4}$.
166. $8\frac{1}{4}$.   **167.** $13\frac{1}{4}$.   **168.** $6\frac{7}{8}$.   **169.** $4\frac{1}{4}$.   **170.** $8\frac{1}{4}$.   **171.** $21\frac{1}{4}$.
172. $10\frac{9}{16}$.   **173.** $18\frac{1}{4}$.   **174.** $18\frac{1}{4}$.   **175.** $18\frac{1}{4}$.   **176.** $5\frac{1}{4}$.   **177.** $20\frac{1}{4}$.
178. 44.   **179.** $96\frac{1}{4}$.   **180.** $94\frac{1}{4}$.   **181.** $49\frac{1}{4}$.   **182.** $59\frac{1}{4}$.   **183.** $24\frac{1}{4}$.
184. $150\frac{1}{4}$.   **185.** $198\frac{4}{9}$.   **186.** 260.   **187.** $\frac{3}{4}$.   **188.** $\frac{4}{7}$.   **189.** $\frac{4}{7}$.
190. $\frac{5}{9}$.   **191.** $\frac{7}{16}$.   **192.** $\frac{7}{13}$.   **193.** $\frac{3}{9}$.   **194.** $\frac{4}{7}$.   **195.** $\frac{11}{17}$.   **196.** $1\frac{4}{9}$.
197. $\frac{4}{9}$.   **198.** $17\frac{1}{4}$.   **199.** $\frac{11}{13}$.   **200.** $1\frac{111}{117}$.   **201.** $\frac{7}{16}$.   **202.** $\frac{4}{7}$.
203. $\frac{7}{13}$.   **204.** $1\frac{223}{1116}$.

Page 212. — **205.** 2.   **206.** $9\frac{1}{4}$.   **207.** $10\frac{11}{13}$.   **208.** $3\frac{11}{17}$.   **209.** $14\frac{1}{4}$.
210. $85\frac{1}{4}$.   **211.** 117.   **212.** 24.   **213.** 2640.   **214.** $\frac{3}{4}$.   **215.** $\frac{7}{15}$.
216. $\frac{7}{13}$.   **217.** $\frac{7}{15}$.   **218.** $\frac{7}{15}$.   **219.** $\frac{7}{17}$.   **220.** $\frac{7}{13}$.   **221.** $\frac{7}{15}$.
222. $\frac{7}{13}$.   **223.** $\frac{7}{13}$.   **224.** $\frac{7}{15}$.   **225.** $\frac{7}{15}$.   **226.** $\frac{7}{15}$.   **227.** $\frac{7}{15}$.   **228.** $\frac{7}{113}$.
229. $\frac{7}{15}$.   **230.** $\frac{10}{15}$.   **231.** $\frac{7}{151}$.   **232.** $\frac{7}{16}$.   **233.** $\frac{7}{13}$.   **234.** $\frac{7}{169}$.
235. $\frac{7}{13}$.   **236.** $\frac{7}{15}$.   **237.** $\frac{7}{13}$.   **238.** $\frac{11}{136}$.   **239.** $550\frac{11}{13}$.   **240.** $355\frac{11}{13}$.
241. $298\frac{11}{13}$.   **242.** $530\frac{7}{13}$.   **243.** $149\frac{7}{16}$.   **244.** $1\frac{7}{13}$.   **245.** 1.   **246.** $\frac{4}{7}$.
247. $1\frac{3}{4}$.   **248.** $\frac{4}{7}$.   **249.** $1\frac{11}{13}$.   **250.** $1\frac{1}{4}$.   **251.** $1\frac{4}{9}$.   **252.** $2\frac{1}{4}$.
253. $2\frac{7}{13}$.   **254.** $1\frac{1}{8}$.   **255.** $\frac{11}{17}$.   **256.** 12.   **257.** $1\frac{1}{4}$.   **258.** 25.
259. 5.   **260.** 14.   **261.** $18\frac{1}{4}$.   **262.** $28\frac{4}{7}$.   **263.** 132.   **264.** 336.
265. $75\frac{4}{7}$.   **266.** $100\frac{1}{4}$.   **267.** $1\frac{11}{16}$.   **268.** $1\frac{7}{13}$.   **269.** $2\frac{1}{4}$.   **270.** $2\frac{1}{4}$.
271. $5\frac{7}{13}$.   **272.** $1\frac{7}{14}$.   **273.** $\frac{4}{7}$.   **274.** $1\frac{11}{17}$.   **275.** $1\frac{11}{13}$.   **276.** $1\frac{7}{13}$.
277. $2\frac{4}{15}$.   **278.** $2\frac{1}{4}$.   **279.** $16\frac{11}{13}$.   **280.** 256.   **281.** $2\frac{43}{105}$.   **282.** 33.
283. 10.   **284.** $8\frac{11}{114}$.   **285.** $5\frac{4}{7}$.

Page 214. — **3.** $\frac{3}{4}$.   **4.** $\frac{1}{4}$.   **5.** $\frac{3}{4}$.   **6.** $\frac{4}{7}$.   **7.** $\frac{4}{7}$.   **8.** $\frac{4}{9}$.   **9.** $\frac{1}{4}$.
10. $\frac{1}{4}$.   **11.** $\frac{1}{4}$ yd.   **13.** Mary's share, 28 qt.; Lucy's share, 21 qt.
14. Frank, $\$.96$; Clay, $\$1.44$.

Page 215. — **1.** $\frac{5}{7}$.   **2.** $\frac{71}{8}$.   **3.** $\$300$.   **4.** $\$4$.   **5.** $98\cancel{c}$.   **6.** $\$5000$.
7. 240 bu.   **8.** $\frac{2}{3} = 2 \times \frac{1}{3}$; $\frac{3}{4} = 3 \times \frac{1}{4}$; $\frac{3}{13} = 3 \times \frac{1}{13}$.   **9.** 6 lb.; 12 lb.;
24 lb.; 30 lb.   **10.** $\frac{1}{4}$; $\frac{1}{4}$; $\frac{1}{4}$; $\frac{3}{4}$.   **11.** $\$35$.   **12.** $\$14$.   **13.** $65\cancel{c}$.
14. $\frac{1}{3}$; $\frac{1}{4}$; 3.   **15.** $\frac{4}{9}$ as much.

Page 216. — **16.** 24 yr.; $\frac{3}{4}$.   **17.** $30\cancel{c}$; $60\cancel{c}$; $\$1.20$.   **18.** 140 rd.
19. 188 mi.; $27\frac{1}{4}$ mi.   **20.** 4 wk.   **21.** 5100.   **22.** 60.   **23.** 24 yd.
24. $\$2.82$.   **25.** $12\frac{1}{4}$.   **26.** $\$1270$.   **27.** $\$7200$.   **28.** $\$1700$.

Page 217. — **29.** $\$39.20$.   **30.** $\frac{1}{4}$.   **31.** $\$1.60$.   **32.** 10 wk.   **33.** 4 hr.
34. $\$2.73$.   **35.** $\$49.74$.   **36.** $\$2.27$.   **37.** $\$1200$; $\$400$.   **38.** $\$53.75$.
39. $\$50.71$.

Page 218. — **40.** $\$1140$.   **41.** $\$8.40$.   **42.** $\$15.04$.   **43.** $\$2.49$.
44. $\frac{1}{4}$.   **45.** $\$1100$.   **46.** $\$400$.   **47.** $\$84$.   **48.** 363 ft.   **49.** 24 times.
50. 20.   **51.** 2514.   **52.** $\$34.80$.   **53.** $15\frac{1}{2}$ da.

Page 219. — 1. .75. 2. .875. 3. .66⅔. 4. .3125. 5. .15. 6. .88⅖.
7. .16. 8. .8. 9. .85⅖. 10. .44⅘. 11. .90⅒. 12. .416⅔. 13. .1875.
14. .4375. 15. .35. 16. ⅛. 17. ⁵⁄₁₂. 18. ⅜. 19. ⁴⁹⁄₉₀. 20. ½. 21. ¹⁄₁₃.
22. ⁵⁄₁₁. 23. ¹⁄₁₃. 24. ⅕. 25. ⅓. 26. ⁵⁄₉. 27. ⁷⁄₁₆. 28. ½. 29. ⁷⁄₉.
30. ⁷⁄₁₇. 31. ⅞. 32. 111.55. 33. 187.64. 34. 66.51. 35. 105.9.
36. 25.823. 37. 45.3407. 38. 35.755. 39. 11.1751. 40. 354.32685.
41. 189.61752.

Page 220. — 42. 343.9706. 43. 2.282566. 44. 984.37936. 45. 1.266.
46. 2.2249. 47. 210.8. 48. 16.2. 49. .5675. 50. .0466. 51. 2.4194.
52. 9.998. 53. 19.724. 54. $33.58. 55. 38.068. 56. 71.125.
57. 113.8909. 58. .101. 59. 100.192. 60. 26.483. 61. .101. 62. 4.3935.
63. 8.2415. 64. .01. 65. .055. 66. .0476. 67. .00081. 68. 3.24.
69. 14.7. 70. 80.214. 71. 4.25. 72. .486. 73. .011055. 74. .1183952.
75. 2.0736. 76. 42.6725. 77. 1203.03208. 78. 1.925. 79. 1.296. 80. 6.
81. 7.936. 82. 31.296. 83. .0428796. 84. .12. 85. 28.65. 86. .000225.
87. 7.762536. 88. 2.75804. 89. 328.252. 90. .12048. 91. .760125.

Page 221. — 92. 911.2. 93. 339. 94. 2759.1. 95. 719.44.
96. .0003424. 97. .030335. 98. 125.472. 99. 42.504. 100. 489.06.
101. 73.8. 102. 2.352. 103. 37.468. 104. 10.7325. 105. 59.2144.
106. 3135.65. 107. 12.054. 108. 404.02944. 109. .61548. 110. 14.8356.
111. .04936. 112. .000608. 113. 1.006008. 114. .00025. 115. 30.04812.
116. .04. 117. .05. 118. .04. 119. .01. 120. 2.2. 121. 2.02.
122. 2.02. 123. 2.01. 124. 1.07. 125. 8.01. 126. 3.01. 127. 8.06.
128. 12.02. 129. 141.6. 130. 6.07. 131. 1.5. 132. 5.06. 133. .1025.
134. .03002. 135. 20.03. 136. .084. 137. .061. 138. .87. 139. .009.
140. .075. 141. 15. 142. 1.8. 143. 2.21. 144. 2.05. 145. 1200.
146. .012. 147. 560. 148. 1.24. 149. 12.5. 150. .0207. 151. .8.
152. 745. 153. 9.7. 154. 248. 155. .018.

Page 222. — 156. .0055. 157. .456. 158. .15. 159. 50.6. 160. 10.25.
161. 3.002. 162. 200.3. 163. .904. 164. .0906. 165. .704. 166. 9.29⅘.
167. .0027. 168. .084. 169. .0704. 170. 1,000,000. 171. 2.5. 172. 2.44.
173. 115.8. 174. 16.4. 175. .0027. 176. 17,500. 177. .033.
178. .000011. 179. 111,000. 180. 100. 181. .0002. 182. 210.
183. .00137. 184. .00023. 185. 125.
1. 47.0625 lb. 2. 23.0875 A. 3. 230.7177. 4. $7154.95.

Page 223. — 5. 866.9375 sq. ft. 6. $20,530.25. 7. 472.44 in.; 13.12¼ yd.
8. $3825. 9. $1200; $3600; $4800. 10. 83 da. 11. $405. 12. 88¼ doz.
13. $18.25. 14. .009. 15. 570³⁄₇ times. 16. 1.25. 17. 18.5 sq. rd.

Page 224. — 18. $20,000. 19. $20. 20. $6.25. 21. $60. 22. $6.25.
23. $10. 24. $45. 25. $200. 26. $300. 27. $200. 28. 2560 lb.
29. 3000 lb. 30. 60 lb. 31. 200 yd. 32. 260 yd. 33. 150 yd.

Page 231. — 2. 49 pt. 3. 67 pt. 4. 38 pt. 5. 45 pt. 6. 7 pt.
7. 3 qt. 8. 4 pt. 9. 1½ pt. 10. 1¼ pt. 11. 7½ pt. 12. 1⅛ qt.
14. 47 gal. 15. 105 gal. 3 qt. 16. 69 gal. 2 qt. 17. 84 gal. 1 qt. 1 pt.
18. 49 qt. 19. 22 pt. 20. 31 qt. 21. 20 pt. 22. 65 qt. 23. 117 gal.
1 qt. 24. 10 gal. 2 qt. 25. 16 gal. 26. 4 gal. 2 qt. 1 pt. 27. 12 gal. 2 qt
28. 200 pt. 29. 6 gal. 1 qt.; $2.

Page 232. — **29.** 3 gal. 3 qt. 1½ pt. ; 1 gal. ½ pt. **31.** 16 da. **32.** 68 ¢.
33. 46⅞ gal. **34.** 71 pt. **35.** $26.97. **36.** 178¼ bbl. ; $446.43.
37. $28.80. **38.** $3.60. **39.** $18.30. **40.** 3 qt. 1 pt.

Page 233. — **1.** 496 pt. **2.** 27 qt. **3.** 161 qt. **4.** 131 qt. **5.** 83 qt.
6. 60 pk. **7.** 3½ pk. **8.** 70 qt. **9.** 6 pk. **10.** 28 qt. **11.** 9 qt.
12. 2 bu. **13.** 4 pk. **14.** 1 bu. **15.** 2 pk. **16.** 4 bu.

Page 234. — **17.** 2½ pk. ; 40 pt. ; $1.60. **18.** 4 bu. 25 qt. **19.** 70 pt.
20. 1 bushel and 13 quarter pecks. **21.** 1 bu. 3½ pk. **22.** 55½½ bu.
23. 34 doz.

Page 236. — **1.** 101 oz. **2.** 6800 lb. **3.** 6448 oz. **4.** 8800 oz.
5. 36 cwt. **6.** 248 oz. **7.** 1825 lb. **8.** 64,080 oz. **9.** 34 lb. **10.** 3 T.
11. 8 lb. **12.** 50 lb. 10 oz. **13.** 9600 lb. **14.** 33,500 lb. **16.** 50 lb.
17. 1750 lb. **18.** 12 oz. **19.** 8 cwt. 75 lb. **20.** 41 ¢. **21.** 80 oz.
22. $4.80. **23.** $3. **24.** $9. **25.** $12. **26.** $4.50. **27.** $1.50.
28. $3.81.

Page 237. — **29.** $671.50. **30.** $59.61. **31.** 1 T. 7 cwt. **32.** 1 T. 10 cwt.
33. 2 long T. 460 lb. **34.** 1½ T. **35.** 45 cwt. ; 2 T. 5 cwt. **36.** 60 oz.
37. 121 lb. **38.** $60. **39.** 3₇²₀ lb. ; $1.30.

Page 238. — **1.** 7½ lb. **2.** 6¼ lb. **3.** 7 lb. **4.** 3¼ lb. **5.** 15 lb.
6. 2 lb. **7.** 6 lb. **8.** 2¾ lb. **9.** 11¼ lb. **10.** 8¼ lb. **11.** 6 lb.
12. 8¼ lb. **13.** 12 lb. **14.** 5¼ lb. **15.** 120 lb. **16.** 240 lb. **17.** 96 lb.
18. 14 lb. **19.** 30 bu.

Page 239. — **20.** 6660 lb. ; 4 loads and 260 lb. over. **21.** 27 loads and
2520 lb. over. **22.** 624,000 lb. ; 8 loads.

Page 240. — **1.** 18¼ ft. **2.** 16¼ ft. **3.** 126 in. **4.** 69 yd. **5.** 69 ft.
6. 16 ft. **7.** 17 ft. **8.** 22¼ ft. **9.** 43 in. **10.** 77 in. **11.** 1604 rd.
12. 240 rd. **13.** 60 yd. **14.** 189 in. **15.** 14¼ ft. **16.** 995 in.
17. 34 ft. **18.** 317 in. **19.** 18 in. **20.** 4 ft. 11 in. **21.** 4 yd. 1 in.
22. 5 yd. 2 ft. **23.** 7 rd. 11½ ft. **24.** 10 rd. 4 yd. **25.** 11 rd. 2½ yd.
26. 1 mi. 200 rd.

Page 241. — **27.** 1 mi. 340 yd. **28** 1 mi. 60 yd. **29.** 128¼ ft.
30. 9768 ft. **31.** 4320 turns. **32.** 88 rd. **33.** $145.20. **34.** 44 mi.
80 rd.

Page 242. — **1.** 186 min. **2.** 123 hr. **3.** 90 da. **4.** 1820 sec.
5. 5¼ da. **6.** 190 min. **7.** 79 hr. **8.** 124 hr. **9.** 45 min. **10.** 3 min.
30 sec. **11.** 4 wk. 2 da. **12.** 6 yr. **13.** 2 da. 2 hr. **14.** 36 sec.

Page 243. — **16.** 32 mi. **17.** $2.95. **18.** 6 min. 18 sec. **19.** 54 mi.
200 rd. **20.** 270 mi. **21.** 192 mi. **22.** $41.60. **23.** 1920 hr.
24. 3¹⁰⁄₁₃ mi. per hr. **25.** 1₄⁵₇ mi. per hr.

Page 245. — **3.** 10 ft. 7 in. **4.** 1 ft. 10 in. **5.** 10 ft. 10 in. **6.** 6 in.
7. 8 ft. 9 in. **8.** 6 ft. 2 in. **9.** 4 ft. 5 in. **10.** 2 ft. **11.** 7 ft. 8 in.
12. 4 ft. 4 in. **13.** 7 yd. **14.** 10 in. **15.** 4 yd. 11 in. **16.** 1 yd. 1 in.
17. 3 yd. 5 in. **18.** 1 yd. 1 ft. 6 in. **19.** 5 yd. 1 ft. **20.** 8 in.
21. 3 ft. 7 in.

Page 245.—**22.** 3 in. **23.** 1 yd. 2 ft. 4 in. **24.** 4 in. **25.** 1 yd. 7 in. **26.** 1 ft. 10 in. **27.** 1 yd. **28.** 2 in. **29.** 10 yd. 2 ft. $5\frac{1}{4}$ in. **30.** $5\frac{3}{4}$ in. **31.** 1 yd. 3 in. **32.** 1 ft. 3.4 in.

Page 246.—**2.** 6 yr. 9 mo. 27 da. **3.** 15 yr. 16 da. **4.** Lee's age, 58 yr. 2 mo. 20 da.; Grant's age, 42 yr. 11 mo. 12 da.; 15 yr. 3 mo. 8 da. older. **5.** 56 yr. 5 mo. 6 da. **6.** 55 yr. 5 mo. 18 da. **7.** 42 yr. 10 mo. 17 da.

Page 252.—**2.** \$32. **3.** $3\frac{3}{5}$ mo. **4.** 18 da. **5.** 18 A. **6.** 90 A. **7.** 72. **8.** 108 horses. **9.** $3\frac{1}{10}$. **10.** 27 lb. **11.** \$1.125. **12.** \$200. **13.** \$50. **14.** 8 words. **15.** 20¢. **16.** 230. **17.** \$3125. **18.** \$60. **19.** 90 A. **20.** \$400. **21.** \$150. **22.** 1209.

Page 253.—**2.** 32%. **3.** 30%. **4.** 30%. **5.** $32\frac{1}{2}$%. **6.** 36%. **7.** $37\frac{1}{2}$%. **8.** 50%. **9.** $31\frac{1}{4}$%. **10.** 8%. **11.** $33\frac{1}{3}$%. **12.** 50%. **13.** 80%. **14.** 80%.

Page 254.—**15.** 50%. **16.** $33\frac{1}{3}$%. **17.** 25%; $12\frac{1}{2}$%. **18.** $8\frac{1}{3}$%. **19.** $6\frac{1}{4}$%. **20.** $8\frac{1}{3}$%. **21.** $\frac{2}{5}$; 40%. **22.** $66\frac{2}{3}$%. **23.** 95%. **24.** 75%; 50%. **25.** 55%; 45%. **26.** 30%. **27.** 50%. **28.** $8\frac{1}{3}$%. **29.** 20%.

Page 255.—**30.** 50%; 200 pupils. **31.** 90%. **32.** 20%. **33.** $66\frac{2}{3}$%; $8\frac{1}{3}$%; $16\frac{2}{3}$%; $8\frac{1}{3}$%. **34.** 25%. **35.** 20%. **36.** $37\frac{1}{2}$%. **37.** 20%. **38.** 20%. **39.** 5%. **40.** 25%.

Page 256.—**41.** 20%. **42.** 58.8%. **43.** $6\frac{2}{3}$%. **44.** 10%. **45.** 20%. **46.** $9\frac{1}{11}$% Dec. **47.** $11\frac{1}{9}$% Inc. **48.** 4% Dec. **49.** 20% Dec. **50.** $16\frac{2}{3}$% Dec. **51.** 20% Dec. **52.** $12\frac{1}{2}$% Inc. **53.** $5\frac{5}{17}$% Dec. **54.** $5\frac{5}{9}$% Dec. **55.** $8\frac{1}{3}$%.

Page 257.—**56.** 1%. **57.** 10%. **58.** 5%. **59.** 10%. **60.** $6\frac{1}{4}$%. **61.** 2%. **62** 2%. **63.** $3\frac{3}{4}$%. **64.** 5%. **65.** 1%. **66.** $5\frac{1}{4}$%. **67.** 2%. **68.** .67%. **69.** 25%. **70.** 20%. **71.** $33\frac{1}{3}$%.

Page 258.—**1.** \$126.56. **2.** \$1145. **3.** \$1248. **4.** \$174. **5.** \$1818. **6.** \$180.52. **7.** \$612.85. **8.** \$251.01. **9.** \$2283.36. **10.** \$6436.44. **11.** \$7777.77. **12.** \$2553.45.

Page 259.—**14.** \$288. **15.** \$267.54. **16.** \$157.03. **17.** \$6.13. **18.** \$4.50. **19.** \$7.65. **20.** \$129.60. **21.** \$4080. **22.** \$1890. **23.** \$1353.75. **24.** \$53.55. **25.** \$291.60; \$21.60 more. **26.** The same.

Page 263.—**1.** \$90. **2.** \$79.70. **3.** \$97.70. **4.** \$70.04. **5.** \$76.40. **6.** \$118.53. **7.** \$532.37. **8.** \$135. **9.** \$1800. **10.** \$456. **11.** \$340.58. **12.** \$31.80; \$24.80. **13.** \$54.36. **14.** \$60.

Page 265.—**2.** \$15. **3.** \$7.50. **4.** \$25.67. **5.** \$18. **6.** \$6. **7.** \$46.07. **8.** \$43.20. **9.** \$173.25. **10.** \$29.43. **11.** \$15.40.

Page 266.—**12.** \$46.20; \$706.20. **13.** \$73.24; \$530.99. **14.** \$54; \$729. **15.** \$192; \$1392. **16.** \$16.48; \$100.98. **17.** \$13.61; \$104.36. **18.** \$92.75; \$622.75. **19.** \$24.50; \$174.50. **20.** \$30.25; \$305.25. **21.** \$16; \$136. **22.** \$16.67; \$641.67. **23.** \$86; \$486. **24.** \$40.10; \$315.10. **25.** \$3. **26.** \$10. **27.** \$19.50. **28.** \$12.48. **29.** \$6.80. **30.** \$7.09. **31.** \$282.50. **32.** \$33.75.

Page 268. — 1. 24,300 sq. ft.   2. 7200 sq. in.   3. 10 A.   4. 260 sq. rd.
5. 26$\frac{54}{121}$ sq. rd.   6. 250,470 sq. ft.   7. 33$\frac{1}{4}$ A.; $2025.   8. 8488 sq. ft.
9. $76.   10. $170.10.   11. $40,425.

Page 269. — 12. $6450.   13. 640 A.; $54,400.   14. 17$\frac{1}{2}$ bu.   15. $500;
16$\frac{2}{3}$⊄.   16. $88.82.   17. 40 rd.   18. (1) 120 sq. rd.; 128 sq. rd.; 175 sq. rd.
(2) 52 rd.; 48 rd.; 64 rd. (3) 2$\frac{121}{160}$ A.   19. (1) 30 rd.; 28 rd.; 38 rd.; 38 rd.
(2) 50 sq. rd.; 45 sq. rd.; 78 sq. rd.; 84 sq. rd. (3) 1$\frac{97}{160}$ A.   20. $63,504.

Page 270. — 21. 20 A.; 1 in. = 40 rd.   22. Oats, 2$\frac{1}{2}$ A.; corn, 5 A.;
pasture 3$\frac{1}{4}$ A.; wheat, 1$\frac{7}{8}$ A.   23. $48.52.   24. $170.

Page 271. — 4. 40 sq. in.   5. 86 sq. in.   6. 225 sq. in.   7. 432 sq. in.
9. 2112 sq. ft.

Page 272. — 10. 224 sq. ft.   11. $25.96.   12. Ends, 35 sq. ft.;
sides, 172 sq. ft.   13. 480 sq. ft.

Page 274 — 3. 628 sq. ft.   4. 120 ft.   5. $37.22.   6. $62.80.
7. 72 times.   8. 288 cu. ft.   9. 166$\frac{2}{3}$ loads; $41.67.

Page 275. — 10. 810,000 cu. in.; 3506$\frac{44}{49}$ gal.   11. $462.   13. $\frac{27}{47}$;
3 lb.   14. 153$\frac{1}{4}$ bu.   15. 4308$\frac{57}{99}$ gal.   16. 13$\frac{1}{2}$ T.   17. 32 T.
18. 3.73+ cu. ft.   19. 54.4 bu.   20. 8$\frac{221}{242}$ bbl.

Page 276. — 1. 540 tiles.   2. 289$\frac{7}{8}$ ft.   3. 277$\frac{1}{2}$ cu. yd.   4. 1350 sq. ft.
5. 1000 bu.   6. $24.   7. $327.68.   8. 7181$\frac{11}{18}$ gal.   9. $168.
10. $16.34.   11. 50 rd.   12. $7.92.   13. 10 rd.

Page 277. — 14. 1728 cakes.   15. $116.67.   16. 15 A.   17. 2048 bu.
18. 4847$\frac{27}{33}$ gal.   19. 150 ft.   20. $96.80.   21. 40 ft.   22. $800.
24. 400 ft.; 292 ft.; 292 ft.; 26 ft.; 216 ft.   25. 1111$\frac{1}{8}$ sq. yd.   26. 511$\frac{1}{8}$
sq. yd.   27. 4$\frac{4}{9}$ sq. yd.   28. 88$\frac{8}{9}$ sq. yd.

Page 278. — 29. 90\frac{10}{19}$.   30. $137.78.   31. 23⊄.   32. $131.74.
33. $23.12.   34. 21544$\frac{4}{7}$ gal.   35. 60 farms.   36. 3000 sods.
37. $62.40.   38. 48 persons.

Page 279. — 1. $40.   2. $10.   4. $196.88.   5. Flora, 8.4 lb.; Bess,
8.17 lb.; Queen, 8.4672 lb.; Daisy, 8.16 lb.   6. $847.50.

Page 280. — 10. 101 kilowatts 207 watts; $7.76.

Page 281. — 12. $7.13.   13. $3.56.   15. 105.6 ft.   16. $\frac{1}{2}$ ft.

Page 282. — 17. 3500 lb.   18. 132 ft.   19. 6.6 ft.   20. 26.4 ft.
21. 49$\frac{1}{2}$ ft.   22. 184.8 ft.   23. 181.5 ft.   24. 33 ft.   25. 7$\frac{1}{2}$ ft.
26. 10 ft.

Page 283. — 1. 110,092.   2. 37.   3. 50.601.   4. 25.061; 125.5;
300.0002.   5. 4742.   6. .0119.   7. 8$\frac{1}{2}$.   8. 55.12.   9. 30⊄.   10. 205 ft.
11. $2160.   12. $88.92.   13. 2$\frac{2}{3}$.   14. .1.   15. $42.

Page 284. — 16. 35$\frac{11}{17}$ mi.   17. 2$\frac{1}{4}$ T.   18. $334.   19. $1080.
20. $100.   21. 14$\frac{2}{7}$%.   22. 8; $1.60; 20.83\frac{1}{3}$.   23. $5625.   24. $4.50.
25. 20 rd.   26. $26.97.   27. $660.   28. 75%.   29. $15,876.

Page 286.— 3. 2, 3, 2, 2, 2, 2. **4.** 2, 2, 5, 5. **5.** 3, 3, 2, 2, 3. **6.** 2, 2, 3, 2, 5. **7.** 5, 5, 5. **8.** 2, 2, 2, 2, 2, 5. **9.** 2, 2, 5, 3, 3. **10.** 7, 2, 3, 5. **11.** 3, 2, 2, 2, 3, 2. **12.** 3, 2, 2, 3, 2, 5. **13.** 2, 2, 2, 2, 3, 3, 3. **14.** 2, 2, 2, 2, , , 2, 3. **15.** 3, 3, 2, 2, 2, 2, 5. **16.** 2, 2, 2, 2, 2, 2, 3, 5. **17.** 3, , 2, 5, **18.** 2, 2, 2, 2, 5, 5. **19.** 2, 2, 2, 3, 5, 5. **20.** 3, 3, 2, 2, 5, 5. **21.** 3, 2, 11, 17. **22.** 11, 3, 5, 7. **23.** 5, 5, 7, 7. **24.** 11, 11, 11. **25.** 5, 5, 7, 8, 3. **26.** 3, 2, 2, 7, 9. **27.** 3, 8, 3, 3, 3, 7. **28.** 3, 7, 2, 3, 17. **29.** 7, 11, 2, 3, 5. **30.** 2, 2, 2, 3, 3, 5, 7. **31.** 7, 11, 2, 5, 5. **32.** 17, 13, 19. **33.** 5, 7, 11, 11. **34.** 5, 7, 7, 7, 3.

Page 287.— 3. 36. **4.** 18. **5.** 17. **6.** 135. **7.** 136. **8.** 190.

Page 289.— 3. 60. **4.** 72. **5.** 42. **6.** 378. **7.** 96. **8.** 756. **9.** 1728. **10.** 2520. **11.** 810. **12.** 240. **13.** 96. **14.** 420. **15.** 945. **16.** 280. **17.** 576. **18.** 2080. **19.** 2100. **20.** 6600. **1.** $3. **2.** $8. **3.** 146 sq. yd. **4.** $29.55. **5.** $181.

Page 290.— 4. 24 bd. ft. **5.** 9 bd. ft. **6.** 45 bd. ft. **7.** 48 bd. ft.

Page 291.— 8. 100 bd. ft. **9.** 360 bd. ft. **10.** $200. **11.** $12. **12.** $260.40. **13.** $400. **14.** 20 cords. **15.** 80 cords ; $440. **16.** $15. **17.** $44.80. **18.** 12½ cords.

Page 292.— 2. $350. **3.** $190. **4.** $150. **5.** $448. **6.** $195. **7.** $192. **8.** $120. **9.** $128. **10.** $140. **11.** $300. **12.** $120. **13.** $80.

Page 293.— 14. $15. **15.** $50. **16.** $10,000. **17.** $3000. **18.** $12,000.